D1734640

Science and Integrated Coastal Management

Goal of this Dahlem Workshop:

To strengthen the effective use and communication of scientific knowledge in the integrated coastal management policy process.

Report of the 85th Dahlem Workshop on
Science and Integrated Coastal Management
Berlin, December 12–17, 1999

Sponsored by: Deutsche Forschungsgemeinschaft

Science and Integrated Coastal Management

Edited by

B. von Bodungen and R.K. Turner

Program Advisory Committee:
B. von Bodungen and R.K. Turner, Chairpersons
J. Dronkers, V. Lee, W. Salomons, and F.V. Wulff

Berlin

Customer Service: dahlem university press
Kaiserswerther Str. 16–18
14195 Berlin, Germany
Tel: +49 (30) 838 55053
Fax: +49 (30) 841 09103
email: dahlemup@zedat.fu-berlin.de

Visit our Home Page at
http://www.dahlem@fu-berlin.de

Die Deutsche Bibliothek - CIP - Einheitsaufnahme
Ein Titeldatensatz für diese Publikation ist bei
Der Deutschen Bibliothek erhältlich

Cover Credits: Figure from "L'Atmosphère: Météorologie Populaire," Camille Flammarion, Paris: Librairie Hachette et C., 1888

Cover design by Rössiger & Swoboda, Berlin, Germany

ISBN 3-934504-02-7

Layout and typeset by Dahlem Konferenzen

Printed and bound by Druckhaus Köthen GmbH, Köthen, Germany

Contents

The Dahlem Konferenzen ix

List of Participants xi

1 **Science and Integrated Coastal Management: An Introduction** 1
B. von Bodungen and R.K. Turner

2 **Eutrophication in the Baltic Sea Area: Integrated Coastal Management Issues** 15
R. Elmgren and U. Larsson

3 **Science and Integrated Drainage Basin Coastal Management: Chesapeake Bay and Mississippi Delta** 37
D.F. Boesch

4 **Transboundary Issues: Consequences for the Wadden Sea** 51
F. Colijn and K. Reise

5 **Eutrophication in the Black Sea and a Basin-wide Approach to Its Control** 71
L.D. Mee

6 **Group Report: Transboundary Issues** 93
T.D. Jickells, Rapporteur
D.F. Boesch, F. Colijn, R. Elmgren, P. Frykblom, L.D. Mee, J.M. Pacyna, M. Voss, and F.V. Wulff

7 **Shoreline and Land-use Development Perspectives in the Context of Long-term Decreasing Functional Diversity in the Delta Region, SW Netherlands** 113
I. de Vries

8 **The Anglian Coast** 121
J. Pethick

9 **The Great Barrier Reef, Australia: Partnerships for Wise Use** 135
C.J. Crossland and R.A. Kenchington

10 **Shoreline Development on the Spanish Coast: Problem Identification and Solutions** 149
R. Sardá

11 **Group Report: Shoreline Development** 165
K. Schwarzer, Rapporteur
C.J. Crossland, A. de Luca Rebello Wagener, I. de Vries, J. Dronkers, E. Penning-Rowsell, K. Reise, R. Sardá, J. Taussik, and M. Wasson

12 **Integrated Coastal Management in Tanzania and Eastern Africa: Addressing Diminishing Resources and a Forgotten People** 191
M.A.K. Ngoile, J. Daffa, and J. Kulekana

13 **River Basin Activities, Impact, and Management of Anthropogenic Trace Metal and Sediment Fluxes to Sepetiba Bay, Southeastern Brazil** 203
L.D. Lacerda, R.V. Marins, C. Barcellos, and B.A. Knoppers

14 **Integrated Coastal Management: The Philippine Experience** 213
L. Talaue-McManus

15 **Integrated Management of the Benguela Current Region: A Framework for Future Development** 229
M.J. O'Toole, L.V. Shannon, V. de Barros Neto, and D.E. Malan

16 **Group Report: Integrated Coastal Management in Developing Countries** 253
C. Richter, Rapporteur
P.R. Burbridge, C. Gätje, B.A. Knoppers, O. Martins, M.A.K. Ngoile, M.J. O'Toole, S. Ramachandran, W. Salomons, and L. Talaue-McManus

17 **River Basins under Anthropocene Conditions** 275
M. Meybeck

18 **The Use of Models in Integrated Resource Management in the Coastal Zone** 295
L.A. Deegan, J. Kremer, T. Webler, and J. Brawley

19 **Governance and Sustainable Fisheries** 307
J.M. McGlade

20 **Inventing Governance Systems That Respond to Coastal Ecosystem Change** 327
S.B. Olsen

21 Group Report: Unifying Concepts for Integrated Coastal Management **341**
K.-C. Emeis, Rapporteur
J.R. Benoit, L.A. Deegan, A.J. Gilbert, V. Lee, J.M. McGlade,
M. Meybeck, S.B. Olsen, and B. von Bodungen

Author Index **365**

Subject Index **367**

The Dahlem Konferenzen

In 1974, the Stifterverband für die Deutsche Wissenschaft[1] in cooperation with the Deutsche Forschungsgemeinschaft[2] founded the *Dahlem Konferenzen.* It was created to promote an interdisciplinary exchange of scientific ideas as well as to stimulate cooperation in research among international scientists. Dahlem Konferenzen proved itself to be an invaluable tool for communication in science, and so, to secure its long-term future, it was integrated into the Freie Universität Berlin in January, 1990.

As has been evident over recent years, scientific research has become highly interdisciplinary. Now, before real progress can be made in any one field, the concepts, methods, and strategies of related fields must be understood and able to be applied. Coordinated research efforts, scientific cooperation, and basic communication between the disciplines and the scientists themselves must be promoted in order for science to advance.

To meet these demands, Dahlem Konferenzen created a special type of forum for communication, now internationally recognized as the *Dahlem Workshop Model.* These workshops are the framework in which coherent discussions between the disciplines take place and are focused around a topic of high priority interest to the disciplines concerned. At a Dahlem Workshop, scientists are able to pose questions and solicit alternative opinions on contentious issues from colleagues from related fields. The overall goal of a workshop is not necessarily to reach a consensus but rather to identify gaps in knowledge, to find new ways of approaching controversial issues, and to define priorities for future research. This philosophy is implemented at every stage of a workshop: from the selection of the theme to its breakdown in the discussion groups, from the writing of the background papers to the formulation of the group reports.

Workshop topics are proposed by leading scientists and are approved by a scientific board, which is advised by qualified referees. Once a topic has been approved, a Program Advisory Committee of scientists meets approximately one year before the workshop to delineate the scientific parameters of the meeting, select participants, and assign them their tasks. Participants are invited on the basis of their scientific standing alone.

Each workshop is organized around four key questions, which are addressed by four discussion groups of approximately ten participants. Lectures or formal presentations are taboo at Dahlem. Instead, concentrated discussion — within a group and between groups — is the

---

[1] The Donors' Association for the Promotion of Sciences and Humanities, a foundation created in 1921 in Berlin and supported by German trade and industry to fund basic research in the sciences.

[2] German Science Foundation.

means by which maximum communication is achieved. To facilitate this discussion, participants prepare the workshop theme prior to the meeting through the "background papers," the themes and authors of which are chosen by the Program Advisory Committee. These papers specifically review a particular aspect of the group's discussion topic as well as function as a springboard to the group discussion, by introducing controversies or unresolved problem areas.

During the workshop week, each group sets its own agenda to cover the discussion topic. Cross-fertilization between groups is both stressed and encouraged. By the end of the week, in a collective effort, each group has prepared a report reflecting the ideas, opinions, and contentious issues of the group as well as identifying directions for future research and problem areas still in need of resolution.

A Dahlem Workshop initiates and facilitates discussion between a certain number — necessarily restricted — of scientists. Because it is imperative that the discussion and communication should continue after a workshop, we present the results to the scientific community at large in the form of this published volume. In it you will find the revised background papers and group reports, as well as an introduction to the workshop theme itself.

The difference between proceedings of many conventional meetings and this workshop report will be easily discernable. Here, the background papers have not only been reviewed by formal referees, they have been revised according to the many comments and suggestions made by *all* participants. In this sense, they are reviewed more thoroughly than scientific articles in most archival journals. In addition, an extensive editorial procedure ensures a coherent volume. I am sure that you, too, will appreciate the tireless efforts of the many reviewers, authors, and editors.

On their behalf, I sincerely hope that the spirit of this workshop as well as the ideas and controversies raised will stimulate you in your work and future endeavors.

Wedigo de Vivanco
Dean of International Affairs and
Director of Dahlem Konferenzen
Freien Universität Berlin
Thielallee 66, 14195 Berlin, Germany

List of Participants with Fields of Research

JEFFREY R. BENOIT Office of Ocean and Coastal Resource Management, NOAA, 1305 East–West Highway, Rm. 10414, Silver Spring, MD 20910, U.S.A.

Protection and conservation of ocean and coastal resources: coastal management, marine protected areas

DONALD F. BOESCH Center for Environmental Science, University of Maryland, P.O. Box 775, Cambridge, MD 21613, U.S.A.

Integration and application of scientific knowledge in ecosystem restoration and management and sustainable resource management

PETER R. BURBRIDGE UNESCO Visiting Professor in Coastal Management at Kiel University and GEOMAR Kiel, Germany, and Dept. of Marine Sciences and Coastal Management, University of Newcastle upon Tyne, Ridley Building, Newcastle upon Tyne NE1 7RU, U.K.

Integrated coastal zone management: policy, enabling mechanisms, concepts, principles, elements of good practice

FRANCISCUS COLIJN Forschungs- und, Technologiezentrum Westküste, Hafentörn, 25761 Büsum, Germany

Phytoplankton ecology; innovation of measuring equipment re phytoplankton; ecological aspects of coastal zone management; coastal ecology

CHRISTOPHER J. CROSSLAND LOICZ International Project Office, Netherlands Institute for Sea Research (NIOZ), P.O. Box 59, 1790 AB Den Burg, Texel, The Netherlands

Processes in coastal ecosystems and nutrient fluxes; coastal zone management; global change in coastal zone — natural and anthropogenic forcing (esp. tourism)

LINDA A. DEEGAN The Ecosystems Center, Marine Biological Laboratory, 7 MBL Street, Woods Hole, MA 02543, U.S.A.

Effects of coastal change on secondary production

ANGELA DE LUCA REBELLO WAGENER Depto. de Quimica, Pontificia Universidade Catolica do Rio de Janeiro, R. Marquès de S. Vicente, 225, 22453–900 Rio de Janeiro, Brazil

Marine and environmental chemistry: cycling of bioassociated elements in the aquatic environment; development of monitoring methodologies; geochronology of organic pollution in tropical estuaries; use of biomarkers in studies of the geochronological variation of carbon storage

IES DE VRIES National Institute for Coastal and Marine Management, RIKZ, P.O. Box 20907, 2500 EX The Hague, The Netherlands

Marine eutrophication and marine ecosystem integrity in relation to land-based sources of nutrients from related drainage basins; analysis and design of land-use dynamics in relation to the integrity of the underlying geo-ecological system

JOB DRONKERS Institute for Coastal and Marine Management, RIKZ, P.O. Box 20907, 2500 EX The Hague, The Netherlands

Coastal physics; coastal management

RAGNAR ELMGREN Dept. of Systems Ecology, University of Stockholm, 10691 Stockholm, Sweden,

Eutrophication of the Baltic Sea: integrated coastal management issues

KAY-CHRISTIAN EMEIS Institut für Ostseeforschung Warnemünde (IOW), an der Universität Rostock, Sektion Marine Geologie, Seestrasse 15, 18119 Warnemünde, Germany

Marine geology, geochemistry

PETER FRYKBLOM Dept. of Economics, Swedish University of Agricultural Sciences, Box 7013, 75007 Uppsala, Sweden

Economics of issues related to waters and coastal zone management

CHRISTIANE GÄTJE Landesamt für den Nationalpark Schleswig-Holsteinisches Wattenmeer, Schlossgarten 1, 25832 Tönning, Germany

Ecosystem research and socioeconomic monitoring in the Schleswig-Holstein Wadden Sea National Park within the framework of the Trilateral Monitoring and Assessment Program (TMAP)

ALISON J. GILBERT Instituut voor Milieuvraagstukken (IVM), Vrije Universiteit, De Boelelaan 1115, 1081 HV Amsterdam, The Netherlands

Ecology economic analysis for the management of (semi-) aquatic ecosystems

TIMOTHY D. JICKELLS School of Environmental Sciences, University of East Anglia, Norwich NR4 7TJ, U.K.

Measurement of nutrient fluxes through estuaries and coastal waters and studies of the factors regulating them; metal fluxes in the past; atmospheric inputs of nutrients and metals; marine environments

BASTIAAN A. KNOPPERS Depto. de Geoquímica, Universidade Federal Fluminense, Outeiro de São João Batista s/n, 24020–007 Niterói, RJ, Brazil.

Contamination of tropical coastal environments, land–sea interaction, marine biogeochemistry

VIRGINIA LEE Coastal Resources Center, Graduate School of Oceanography, University of Rhode Island, Narragansett, RI 02882–1197, U.S.A

Coastal management and policy for U.S.A.

OLASUMBO MARTINS GKSS Research Centre, Max-Planck-Straße, 21502 Geesthacht, Germany

River basin activities and pressures at the coastal zone of developing countries

JACQUELINE M. MCGLADE Dept. of Mathematics & CoMPLEX, University College of London, Gower St., London WC1E 6BT, U.K.

Developing a new concept for the governance of natural resources; developing software for integrated coastal management

LAURENCE D. MEE Plymouth Environment Research Centre, Environmental Policy Unit, University of Plymouth, Drake Circus, Plymouth PL4 8AA, U.K.

Cross disciplinary environmental assessment and policy development in transboundary international waters; current focus on eutrophication in the Black Sea and on new generation agrochemicals in the Tropics

MICHEL MEYBECK SISYPHE, UPMC/CNRS, Université de Paris 6, 4, place Jussieu, 75232 Paris Cedex 05, France

River basin water quality issues at local (Seine River) and global scale; personal studies on nutrients and carbon, and on heavy metals

MAGNUS A.K. NGOILE National Environment Management Council (NEMC), P.O. Box 63154, Dar es Salaam, Tanzania

Development of integrated coastal management, linking science to policy and management, in Eastern Africa; planning and implementation of management programs; community-based ICM and marine protected areas; general environmental management issues

STEPHEN B. OLSEN Coastal Resources Center, Graduate School of Oceanography, University of Rhode Island, Narragansett, RI 02882–1197, U.S.A.

Formulation of a common methodology for learning from the practice of coastal management

MICHAEL J. O'TOOLE Kinvara, Co. Galway, Ireland

Coordination of Benguela Current Large Marine Ecosystem Programme (BCLME); biological oceanography

JOZEF M. PACYNA Center for Ecological Economics, Norwegian Institute for Air Research, P.O. Box 100, 2007 Kjeller, Norway

Socioeconomic aspects of fluxes of chemicals into the aquatic ecosystems; cost-benefit analysis of strategies for pollution abatement

EDMUND PENNING-ROWSELL Flood Hazard Research Centre, Middlesex University, Queensway, Enfield EN3 4SF, U.K.

Economics of coastal defence

SUNDARARAJAN RAMACHANDRAN Institute for Ocean Management, Anna University, Chennai 600 025, India

Integrated coastal and marine area management; remote sensing and GIS applications

KARSTEN REISE Wattenmeerstation Sylt, Alfred-Wegener-Institut für Polar- und Meeresforschung, Hafenstrasse 43, 25992 List/Sylt, Germany

Coastal ecology; Wadden Sea ecology; community interactions and long-term changes

CLAUDIO RICHTER Zentrum für Marine Tropenökologie, Fahrenheitstraße 1, 28359 Bremen, Germany

Research management in Red Sea area; coral reef ecology, pelagic–benthic coupling in coral reefs

WIM SALOMONS GKSS Research Centre, Max-Planck-Strasse, 21502 Geesthacht, Germany

Coastal zone management

RAFAEL SARDÁ CEAB-CSIC, Centre d'Estudis Avançats de Blanes (CSIC), Camí de Santa Barbara, s/n, 17300 Blanes, Girona, Spain

Marine biology, coastal zone management

KLAUS SCHWARZER Institut für Geowissenschaften, der Universität Kiel, Olshausenstrasse 40, 24118 Kiel, Germany

Coastal evolution; sediment transport; nearshore morphodynamics; coastal defense

LIANA TALAUE-MCMANUS Marine Science Institute, College of Science, University of the Philippines, Diliman, Quezon City 1101, Philippines and Division of Marine Affairs, Rosenstiel School of Marine and Atmospheric Science, University of Miami, 4600 Rickenbacker Causeway, Miami, FL 33149 U.S.A.

Community-based coastal management and local governance; integrated economic and biogeochemical modeling; planktonic production along gradients of human impact

JANE TAUSSIK Dept. of Land and Construction Management University of Portsmouth, Portland Building, Portland Street, Portsmouth PO1 3AH, U.K. Mailing address: 4 Langstone Ave., Havant, Hants PO9 1RU, U.K.

The contribution of town and country planning to coastal zone management

BODO VON BODUNGEN Institut für Ostseeforschung Warnemünde (IOW), an der Universität Rostock, Postfach 301161, Sektion Marine Geologie, Seestrasse 15, 18119 Rostock, Germany

Biological oceanography; marine biogeochemistry

MAREN VOSS Institut für Ostseeforschung Warnemünde (IOW) an der Universität Rostock, Postfach 30 11 61, Sektion Biologische Meereskunde, Seestrasse 15, 18119 Rostock, Germany

Nitrogen cycling in the coastal oceans and Baltic Sea; impacts of riverine eutrophication (by means of stable isotope measurements)

MERRILYN WASSON Ecosystems Dynamics Group, RSBS, Institute of Advanced Studies, Australian National University, Canberra, A.C.T. 0200, Australia

Institutions analysis: conflict/integration of international, national, and local institutions which impact on coastal ecosystems (Asia–Pacific); institutional conflict and integration between Southeast Asian rainforest users and coastal communities

FREDRIC V. WULFF Dept. of Systems Ecology, University of Stockholm, 10691 Stockholm, Sweden

Marine systems ecology, land–ocean interations in the coastal zone — global climate research program of IGBP; models of biogeochemical cycles and processes im marine systems; Baltic Sea integrated research

1

Science and Integrated Coastal Management

An Introduction

B. VON BODUNGEN[1] and R.K. TURNER[2]

[1]Institut für Ostseeforschung Warnemünde (IOW), an der Universität Rostock, Postfach 30 11 61, 18112 Rostock, Germany
[2]CSERGE, School of Environmental Sciences, University of East Anglia, Norwich NR4 7TJ, U.K.

THE PROBLEM

Today more than 60% of the world's population lives within a narrow strip of land about 100 km wide along the world's seashore and much more in the entire drainage basins of the coastal seas. Most of the megacities are located near the sea, and urbanization in the coastal zone, and thus population density, is expected to increase in the future. It appears that coastal development together with ongoing protection measures have grown out of control, and the consequent degradation or destruction of the coastal environment continues to increase. It has been estimated, for example, that within the next sixty years, erosion could destroy up to 85,000 houses (not including new development) along the 10,000 miles of U.S. ocean and Great Lake shorelines. The estimated economic cost of this property loss alone is around US$ 410 million per year (Dunn et al. 2000). The problem that we face, therefore, is how can we regain control and mitigate resource degradation to conserve environmental systems and the socioeconomic activity that depends upon them.

BACKGROUND

Three particular characteristics of coastal zones deserve attention: the extreme variability present in coastal systems, the highly diverse nature of such systems, and their valuable

Science and Integrated Coastal Management
Edited by B. von Bodungen and R.K. Turner

multifunctionality. The highly diverse and variable environment of the coastal zones is shaped by various processes in the land–ocean interface, such as waves and currents, sediment transport, chemical and biological modifications, and their interactions with the coastal structure. These interactions result in a dynamic equilibrium of the morphology and ecology in the coastal zones and encompass a wide range of spatial and temporal scales. Naturally fertilized by land runoff and atmospheric input, coastal oceans are the most productive realms in the marine biome, yielding high biomass in a large variety of plants and animals.

The rich diversity in "phenotypes" of coastal ocean ecosystems provides a great number of resources for human exploitation, among which food provision and permanent settlement date back to the very early phases of human development. As human societies have increasingly populated the coastal areas, exploitation of resources has become more intense and diverse. In addition to fishing, mining, trade, coastal engineering, and recreation, pressures in the coastal zone which impact coastal seas are linked to sewage and waste disposal due to urbanization, agriculture, and industrialization inland in the drainage basin and are significant. Many effects in the latter categories are in principle reversible within human lifetimes, while overfishing, mining, changes in land use (i.e., wetlands to arable land), and coastal engineering may only be remediated on a long term basis and at a very high cost. Coastal areas are therefore important economic zones supporting billions of livelihoods through flows of income derived from the utilization of the *in situ* natural capital stock and through global trading links. Simultaneously, coastal areas are sociocultural entities, with specific historical conditions and symbolic significances, as well as institutional domains with administrative boundaries that can cross national jurisdictions and which are not coincident with the scales and susceptibility of biogeochemical and physical processes.

Forecasts of economic and demographic growth and development predict dramatic increases in the habitation of coastal areas and in the use of the land–sea interface. These forecasts invariably imply considerable strain on natural resources in the coastal zone, both in the terrestrial and the adjoining marine realm. The added strain may coincide with anthropogenic and natural changes in climate and sea level. In a relatively short time, many coastal zone environments could lose their natural appearance, and their carrying capacity for human exploitation would thus diminish progressively. Intervention in this coevolutionary, jointly determined, ecological and socioeconomic system will need to be carefully undertaken and will require flexible and adaptive projects, policies, programs, and institutions. Management agencies will need to find better ways to manage the causes and consequences of the environmental change process across a range of coastal situations and the connected drainage basins. Given the generic policy goal of sustainable development, management agencies seeking to utilize coastal ecosystems sustainably should be giving a high priority to the maintenance of systems' resilience, i.e., their ability to cope with stress and shock. Such an integrated coastal strategy management, in turn, needs to be based on as good an appreciation of the systems functioning and outputs of economic and sociocultural goods and services, as is feasible.

A key objective is to retain as much coastal functional diversity as is practicable. The management strategy will require the adoption of a relatively wide perspective, in order to understand and potentially manage larger-scale (landscape) ecological processes and relevant environmental and socioeconomic driving forces more effectively. Properties of structures and processes, both natural and socioeconomic, must become subsumed with the change in both spatial and temporal scales. Climate change (from Milankovich cycles to centennial scales), and the accumulation of greenhouse gases (at anthropogenic scales) impact the nature of coastal waters, their drainage basins and the way people make their living in these changing systems. Social and economic parameters also change with the process of market globalization (< decadal scales). This has national macroeconomic consequences, which together with the actions of transnational corporations and institutions will further impinge on use of resources and services from the coastal systems (both littoral and inland).

THE SOLUTION: INTEGRATED COASTAL MANAGEMENT

Most managed ecosystems are complex and their hierarchically organized nature is poorly understood. Coastal and related drainage basin systems and processes pose a particularly complex challenge because of the spatial scales and the degree of complexity and variability in the systems that are involved. In an effort to exert some "control", i.e., reduce risk and overcome uncertainty in the coastal environment, persistent human intervention has in many ways only resulted in a state of permanent disequilibrium or undesired new states of equilibrium. Such a set of environmental conditions, driven by human reclamation and continued protection of intertidal land for economic reasons (and more recently on nature conservation grounds), is arguably more risky to humans and not less so. In the light of these difficulties, capturing the range of relevant impacts on natural and human systems under different management options is and remains a formidable challenge. An interdisciplinary scientific effort is needed to develop methodologies for better understanding and detection of ecosystem change, as well as evaluation of different ecological functions. Modeling work, monitoring of robust indicators for change, and scientific experimentation all need to be integrated more effectively.

Given the current high level of uncertainty and ignorance, the manifold socioeconomic and cultural value, as well as the pressure on coastal systems by conflicting stakeholders interests, many analysts have been advocating a much more integrated and holistic approach to coastal management (Salomons et al. 1999). This steering mechanism should be underpinned by the following interrelated sustainability principles:

- economic and ecological efficiency and the cost-benefit principle (including the "polluter pays" principle), which addresses the practical need to find long-term, cost-effective resource allocation options within the ever-present problem of resource scarcity;
- equity and fairness principle, which encompasses a number of requirements such as the need for more "civic science" in which scientists actively participate in the communication and use of science in the political process as well as more inclusionary processes to engage all relevant coastal stakeholders together with the placement of power and responsibility for planning and decision making at the lowest feasible level of governance (subsidiarity principle);
- the precautionary principle, which gives appropriate recognition to the fact that coastal science and management is and will continue to be conditioned by data and knowledge gaps as well as decision-making systems able to operate in and to adapt to the context of this uncertainty.

As a future goal, integrated coastal management (ICM) is a continuous, adaptive, day-to-day process that consists of a set of tasks, typically carried out by several or many public and private entities (Bower and Turner 1998). The tasks together produce a mix of products, services and other gains/losses of sociocultural significance from the available coastal resources. In principle, the core objective of coastal zone management is the production of a "socially desirable" mix of coastal environmental system states, products, and services. In practice, this mix is subject to intense stakeholder debate and is likely to change over time with changing demands, changing knowledge, and changing pressures.

A future, more integrated, coastal management process should include:

- integration of programs and plans for economic development, environmental quality management, and ICM;
- integration of ICM with programs for such sectors as fisheries, energy, transportation, water resources management, disposal of wastes, tourism, and natural hazards management;
- integration of responsibilities for various tasks of ICM among the levels of government — local, state/provincial, regional, national, international — and between the public and private sectors;
- integration of all elements of management, from planning and design through implementation (i.e., construction and installation), operation, maintenance, and feedback from monitoring and evaluation overtime;
- integration among disciplines, e.g., ecology, geomorphology, marine biology, economics, engineering, political science, law; and
- integration of the management resources of the agencies and entities involved.

In summary, the ICM process must aim to unite government and the community, science and management, as well as sectoral and public interests. It should *inter alia* improve the quality of life of human communities who depend on coastal resources while maintaining the biological diversity and productivity of coastal ecosystems (GESAMP 1996: Figure 1.1), and therefore the functioning of nature. Clearly, this is a formidable task, one that will only be achieved incrementally over time.

The formulation and implementation of ICM will depend on advances in transdisciplinary research and knowledge. The ultimate goal of integrated coastal zone management is to produce a set of products and services that allows the maximum net benefits for the society over as long a period of time as possible. A consensus on what constitutes a net societal benefit and the values that are implied is most likely to emerge from changes in the interests and priorities of society, as interpreted by political institutions and reflected in legislation, policy statements, principles, rules, regulations, and last but not least from advancing knowledge about ecosystem functioning and behavior. Besides regional and national economic and political activities, cross-border impacts, trade policies and international conventions also have to be considered in this context.

A prerequisite for institutional advances is a more rigorous use of current knowledge and an increase in scientific knowledge about coastal problems and their nature as well as methodologies that will help to define, analyze, and mitigate problems. For science, a novel and challenging area is emerging, where the need is clear but the appropriate approach is less clear. This Dahlem Workshop provided a platform for assembling the required expertise in environmental, cultural, social, and economic science to discuss and define a transdisciplinary approach and new scientific products in the framework of integrated coastal zone management.

GOALS AND RATIONALE FOR THE DAHLEM WORKSHOP

The overall goal of our Dahlem Workshop was to strengthen the effective use and communication of scientific knowledge in the integrated coastal management policy process. To meet this objective, it was perceived that there was a need to analyze the scientific requirements of ICM and to develop a strategy for a transdisciplinary approach to identify the problems and their nature, to find solutions, and to formulate products that could be used as guidelines for valuation, assessment, and policy making. A further step is to transfer such products by more intensified communication. "Science cannot provide the solutions, but it can help understand the consequences of different choices" (Lubchenco 1998). Thus, transfer of relevant scientific knowledge must be understandable to nonscientists involved in ICM.

The analysis of the scientific needs had to take into account the fact that present knowledge about the natural dynamics of the coastal environment and socioeconomic processes is limited, as is the knowledge about the interaction between these processes. Together with the natural dynamic equilibrium between coastal ecological systems and land–ocean processes, the balance between socioeconomic and natural processes is the new comprehensive scientific context for ICM. Thus, ICM cannot just be based on stringent scientific predictions but will have to cope with both social and scientific uncertainties. Furthermore, traditional values and perceptions existing on local to regional scales serve more often as a basis stakeholder decisions than do facts and figures. It was realized that there was an urgent need for definitions of the new role of "civic science" and identification of important scientific tasks to be conducted in ICM, accepting that a certain degree of uncertainty will always be present. A further need of equal importance is to decide whether uncertainty should be communicated to all interested parties and by what means this can be done.

The combined effects of socioeconomic and environmental changes clearly require an overall framework to investigate the interaction between environmental, social, economic, and institutional subsystems as well as to identify crucial processes and interactions. Difficulties inhibiting the formulation and implementation of an acceptable framework and its deployment are manifold: the diversity in phenotypes of coastal ecosystems and thus in their functional value, regional, and national differences in socioeconomic developmental stages, the pace of development and cultural constraints on social and environmental attitudes. Another major analytical challenge to overcome is the different temporal and spatial scales on which these subsystems react and interact, and thus the different required scales of prediction. Considerable changes in socioeconomic development can emerge within relative small spatial scales at the shoreline, for example, whereas the reaction of the marine system is influenced by changes and systems operating much larger scales. Not every coastal problem will require a fully integrated approach. However, generic strategies focused around a core of sound interdisciplinary science must be the basis for any future adaptive and inclusionary coastal management strategy.

A FRAMEWORK FOR MULTIDISCIPLINARY RESEARCH

Some analysts have questioned the entire rationale of ICM. According to Nichols (1999) ICM is actually an attempt to reorganize coastal spaces and political systems for the purpose of facilitating investment penetration by governments and/or transnational corporations. The consequence (particularly in developing countries) is the political and spatial marginalization of pre-existing resources users. To address this equity issue and others, ICM has to be more than just a process by which efficient utilization of coastal resources is promoted. Olsen et al. (1997) have strongly argued that the fundamental challenge of coastal management is one of governance (objective, process, and structures) and not of technology transfer or refined scientific knowledge alone. They recommend a learning-based approach to coastal management, which assumes that such intervention is a young endeavor inevitably beset by uncertainties, instability, and rapid rates of change. It follows that progress towards effective coastal management and sustainable forms of coastal development will only come incrementally, through analysis and experience learning over decades. A learning-based approach calls for framing coastal management initiatives as experiments and subjecting them to formal scientific testing analysis.

According to GESAMP (1996) there are five consecutive stages forming an ongoing, interactive ICM process. The process itself may go through a number of cycles before the program is sufficiently refined to produce effective results (Figure 1.1). Thus in Stage 1 of the GESAMP cycle, natural and social scientists together need to compare problem issues in the light of their different methodologies, models, and value systems (the science challenge for ICM). A consensus on a common set of pressures/problems issues needs to be established. Any gaps in scientific knowledge, their likely consequences for ICM, and the practicable possibilities for their mitigation within an acceptable time frame also need to be addressed.

The GESAMP cycle offers an excellent tool to evaluate the contribution of science as well as the most urgent needs for society that should be addressed by analysts according to their importance in the ICM policy cycle. In order to grasp the many issues, problems, and disagreements surrounding the scientific analysis, valuation, and management of coastal and related drainage basin resources, a more detailed, practical framework may be adopted. This is the organizational and auditing Drivers-Pressures-State-Impact-Response (D-P-S-I-R) approach, which although simple is flexible enough to be conceptually valid across a range of spatial scales. It also serves to highlight the dynamic characteristics of ecosystem and socioeconomic systems changes, involving multiple feedbacks with a coevolutionary process of change (Figure 1.2).

For any given coastal area (defined to encompass the relevant drainage area), there exists a spatial distribution of socioeconomic activities and related land uses: urban, industrial mining, agriculture, forestry, aquaculture, fisheries, commercial, and transportation. This spatial distribution of human activities reflects the final demand for a variety of goods and services within the defined area and from outside the area. Environmental pressure builds up via these socioeconomic driving forces and is augmented by natural systems variability, which leads to changes in environmental systems states and finally to the loss of goods and services.

Production and consumption activities result in different types and quantities of residuals as well as goods and services measured in gross national product (GNP) terms. Thus the concern might be, for example, the role and extent of changes in C, N, P, and sediment fluxes as a

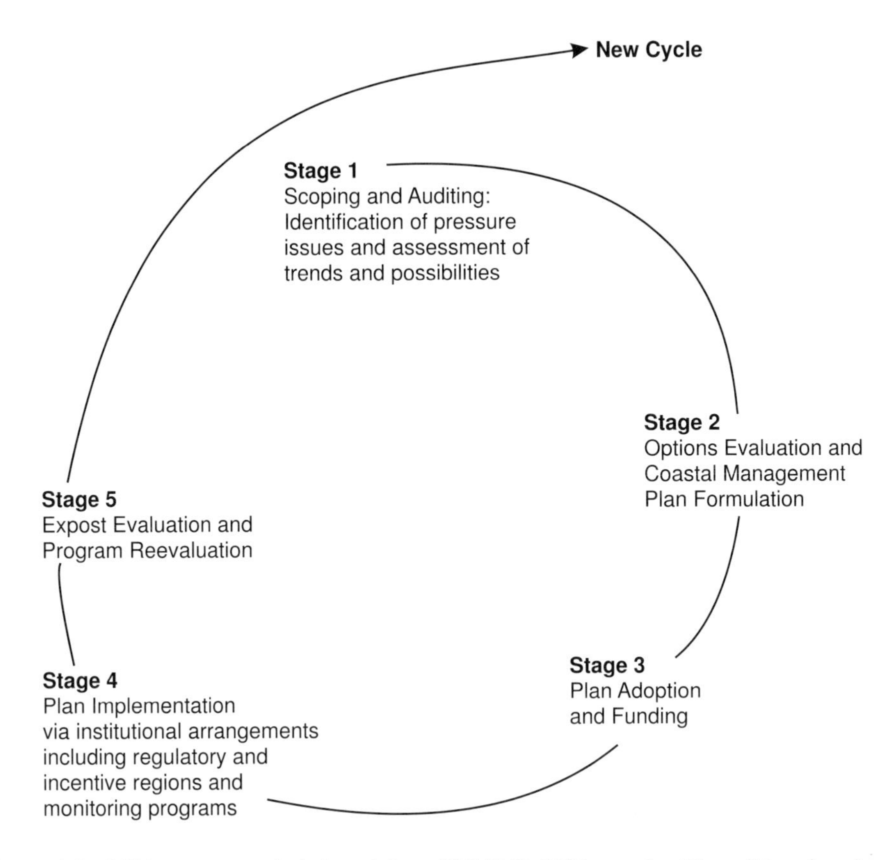

Figure 1.1 ICM program cycle (adapted from GESAMP 1996; see also Olsen, this volume).

result of land-use change and other activities. Conceptually what we have are a multiplicity of input–output (I–O) relationships, where outputs are joint products (combinations of goods and services and nonproduct outputs or residuals, which if not recycled become waste emitted/discharged into the ambient environment). I–O relationships will operate at the individual industrial process/plant level, through population settlements, agricultural cropping regimes/practices, and up to regional drainage basin scale. These residual estimates will then serve as the input to the natural science models, such as nutrient budgets. Environmental processes will transform the time and spatial pattern of the discharged/emitted residuals into a consequent short-run and long-run time and spatial ambient environmental quality patterns.

These state environmental changes impact on human and nonhuman receptors, resulting in a number of perceived social welfare changes (benefits and costs). Such welfare changes provide the stimulus for management action, which depends on the institutional structure, culture/value system, and competing demands for scarce resources and for other goods and services in the coastal zone. Within its analytical framework, an integrated (modeling) approach will need to encompass the socioeconomic, biogeochemical, and physical drivers that generate the spatially distributed economic activities and related ambient environmental quality to provide information on future environmental states.

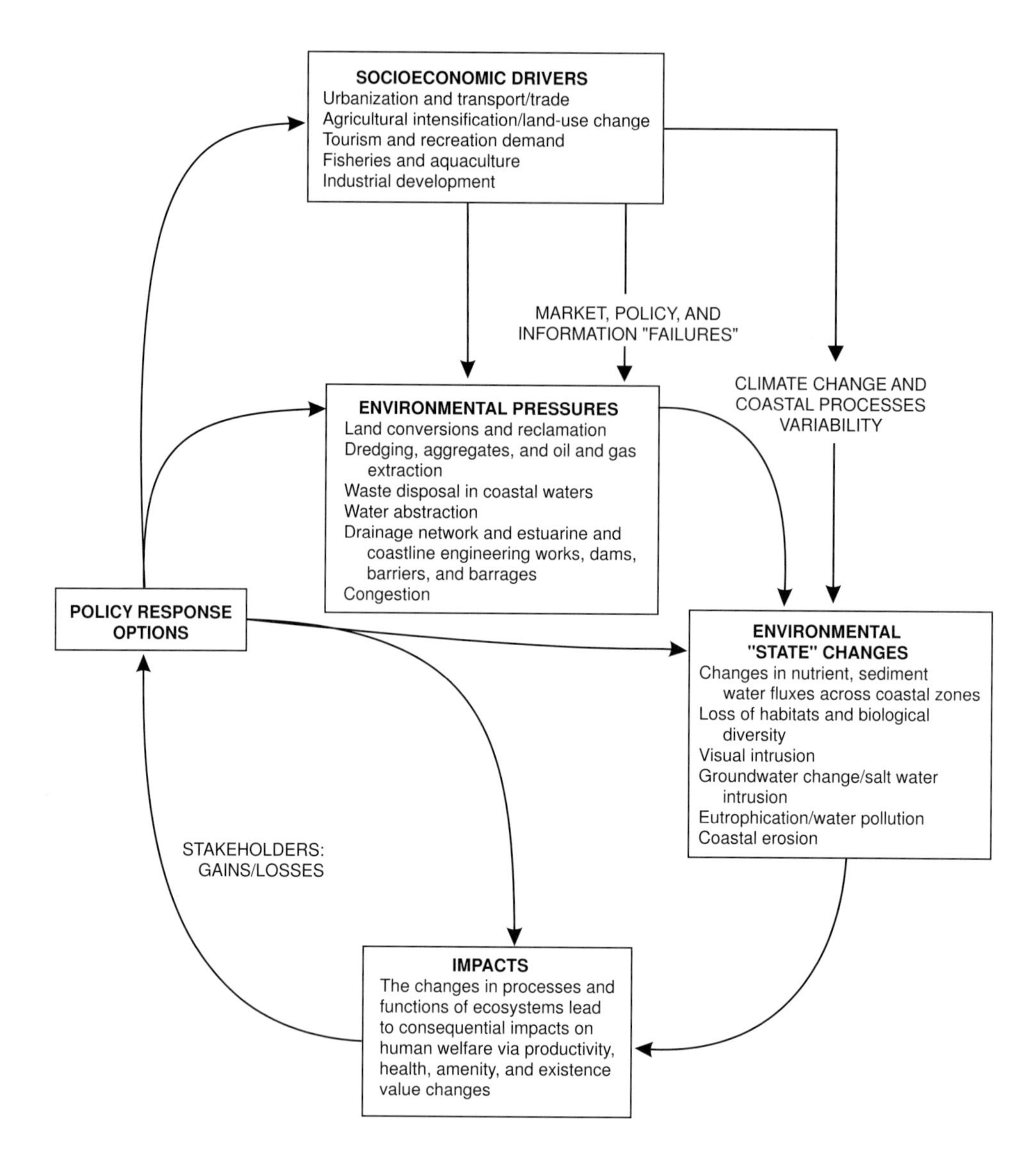

Figure 1.2 D-P-S-I-R framework: Continuous feedback process in coastal areas (Turner, Lorenzoni et al. 1998).

At the core of this interdisciplinary analytical framework is a conceptual model, based on the concept of functional diversity, which links ecosystem processes, composition, and functions with outputs of goods and services, and can ultimately be assigned monetary economic and/or other values (Figure 1.3). A management strategy based on the sustainable utilization of coastal resources principle should have at its core the objective of ecosystem integrity maintenance, i.e., the maintenance of system components, interactions among them and the

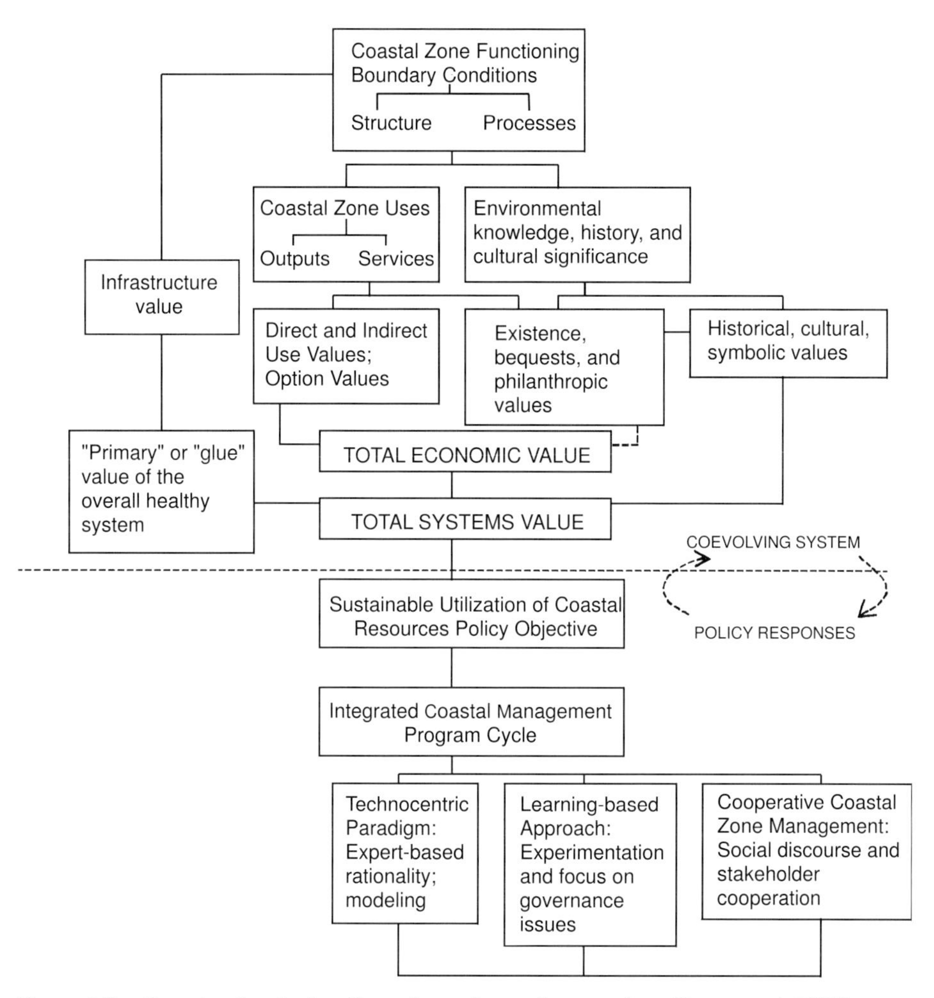

Figure 1.3. Functional and other dimensions of coastal zone values (Turner et al. 2000).

resultant dynamic behavior of the system. Functional diversity can then be defined and possible changes in functional diversity evaluated by the variety of responses to environmental change, in particular the variety of spatial and temporal scales with which organisms react to each other and to the environment. Marine and terrestrial ecosystems differ significantly in their functional responses to environmental change, and this will have practical implications for management strategies. Thus, although marine systems may be much more sensitive to changes in their environment, they may also be much more resilient (i.e., more adaptable in terms of recovery response to stress and shock). The functional diversity concept encourages analysts to take a wider perspective and examine changes in large-scale ecological processes, together with the relevant environmental and socioeconomic driving forces. The focus is then on the ability of interdependent ecological–economic systems to maintain functionality under a range of stress and shock conditions.

In Stage 2 of the GESAMP cycle (Figure 1.1), the characteristics and conditions of coastal systems that cause concern or otherwise warrant attention need to be analyzed. The scale of any habitat destruction needs to be determined together with the supporting natural processes, their linkages to habitats, and their recovery times. At this stage, the concepts of ecological integrity and functional diversity need to be operationalized. Given the overall policy objective of sustainable utilization of coastal resources, the pressure to state and stage changes to human welfare impacts in the D-P-S-I-R approach need to quantified and evaluated as comprehensively as is practical (Figure 1.3). The particularly difficult questions of the mismatch of spatial and temporal scales need to be tackled in a pragmatic but rigorous fashion, as the analysis moves from natural systems dynamics to the socioeconomic and politico-cultural realms. Keystone processes and functions, when and if they are identified, may need to be subject to a "no net loss" regulation, the cost-benefit implications of which will require examination. Multi-criteria decision support systems will be required to tackle the multiple use conflict situations that will almost inevitably arise. Futures scenario analyses based on growth rate projections and/or ecosystem management strategies can play an important role in this stage.

In Stages 3 and 4 of the cycle, the formal adoption of an ICM plan will require a reaffirmation of the cost-benefit and decision analysis work by new institutional arrangements. Monitoring of the rate and extent of change in the coastal area will be essential, as will enforcement systems (the governance challenge of ICM).

In Stage 5, natural and social scientists should evaluate the relevance, reliability, and cost-effectiveness of scientific information generated by research and monitoring and advise on the suitability of control data. These results should then be compared to a scenario without ICM (GESAMP 1996; Bower and Turner 1998).

Progress through the ICM policy cycle will also be conditioned by the degree to which "accountability" and "trust" issues are successfully tackled. No process of ICM can produce legitimate answers (and effective solutions) to the challenges posed without meaningful public participation. The public must be incorporated in a proactive, participatory, and conflict minimizing fashion. Davos (1998) believes that if ICM is crucially dependent on the voluntary cooperation of stakeholders, this raises doubts about the value of positivistic or normative ICM prescriptions in the absence of consensus. He argues that the alternative is to pursue a "cooperative coastal zone management" approach, which would rely on social discourse as its defining property. Such discourse also needs a guiding framework to facilitate the achievement of cooperative collective decisions. There is a need to establish "windows of opportunity" where policy, politics, and participants can operate together to set the sustainable resource utilization agenda and to implement it effectively (Davos 1998).

WORKING GROUP STRUCTURE AND DISCUSSION FOCI

Drawing on the GESAMP–ICM policy cycle and the D-P-S-I-R framework, background papers and discussions in the four working groups addressed the following themes/topics at this workshop:

- transboundary issues,
- shoreline development,

- integrated coastal management in developing countries,
- unifying concepts for ICM.

By addressing a variety of environmental pressures (such as land-use change in the drainage basin, habitat degradation, conflicting developments of shorelines, the role of climate change, and the persistence of resource overexploitation), the discussion groups analyzed the scientific requirements of ICM. They also sought to develop a transdisciplinary approach to identify problems, solutions, and formulate products for use in valuations, assessment, and policy meetings.

The background papers provide information about the stages of ICM in a variety of case studies. In more general terms, they discuss how to improve the use of science in the various stages of the ICM process.

In background papers and group reports, the following issues were addressed:

- methods for communicating the state of the coastal environment to the public and decision makers, including the issues of responsibility, persuasiveness, and trust (see, e.g., Boesch, de Vries, and McGlade, all this volume),
- illustration of the potential magnitude and socioeconomic consequences of anthropogenic-induced change in coastal systems (e.g., Elmgren and Larsson, Boesch, Mee, Colijn and Reise, Sarda, Pethick, Meybeck, and Deegan et al., all this volume),
- estimation of the consequences and costs of less integrated approaches compared to the possible benefits of implementing an ICM policy cycle (e.g., Crossland and Kenchington, and O'Toole et al., both this volume),
- forecasting ecosystem changes and their long-term consequences (e.g., de Vries, Meybeck, and Olsen, all this volume),
- gaining a better scientific understanding of feedbacks for the adaptive approach to management (Boesch, Olsen, Deegan et al., all this volume)
- monitoring requirements for the assessment of management results across different spatial scales and time spans.

SOME GENERAL IMPRESSIONS FROM THE DISCUSSIONS

Detailed summaries and outcomes from the four discussion groups are presented in the group reports (Jickells et al.; Schwarzer et al.; Richter et al.; Emeis et al., all this volume). Here, we present some general impressions from the workshop, which in actuality may be indicative of the ICM process itself.

As a whole, workshop participants represented a wide range of high quality expertise in the field of natural and socioeconomic sciences as well as in policy making. Initial discussions were characterized, however, by a high level of generality. This was probably due to the diversity of the different perceptions of ICM. Some participants considered ICM as an organizational endeavor, in which the core objective is to organize stakeholders and define rules for good management practices. In this sense, ICM is regarded as process management that needs to be made more inclusionary in order to deal with competing interests and power relationships. Another group of participants looked at ICM from a technocratic point of view. Here, the aim was to optimize resource exploitation with concomitant conservation of the environment. Solutions for decoupling economy from ecology appeared to be paramount from this perspective.

Such diversity in basic approach led to different opinions on how to achieve the conference objectives – improved effectiveness and communication of science in ICM. The former group put much more emphasis on the development of appropriate communication, whereas the latter focused more on improving the effectiveness of science-based solutions in ICM. This difference in foci probably reflects the existing ambiguity about the role of science in ICM, which often creates a division between coastal managers and coastal scientists. Such polarization was also reflected in the different perception of what was the limiting factor to more efficient progress in ICM: the low capacity to practice coastal governance at local, national, and regional levels or the insufficient knowledge about the functioning of the natural environment. It appeared that most participants came with a preconceived notion that the latter was the reason for slow progress. Only a few background papers defined the coastal ecosystem as comprising both the "natural environment" and the human societies living therein (Ngoile et al. and Olsen, both this volume). Nevertheless, by the end of the workshop, there was much more agreement that this definition was most appropriate and that the lack in governance capacity may be as, or more, serious for ICM than the gaps in scientific knowledge (Emeis et al., this volume).

This conclusion was further underlined by the recognition that in reality important stakeholders are often disempowered. Simple participation itself does not necessarily resolve this problem as power relationships can be complex and often informal alliances control how decisions are made (Lacerda et al.; Talaue-McManus; Ngoile et al., Richter et al., this volume).

Thus, a great deal of discussion time was devoted to attempts to reduce the difference between the various mind-sets, or at least to make a strong case that a considerable change in the more narrow way of thinking (mind-sets) was overdue. Further points of emphasis in the debate highlighted the issues of uncertainty, the experimental character of the ICM policy cycle, and the lack of incentives for scientists to join the policy cycle in all phases (see also Healey and Hennessy 1994; Emeis et al., Jickells et al.; Richter et al., and Schwarzer et al., all this volume).

The discussions about uncertainty and whether and/or how to convey it were illuminated through an assessment of different categories of uncertainty. McGlade (this volume) suggests two fundamentally different kinds of uncertainty: measurable uncertainty and ignorance. Measurable uncertainty is derived from "error bars," imprecision, and averaging over space and time, while ignorance is based on gaps in knowledge, i.e., things we do not know about or understand. Further uncertainty stems from complex model predictions (Boesch and Deegan et al., both this volume). In models, uncertainty stems from errors in measurements and understanding. Ignorance and errors may be partly reduced by the acceptance and continuous evaluation of differing opinions and by increasing appreciation of indigenous/traditional ecological knowledge (see also Berkes et al. 2000).

The ambitious but key requirement for a future, more sustainable coastal zone management is that both scientists and managers need to be involved in a continuous interactive process, such that scientists gain a better appreciation of policy formulation and implementation while managers/users better understand the functioning and variability of natural systems and the consequences of socioeconomic activities. Fisheries policies in most countries have reflected the schism between science and users. It has not been straightforward for scientists to relate their science to the various stages in the policy cycle, and for policy makers to recognize what science is needed or what scientific results have to be incorporated at which time in the policy cycle. However, if science and management proceed as largely unrelated endeavors, science and scientists may be directly involved in some stages of ICM but excluded from others. Traditionally, science has had most influence in the "issue identification/assessment" and "program preparation" stages, whereas its participation has been weak to nonexistent in later stages. No real-world examples of complete integration throughout the whole policy cycle came to light during the discussions. The cases of the management of Chesapeake Bay and the Baltic Sea come closest to this situation, but in the majority of cases, the policy cycle has not been moving beyond its early stages.

Many scientists, particularly natural scientists, are hesitant to communicate uncertainty to the public or to policy makers, as they fear that their advice will not be taken seriously. However, conference participants clearly recognized that the problem of uncertainty should not be exclusively used to ask for new science, rather that scientists should give advice based on what is known, even when this may be minimal. From the discussions it also emerged that efforts to constrain the communication of uncertainty are not justified. A striking example is the recent measures belatedly taken by governments and accepted by the public to reduce the risk of BSE, which are based on little knowledge on the pathways of infection from cattle to humans and on little quantitative information of the risks taken by consuming beef.

Most participants agreed that ICM should adopt subsidiarity and precautionary principles and should choose adaptive management as the natural way to promote ICM. In this context, scientists may be more willing to participate in all stages of the public policy cycle, as they can expect that the scientific method, testing hypothesis, will be applied more frequently. Each ICM policy cycle would appear as a part of an ongoing experiment to achieve a sustainable use of the natural resources. Choices to be taken can be underpinned with risk assessments, and predictions can be presented in "what if" scenario models (Boesch, Deegan et al., and Emeis et al., all this volume).

A largely unresolved problem which inhibits the fuller engagement of scientists in ICM is related to measurement of scientific success by different standards: peer review sets the standard within the science community. The wider practical value of science is considerably undervalued, and there is a danger that striving for scientific excellence without reference to social benefit and vice versa may lead to an ineffective use of science in ICM. The review system should be broadened and institutions should set aside funds for awards for excellent fundamental science together with its appropriate application (Richter et al., this volume).

It was acknowledged that the discussions brought new insights into the problems of using and communicating science in ICM. The background papers present the failures and successes of ICM (in many case studies) while the group reports compile sets of recommendations designed to overcome these problems. It was also recognized that a mixed group, like the one assembled for this Dahlem Workshop, is needed to focus on more specific unresolved issues in ICM, such as indicators and indices of coastal change (Olsen, this volume), as well as estimations of and use of background values to meet the requirements of the various legal directives set by national and international bodies. Furthermore, in the future it would be useful to bring together periodically (5–10 years) a mixed group of experts to focus more narrowly on a particular societally important issue (e.g., eutrophication of coastal waters, land use, shoreline destruction) to evaluate what is and is not known, as well as the implications of that state of knowledge for society. The usefulness of such an approach has been shown impressively by the activities of the International Panel for Climate Change on global warming.

ACKNOWLEDGMENTS

We would like to thank the Deutsche Forschungsgemeinschaft, the Freie Universität Berlin, the organizers, and staff of Dahlem Konferenzen for the opportunity they gave participants to discuss transdisciplinary approaches aimed at meeting the great challenge of ICM. The facilities and ambience provided by Dahlem were exceptional and very much appreciated. We also wish to thank all participants for the lively discussions, valuable contributions, and the open minded and collegiate mood in which the conference was held.

REFERENCES

Berkes, F., J. Colding, and C. Folke. 2000. Rediscovery of traditional ecological knowledge as adaptive management. *Ecol. Appl.* **10**:1251–1262.

Bower, B., and R.K. Turner. 1998. Characterising and analysing benefits from integrated coastal management. *Ocean Coast. Manag.* **38**:41–66.

Davos, C.A. 1998. Sustaining co-operation for coastal sustainability. *J. Env. Manag.* **52**:379–387.

Dunn, S., R. Friedman, and S. Baish. 2000. Coastal erosion: Evaluating the risk. *Environment* **42**:36–45.

GESAMP. 1996. The contributions of science to integrated coastal management. Reports and Studies No. 61. Rome: FAO (UN Food and Agriculture Organisation).

Healy, M.C., and T.M. Hennessy. 1994. The utilization of scientific information in the management of estuarine ecosystems. *Ocean Coast. Manag.* **23**:167–191.

Lubchenco, J. 1998. Entering the century of the environment: A new social contract for science. *Science* **279**:491–497.

Nichols, K. 1999. Coming to terms with integrated coastal management: Problems of meaning and method in a new area of reserve regulation. *Prof. Geogr.* **51**:388–399.

Olsen, S., J. Tobey, and M. Kerr. 1997. A common framework for learning from ICM experience. *Ocean Coast. Manag.* **37**:155–174.

Salomons, W., et al., eds. 1999. Perspectives on Integrated Coastal Zone Management. Berlin: Springer.

Turner, R.K., I.J. Bateman, and W.N. Adger, eds. 2001. Economics of Coastal and Water Resources: Valuing Environmental Functions. Dordrecht: Kluwer, in press.

Turner, R.K., et al. 1998. Coastal Management for sustainable development: Analysing environmental and socio-economic changes on the UK coast. *Geogr. J.* **164**:269–281.

2

Eutrophication in the Baltic Sea Area

Integrated Coastal Management Issues

R. ELMGREN and U. LARSSON

Dept. of Systems Ecology, Stockholm University, 10691 Stockholm, Sweden

ABSTRACT

The development of eutrophication in the Baltic Sea area is discussed in relation to the management of nutrient discharges to the coast. Since the open sea itself is eutrophicated in much of the area, regional as well as local effects must be taken into account when managing nutrient discharges to the Baltic coastal zone. International efforts to reduce Baltic eutrophication are described. Market economy states have been quite successful in reducing phosphorus and in some cases also nitrogen from discharges of treated municipal sewage. So far, efforts to reduce the total nitrogen load substantially have failed because the difficulty of the task has been underestimated, particularly for agricultural inputs and atmospheric deposition. Renewed efforts are underway, but progress is likely to be slow.

The Himmerfjärden Bay system in Sweden is presented as a local example of adaptive management. Its nitrogen load has been dominated by the discharge of treated municipal sewage, which has recently been reduced from 900 to about 100 tons of nitrogen per year. Getting local stakeholders in Sweden involved in nutrient management has been difficult, since emission standards are now decided in Brussels — far from local control — whereas earlier they were set by the Swedish Environmental Protection Agency in Stockholm.The future EU Water Framework Directive offers an opportunity to create a drainage basin-based management system, with stakeholder involvement and cost optimization through bubble policies. Finally, areas are highlighted in which further progress in cross-disciplinary studies and in the natural sciences are needed to support such future management initiatives.

INTRODUCTION

In the following, we present an outline of the Baltic eutrophication problem and of the several decades of policy efforts to counteract it. Our focus is on coastal management in Sweden, the

Science and Integrated Coastal Management
Edited by B. von Bodungen and R.K. Turner

country with the longest Baltic coastline, which we exemplify by the case of Himmerfjärden, a coastal bay receiving treated sewage from almost 250,000 persons. We also discuss developments in other coastal states. Our perspective is that of natural scientists; however, our work, lately within the "Sustainable Coastal Zone Management" (SUCOZOMA) research program of the Swedish Foundation for Strategic Environmental Research (MISTRA), has made us aware that on its own, natural science is powerless to protect coastal waters. This necessitates an effective management system, the design of which requires interdisciplinary input from the socioeconomic, legal, and the natural sciences (Gren et al. 2000).

THE BALTIC SEA

The Baltic Sea is one of the largest brackish water areas on Earth, some 377,000 km^2 (415,000 km^2 including the Danish Straits and the Kattegat), with a fourfold larger drainage area, extensive industrial activity, and a population of 85 million (Table 2.1). It consists of a series of basins joined by narrow sounds or by wide, shallow connections (Figure 2.1). A mean depth of about 56 m gives it a volume of 21,000 km^3. It extends from 54°N to near 66°N, almost to the polar circle. Its positive water balance (Table 2.2) and long residence time provide a rather stable surface salinity, increasing from 2–3 in its innermost reaches, to 5–6 in the southern Gulf of Bothnia, 6–8 in the Baltic Sea proper, and up to 15–25 in the Kattegat. The deep water is saltier, in the Baltic proper up to 13–20, and is generally separated from the surface water by a permanent halocline. The surface layer is mixed in the cold season, but in the summer a thermocline develops at a depth of 15–25 m. Since tides are negligible, wind mixing and the thermal balance govern the formation and breakdown of the thermocline. The

Table 2.1 Drainage basin of the Baltic Sea area: National areas and their respective populations (from Sweitzer et al. 1996).

Country	Area in basin, 1000 km^2	Population, millions
Belarus	90	4.0
Czech Republic	8	1.6
Denmark	33	4.5
Estonia	46	1.6
Finland	308	5.0
Germany	28	3.1
Latvia	65	2.7
Lithuania	66	3.7
Norway	14	<0.1
Poland	316	38.1
Russia	328	10.2
Slovakia	2	0.2
Sweden	426	8.5
Ukraine	14	1.8
Sum Total	1745	84.9

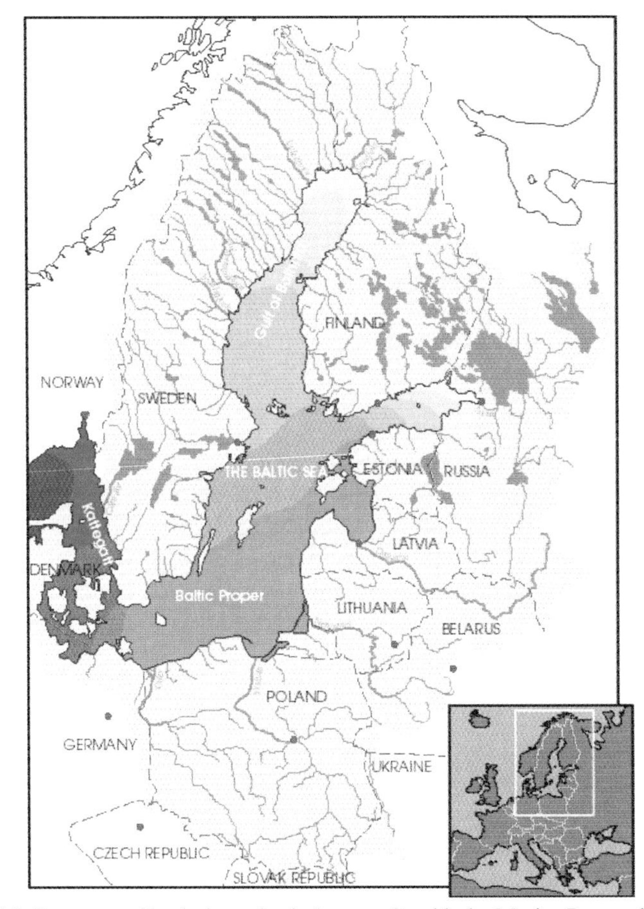

Figure 2.1 The Baltic Sea area and its drainage basin (source: Stockholm Marine Research Centre).

bottom waters of the Baltic proper are replaced intermittently by large inflows of denser water through the Danish Straits. The intervening stagnation periods are normally some 3–4 years (recorded maximum 16), during which oxygen is used up and the deep waters tend to become anoxic. It has recently been suggested that damming of Baltic rivers for hydropower may have made major inflows of high-salinity water less likely by changing the seasonal pulse of freshwater input (Schinke and Matthäus 1998).

Table 2.2 Baltic water balance (after sources in Ehlin 1981).

	$km^3\ a^{-1}$
River inflow	~440
Seawater inflow	~470
Precipitation minus evaporation	~50
Seawater outflow	~960

The Gulf of Bothnia receives a large input of freshwater, hence its low salinity. The natural poverty of nutrients in the freshwater input and the shorter productive season in the north (4–5 months vs. 8–9 in the south) lead to a marked, natural increase in all aspects of productivity from north to south in the Baltic Sea. The Kattegat, which borders the nutrient-rich North Sea, is even more productive than the southern Baltic proper (Richardson and Christoffersen 1992).

Only a small selection of marine and freshwater organisms, and a few brackish-water specialists, can live in the low salinity waters of the Baltic Sea, giving it a very species-poor assemblage of organisms (Elmgren and Hill 1997). The low biodiversity is further accentuated in the deep, nowadays extensively oxygen-deficient areas, where few macroscopic species survive (Cederwall and Elmgren 1990).

The Baltic Coastal Zone

Extensive rocky or moraine-dominated archipelagos are typical of the coasts of Sweden, Finland, and Estonia. The other coastal states tend to have straighter coastlines, with coastal cliffs or sandy beaches, and often with lagoons and bays into which major rivers empty. Archipelagos, lagoons, and bays have limited water exchange, which means that the local effect of pollutant emissions may be quite different from the regional effect in the open Baltic outside of these areas.

Himmerfjärden Bay

The Himmerfjärden Bay system (Figure 2.2) has an area of 232 km^2, a mean depth of 15 m, and is situated on the northwestern coast of the Baltic proper, at 59°N. Total freshwater input is 19 $m^3 s^{-1}$ and average yearly precipitation is ~460 mm, giving Himmerfjärden proper, the fjord-like central bay, a surface salinity that is ~0.5 lower than the 6–7 of the open Baltic proper (Elmgren and Larsson 1997). Water exchange is limited, particularly below the summer thermocline and under ice cover (Engqvist 1990). The drainage area is mainly forested, with a good deal of arable land.

BALTIC EUTROPHICATION

Discovery of the Problem

Eutrophication has been a problem near some of the larger Baltic cities since the early 20th century (Kiel: Rheinheimer 1999; Stockholm: Cronström 1986; Helsinki: Häyrén 1921). A suspicion that anthropogenic eutrophication might also be affecting the open Baltic Sea came only with Fonselius's (1969, 1972) studies of oxygen deficiency in the deep basins of the Baltic proper. Fonselius showed that the phosphorus concentration of the Baltic proper had increased since the 1930s and suggested that pollution by nutrients and oxygen-consuming wastes could be at least partly responsible for the recorded decline in deep-water oxygen. Fonselius also pointed out that climatic variations could potentially explain most of the changes. An unusually long period of stagnation would give rise to anoxia in the deep water, during which phosphate would be released from the sediments. When mixed into the surface

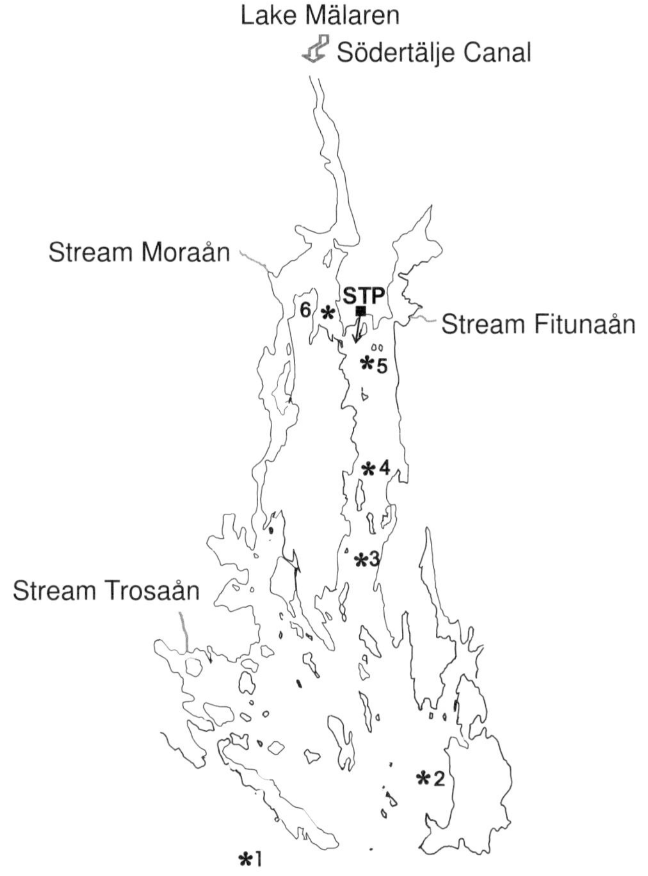

Figure 2.2 Map of Himmerfjärden Bay, with sewage treatment plant (STP), monitored streams, and inflow from Lake Mälaren and sampling stations (1–6). Simplified after Elmgren and Larsson (1997).

layer, this phosphate would stimulate production, leading to greater sedimentation of oxygen-consuming organic matter and increased deep-water anoxia, more release of phosphate, and so on. The high production resulting from this positive feedback might be expected to persist until terminated by another climate fluctuation. This idea gained in credibility when Hallberg (1974) reported evidence from sediment cores of anoxic periods centuries ago in the eastern Gotland basin. In this scenario, which is based on phosphorus controlling phytoplankton production in the Baltic proper, large fluctuations in productivity and deep-water oxygen deficiency could result from natural climatic variability, without anthropogenic influence. Thus even a large increase in oxygen deficiency and phytoplankton production would not necessarily require society to take countermeasures. It might be an entirely natural phenomenon that would correct itself in due time, without human intervention. This climate-driven scenario lost in appeal when it was shown that phytoplankton in the Baltic proper is largely nitrogen limited (Granéli et al. 1990) and that, on average, the sediments are nutrients sinks, not sources (Larsson et al. 1985).

Nature of the Problem

It is now widely agreed that a series of eutrophication-related ecological changes in the Baltic Sea and the Kattegat have been caused primarily by increased anthropogenic nutrient inputs, mainly after World War II (Baltic: Larsson et al. 1985; Larsson 1986; Elmgren 1989; Jansson and Dahlberg 1999; Kattegat: Richardson 1996). The major effects can be summarized as:

- Reduced water transparency, which has reduced submersed vegetation, affected fish stocks, and lowered the value of the coast for tourism, recreation, and nature conservation.
- Increase of toxic or noxious algal blooms. Cyanobacterial blooms, in the open sea particularly of the toxic, nitrogen-fixing genus *Nodularia,* are the main problem; however, a number of fish kills by the prymnesiophyte *Prymnesium parvum* have been reported from the Baltic proper coastal zone. The former is stimulated by low inorganic N/P ratios; the latter is thought to be favored by high N/P ratios.
- Increased areas of oxygen-deficient bottom waters. While not easily perceived at the surface, except in extreme cases, oxygen deficiency will kill fish and invertebrates and change nutrient cycling in ways that mostly enhance eutrophication (positive feedback).
- Changes in fish stocks have been both positive, due to increased food supply (e.g., pike perch in Baltic archipelagos) and negative (e.g., oxygen deficiency reducing Baltic cod recruitment) (Vallin et al. 1999).

BALTIC NUTRIENT LOAD

Quantities and Transport Routes

Loads of both nitrogen and phosphorus on the Baltic Sea increased greatly in the 20th century (Larsson et al. 1985). A recent study (Stålnacke et al. 1999a) concluded that most of the load increase must have occurred before 1970, and that since 1980, variability in the riverine loads is primarily the result of variations in runoff. Concentrations of total phosphorus and total nitrogen in the open Baltic proper surface water likewise increased until the 1980s and have since fluctuated with no clear trend (HELCOM 1996; Larsson and Andersson 1999). In Table 2.3 we compile estimates of riverine load, coastal point sources, and atmospheric deposition to obtain total external nutrient loads on the Baltic. Published atmospheric deposition estimates for the Baltic Sea include only inorganic nitrogen. We have added 25% to these values in an effort to include the deposition of organic nitrogen (cf. Cornell et al. 1995). We used 1% of nitrogen input as an estimate of atmospheric P input to the Baltic. The total external nitrogen load thus estimated is some 50 times the phosphorus load, as a molar ratio, far higher than the so-called Redfield N/P ratio of 16 (Redfield et al. 1963). Even if more nitrogen than phosphorus is in organic, relatively refractory form, this massive nitrogen excess has led some authors to conclude that the Baltic Sea must be phosphorus limited. Nitrogen-fixing cyanobacteria add a further 200,000–400,000 tons of combined nitrogen per year (Larsson et al. 1999). Yet budget calculations (Larsson et al. 1985; Wulff and Stigebrandt 1989) show that a huge nitrogen sink must exist within the Baltic Sea, and measurements of denitrification and of N_2 and N_2O agree in suggesting bacterial denitrification to be the main

Table 2.3 Baltic nutrient load, thousand tons per year of nitrogen and phosphorus.

Area	Riverine Load (natural and anthropogenic)	Coastal Point Sources	Atmospheric Load (natural and anthropogenic)	Total
Gulf of Bothnia	N: 100	N: 10	N: 48	N: 158
	P: 5	P: 1	P: <1	P: 6
Baltic proper	N: 363	N: 27	N: 185	N: 575
	P: 23	P: 4	P: 2	P: 29
Gulf of Finland	N: 126	N: 31	N: 23	N: 180
	P: 6	P: 4	P: <1	P: 10
Gulf of Riga	N: 113	N: 5	N: 11	N: 129
	P: 2	P: 1	P: <1	P: 3
Danish Straits plus Kattegat	N: 124	N: 21	N: 62	N: 207
	P: 4	P: 3	P: 1	P: 8
Baltic Sea, Area Total	N: 826	N: 94	N: 329	N: 1249
	P: 41	P: 12	P: 3	P: 56
Sources:	Stålnacke et al. (1999a)	HELCOM (1993)	HELCOM (1997)*	

*(average 1991–1995) × 1.25; Gulf of Riga load = North Baltic proper load; P = 1% of N load

sink (Rönner 1985). Efficient nitrogen removal through denitrification is the only plausible explanation for the predominant nitrogen limitation of Baltic phytoplankton biomass (Granéli et al. 1990), in spite of the apparent excess of nitrogen in the load.

Root Causes

Which human activities are the main causes of the eutrophication of the Baltic Sea? The single largest source of phosphorus is municipal sewage (including individual households), released either directly at the coast or into inland waters. Agriculture is second in importance, and industry (mainly the fertilizer and paper industries) a distant third. Aquaculture can be locally important in coastal areas but is small as a source for the whole Baltic. For nitrogen, long-range atmospheric transport, partly from outside the drainage basin, is far more important than for phosphorus. Agriculture is by far the largest nitrogen source: directly by runoff and drainage from fields; indirectly to the air by emissions of ammonia from fertilizers, manure, and farm animals, and of nitrogen oxides from farm machinery. Combustion in domestic heating, transport, and industry is the second largest source, releasing large quantities of nitrogen oxides. Sea transport has long been a neglected source, since no country held itself responsible for such emissions. Municipal sewage is also a considerable nitrogen source for the Baltic as a whole, and particularly important near major cities. Industry emissions into water can be locally important, as can aquaculture (HELCOM 1993; Enell and Fejes 1995; Jansson and Dahlberg 1999).

COUNTERMEASURES

International Conventions

The first serious international efforts to protect the Baltic Sea environment, apart from its fish stocks, were inspired more by fear of toxic pollutants than by the eutrophication problem. This led in 1974 to the signing of the first Helsinki Convention for the Protection of the Marine Environment of the Baltic Sea Area (ratified 1980). Since the Paris Convention covered the North Sea and Skagerrak, the Helsinki Convention had to include the Kattegat to avoid an unregulated gap between convention areas. This administrative decision has led to the Baltic Sea nowadays often being taken to include the Kattegat, even though this is wrong both historically and in strict geographical terms. The original Helsinki Convention expressly excluded near-coastal waters, but a renegotiation of the convention in 1992 included them (Fitzmaurice 1993). In 1988 the Helsinki Commission (HELCOM) adopted the Declaration for the Protection of the Baltic Marine Environment, in which the Baltic coastal states pledged themselves to halve their anthropogenic discharges of nitrogen and phosphorus between 1987 and 1995. At the Ronneby Conference in 1990, all governments of Baltic coastal states agreed on the goal of "restoring the ecological balance of the Baltic Sea." The HELCOM 50% nutrient reduction goal was now seen as a first step only, with further reductions to follow. This conference also initiated the elaboration of the "Joint Comprehensive Programme" (JCP) aimed, among other goals, at mitigating 132 pollution "hot spots" in the Baltic region at a total cost of 18×10^9 ECU over twenty years.

Since Sweden and Finland joined the European Union (EU) in 1995, most of the coastal area covered by the Helsinki Convention is also subject to EU directives. This includes the Urban Wastewater Treatment Directive (UWWTD), directives on the quality of waters for bathing, fishery, and mussel culture, the Nitrates Directive, and the upcoming EU Water Framework Directive. In areas described as sensitive to eutrophication, the UWWTD sets emission standards for all sewage treatment plants above a minimum size. These standards apply even where emissions from the sewage treatment plant are dwarfed by the discharge of a major river nearby. All Baltic coastal states still outside the EU, except Russia, have applied for membership.

National Management

National cleanup efforts in market economy countries in the Baltic Sea area were initially focused around major cities. Gross fecal pollution was often eliminated early in the 20th century, but nutrient removal efforts started only in the 1970s. In Sweden, phosphorus removal had already been successful for lakes, and was now implemented around Stockholm to reduce rampant eutrophication of the beautiful archipelago that connects the city to the Baltic. A thorough pre-implementation study (Wærn and Pekkari 1973; Wærn and Hübinette 1973) predicted that phosphorus removal would reduce algal biomass in the inner archipelago but would also lead to increased export of nitrogen to the nitrogen-limited outer archipelago (Karlgren and Ljungström 1975). As predicted, phosphorus removal eliminated blooms of cyanobacteria, some of which were nitrogen fixers, from the inner archipelago, while indications of increased phytoplankton biomass in the spring were seen in the middle archipelago (Brattberg 1986).

While Swedish policy concentrated on phosphorus reduction until the late 1980s, the Danes realized early on that nitrogen was often the limiting nutrient in their coastal waters. A series of studies in the 1970s characterized their coastal bays and fjords as limited either by nitrogen or phosphorus. The Danes then regarded open Danish waters as free from eutrophication (Ærtebjerg 1980) and therefore only recommended removal of the locally limiting nutrient, often nitrogen in one coastal bay, and phosphorus in the next. The nonlimiting nutrient, i.e., that present in the greatest supply in relation to plant needs, was not seen as important. From 1980 onwards, a series of major phytoplankton blooms, accompanied by benthic oxygen deficiency and fish kills, in open Danish waters in the straits at the entrance to the Baltic and in the Kattegat, demonstrated the need to consider also regional effects. These blooms, which also affected adjacent Swedish waters, became the starting signal for major research programs on marine eutrophication in Denmark (Jørgensen and Richardson 1996) and Sweden (Elmgren et al. 1986; Rosenberg et al. 1990).

Present Swedish Policy

The management of coastal sewage treatment in Sweden is based mainly on the UWWTD, as a result of Sweden having declared all its coastal waters south of the border between the Baltic proper and the Gulf of Bothnia as eutrophication sensitive. The UWWTD prescribes upper limits to the concentrations of nitrogen and phosphorus that can be discharged from sewage treatment plants larger than 10,000 person-equivalents. However, it sets no limit on the total quantities discharged and gives no guidance on how to handle nonpoint sources of nutrients, which nowadays increasingly dominate the local loads of both nitrogen and phosphorus to the Swedish coastal zone. Implementation of the EU regulations has so far left little scope for the meaningful involvement of local stakeholders in the management of coastal nutrient discharges.

Local regulations can, in principle, set lower nutrient emission standards than the UWWTD and also set limits in terms of total element mass emitted. The Himmerfjärden sewage treatment plant is the only one in Sweden that has been given some latitude in emission standards, so that the plant can be operated in an adaptive management mode, with the aim of optimizing environmental conditions in receiving waters (Larsson 1997). The idea at present is to remove ~90% of both nitrogen and phosphorus and to monitor receiving waters closely. When there are indications that a bloom of toxic cyanobacteria may occur, more nitrogen can be released to prevent or reduce the bloom. This ruling was obtained only after a lengthy political and legal process, involving also local nongovernmental environmental organizations. The Himmerfjärden case is thus of interest both from an ecological and a management point of view.

OUTLOOK FOR THE FUTURE

What Are Our Goals?

After decades of efforts to reduce its eutrophication, there is still no international agreement on environmental quality goals for the restoration of the Baltic Sea. There is an often voiced political desire to restore the area to an earlier, less polluted state, but no clear formulation of the end goal. Currently, it is thus not possible to say when or if the goals have been attained. It is also not possible to say which goal has priority, should a conflict become obvious. When phosphorus was reduced in Lake Erie, and water transparency improved greatly, there were protests from fishermen as the walleye population dropped by over half (Pelley 1998). Likewise in the Baltic Sea area, some fish stocks may decrease as nutrient input is reduced, but it is hoped that, at least initially, the decline will be in species of low economic value, such as cyprinids and clupeids. Fears have also been voiced that reducing Baltic eutrophication might yield higher concentrations of toxic pollutants in its biota and fish catches (Gunnarsson et al. 1995). The preferred plan of action is thus likely to depend critically on what goals we want to reach — a discussion that has hardly started in the international arena. Sweden has, however, decided on a set of national environmental goals and subgoals, some of which apply directly to coastal marine eutrophication, namely:

1. *A balanced marine environment and sustainable coastal areas and archipelagos.* Biotopes important for biodiversity should be protected. Animals and plants should maintain healthy, sustainable populations in their natural distribution areas.
2. *No eutrophication.* Swedish coastal and open marine waters should have good ecological status, according to the definition of the EU Water Framework Directive.
3. *A toxin-free environment.* Levels of substances that occur naturally in the environment must be close to background levels and levels of human-made substances must be close to zero.

While useful as a starting point for discussion, these goals are not yet well enough defined, even with further exemplification, to serve as a clear yardstick against which to measure our progress in protecting the Baltic marine environment. In Himmerfjärden, the management goal is to prevent blooms of nitrogen-fixing cyanobacteria, while minimizing both nutrient exports to the bay and the bottom area affected by oxygen deficiency (Larsson 1997).

Which Nutrient Should Be Removed?

Since both nitrogen and phosphorus concentrations have undoubtedly increased in most of the Baltic Sea proper, Danish Straits, and the Kattegat, the internationally agreed upon strategy has been to reduce loads of both elements. This may seem prudent and risk-free, but it could be economically inefficient, and thus indirectly harm the environment by diverting funds and efforts from more effective countermeasures. In the northern Gulf of Bothnia, nitrogen concentrations have increased, but no eutrophication has resulted, since phytoplankton biomass is phosphorus-limited (Granéli et al. 1990). In nitrogen-limited Tampa Bay, Florida, nitrogen removal in a sewage treatment plant has lowered nitrogen levels and improved water transparency; this has resulted in the recovery of local seagrass beds, even though phosphorus was not removed and phosphate concentrations remained high (Johansson and Lewis 1992). In both cases, investment in removal of the nonlimiting nutrient would not have been cost-effective from a local or regional perspective. In the Danish experience recounted above, locally cost-effective nutrient removal turned out to be ineffective on a regional scale. This may often be the case in estuaries, which are commonly phosphorus-limited in the low-salinity range, and nitrogen-limited in the more marine sections (e.g., Chesapeake Bay, Fisher et al. 1999). In such cases, removal of the locally limiting nutrient will have the side effect of leaving a greater surplus of the nonlimiting nutrient that can be exported to areas where it is limiting, and will increase production and biomass. *It is thus clear that both local and regional effects must be taken into account when managing nutrient discharges to the Baltic coastal zone, and that these are best treated on a case-by-case basis.*

In the Baltic Sea, blooms of nitrogen-fixing cyanobacteria are a further complication. Such blooms, which occur regularly in the Baltic proper each summer, sometimes extending into autumn, have their own inherent source of combined nitrogen and at their peak often show indications of phosphorus limitation (Larsson et al. 1999). Addition of phosphorus thus seems likely to enhance such blooms, which can be toxic. This is the rationale for recommending that phosphorus should be reduced in the load to the Baltic proper, and not only nitrogen, the main limiting nutrient (e.g., Granéli et al. 1990). Stronger assumptions are sometimes made that any further phosphorus increase will give a proportionally greater biomass of nitrogen-fixers, and that any reduction of nitrogen loads on the water body will be negated by a compensating increase in nitrogen fixation, making nitrogen reduction ineffective (e.g., Hellström 1996). A biogeochemical model incorporating such assumptions predicts that nitrogen removal will still reduce algal biomass on a local scale in the coastal zone, but not in the open Baltic proper, where nitrogen fixation will replace the nitrogen removed (Savchuk and Wulff 1999). In the Kattegat, the higher salinity seems to inhibit cyanobacterial blooms (Stal et al. 1999).

There is, however, little support for these stronger assumptions in empirical science and practical experience from the Baltic area. Thus, iron rather than phosphorus has been reported to limit nitrogen fixation in the Baltic Sea (Stal et al. 1999). Iron tends to be effectively sequestered by sulfide in marine sediments (Caraco et al. 1990; Gunnars and Blomqvist 1997). The sulfate in brackish waters is reported to limit nitrogenase activity in nitrogen-fixing Baltic cyanobacteria (Stal et al. 1999). Basic biogeochemical differences between fresh and marine waters are thus likely to help limit the development of blooms of nitrogen-fixing cyanobacteria in brackish waters (see also Howarth et al. 1999). In addition, cultures of *Aphanizomenon*, the dominant nitrogen-fixing cyanobacterial genus in the Baltic, produce less biomass per unit of phosphorus when grown diazotrophically than when grown with a sufficiency of fixed nitrogen (De Nobel et al. 1997). Maintaining a basic nitrogen limitation should then potentially give a smaller export of oxygen-consuming organic matter to bottom waters.

Taken together, the presence of the cyanobacterial pigment zeaxanthin and of low stable nitrogen isotope ratios in deep sediment cores from the Baltic proper indicates that blooms of nitrogen-fixing cyanobacteria are almost as old as the Baltic Sea itself (Bianchi et al. 2000). Given that the present management goal is to return the Baltic to a more natural state, this requires maintaining the basic nitrogen limitation that causes nitrogen-fixing cyanobacterial blooms. Finally, the more strongly nitrogen-limited the spring phytoplankton bloom in the Baltic proper is, the smaller the export of potentially oxygen-consuming organic matter to the deep waters will be. This should be helpful in reducing deep-water oxygen deficiency, even if total annual net production is maintained by a compensating increase in the summer, since less of the organic matter produced by phytoplankton sinks out during the summer (Heiskanen and Tallberg 1999). There are thus good reasons to continue trying to reduce the loads of both nitrogen and phosphorus in the Baltic Sea.

Adaptive Management of Himmerfjärden Bay

The Himmerfjärden sewage treatment plant started operating in 1974. The population served increased gradually from 90,000 to near 250,000 today. Since 1976, we have studied the ecological effects of the nutrients discharged from the plant. The load changed with the population served, but also through an experimental three- to fourfold increase in P release for one year (1983/84), and the stepwise introduction of nitrogen removal, reaching ~90% by 1998. During 1976–1994, the annual load from the sewage treatment plant varied between 500 and 900 tons N, and between 6 and 31 tons P. The total N load from land and air to the study area was 1050–1730 tons and the total P load 27–72 tons (Larsson, pers. comm., slightly modified from Skärlund 1997, pp. 28–29). Sewage was thus the dominant source of nitrogen, while land runoff dominated the phosphorus load. There was a good correlation ($r^2 = 0.69$) between load from the treatment plant and concentration in inner Himmerfjärden for total nitrogen (Larsson 1997, p. 166), but not for total phosphorus. Over 18 years, both Secchi depth and chlorophyll *a* concentrations in the study area correlated well with total nitrogen concentrations, as annual means (Figures 2.3 and 2.4). Weaker correlations were found with total phosphorus (exemplified by Figure 2.5), probably because total N and total P covary, but also since the inner bay was P limited for part of the year. The spring bloom maximum chlorophyll *a* concentration correlated with the inorganic nitrogen concentration available (Figure 2.6), a correlation that improved when suspected cases of P limitation were excluded (Figure 2.7). Overall, these observations strongly indicated that removing nitrogen would reduce local algal biomass, in agreement with algal growth potential tests (Granéli et al. 1990). To the extent that phosphorus removal would reduce plant production, it would leave more nitrogen available for export to nitrogen-limited areas outside, where algal growth would instead be stimulated. Algal production would thus be moved about, rather than reduced.

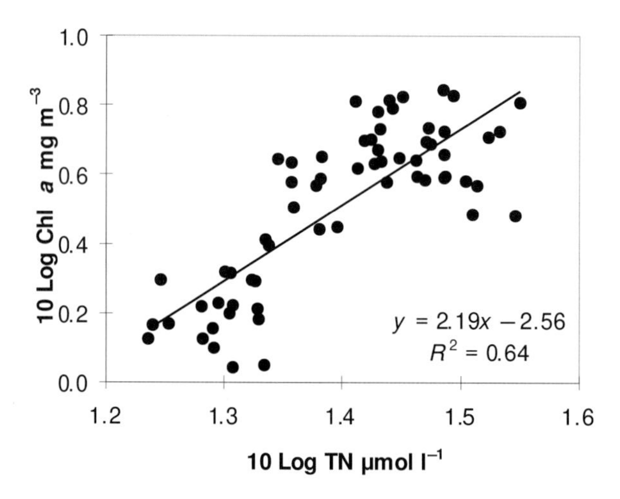

Figure 2.3 Yearly mean chlorophyll *a* concentration in the trophogenic layer as a function of total nitrogen concentration (TN) at five stations (B1, H2–H5 in Figure 2.2) in 1977–1994 (after Elmgren and Larsson [1997], with two more years of unpublished data added).

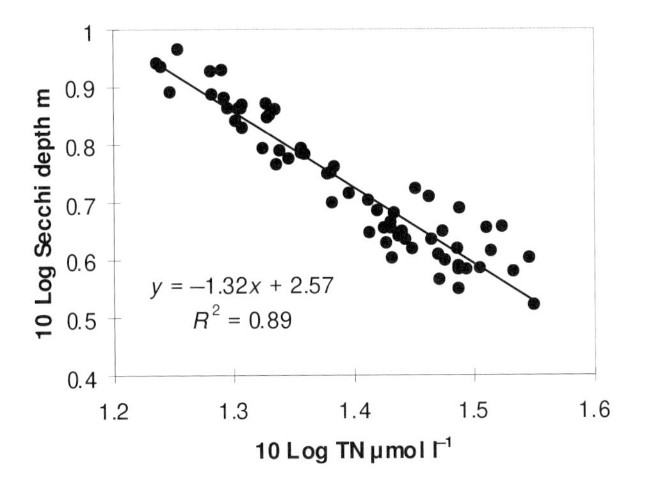

Figure 2.4 Yearly mean Secchi depth as a function of total nitrogen concentration (TN) at five stations (B1, H2–H5 in Figure 2.2) in 1977–1994 (after Elmgren and Larsson [1997], with two more years of unpublished data added).

Before the sewage treatment plant was fitted with fluidized beds for denitrification (Larsson 1997), empirical relations, like the above, in combination with a year-long experiment (1983–1984) with increased phosphorus release, were used to predict effects of further nitrogen removal in Himmerfjärden. A removal of 70% was predicted to decrease total nitrogen by 23% and chlorophyll *a* by 38% as annual average, and by 14% and 30%, respectively, as summer means. Secchi depth was predicted to increase by 37% over the year and by 34% in summer. More than 70% removal was predicted to stimulate the growth of previously rare nitrogen-fixing cyanobacteria. Results from the first two full years (1998–1999; Larsson and

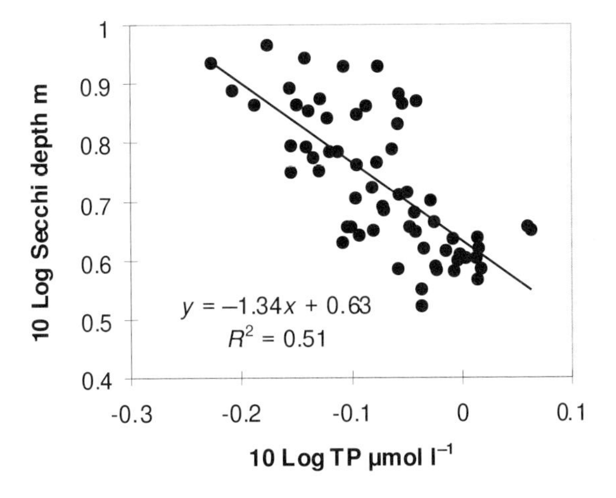

Figure 2.5 Yearly mean Secchi depth as a function of total phosphorus concentration (TP) at five stations (B1, H2–H5 in Figure 2.2) in 1977–1994.

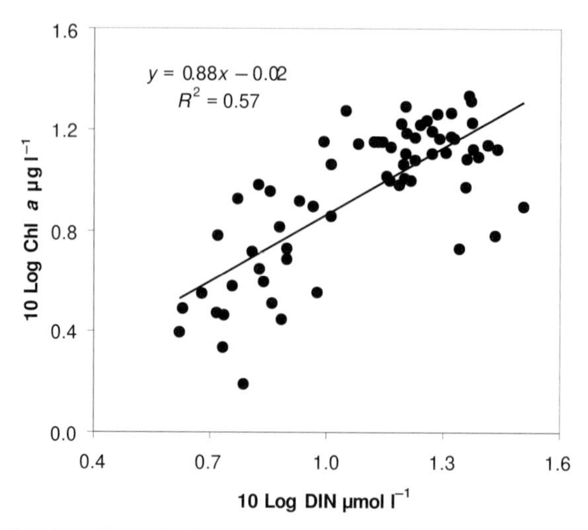

Figure 2.6 Maximal spring chlorophyll *a* concentration as a function of maximum inorganic nitrogen concentration (DIN) in January/February, before the start of the spring bloom. Five stations (B1, H2–H5 in Figure 2.2) in 1977–1994 (after Elmgren and Larsson [1997], with two more years of unpublished data added).

Elmgren, unpublished data from station 4) of ~90% nitrogen removal show that the levels predicted for 70% removal have almost been reached for total nitrogen (–19% annual mean, –12% July–August mean) and chlorophyll *a* (–31% annual mean, –20% July–August mean). Secchi depth, on the other hand, changed little (+1% annual mean, –15% July–August mean). As expected, biomass of nitrogen-fixing cyanobacteria increased in the summer, but nontoxic

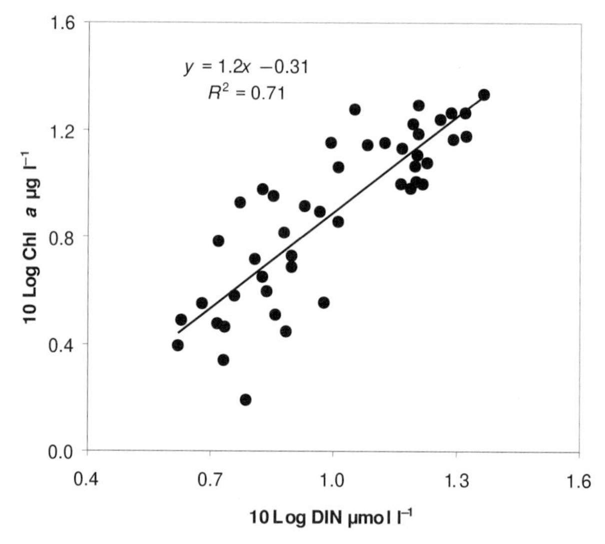

Figure 2.7 As in Figure 2.6, but excluding data from stations with inorganic N/P atom ratios above 22 before the start of the spring bloom.

species dominated. A longer period still is needed for firm conclusions, since it may take some time for full effects to develop (cf. Tampa Bay, where it took several years for effects of nitrogen reduction to be seen; see Johansson and Lewis 1992).

Which Sources Can Be Reduced, and How?

Swedish efforts to reduce nutrient input to the Baltic Sea have been successful for sewage treatment plants and water pollution by industry, where an existing management apparatus could be brought to bear on the new problem. These point sources were important for their P discharges, which have also been successfully reduced in Denmark, Germany, and Finland. For agriculture and air pollution, progress has been slow, in some areas almost nonexistent, and much the same is true for sewage from individual and small groups of households. As a result, no HELCOM country has fully complied with the EU Nitrates Directive (Iversen et al. 1998) or reached the 50% goal for anthropogenic nitrogen reduction. By 1995, Sweden claimed a reduction of nitrogen loss from agriculture of 29% (SEPA 1997) with Denmark claiming 17% (Iversen et al. 1998). Reduction was calculated as leakage from the root zone, but few statistically significant decreases of concentrations in the rivers have yet been reported (Sweden: Stålnacke et al. 1999b).

In Poland and the Baltic states, fertilizer use in agriculture dropped dramatically after the collapse of the Soviet Union (Nehring et al. 1995), yet there are few signs of decreased nutrient concentrations in rivers (Löfgren et al. 1999). The reason is twofold: fallow fields leak more than farmed fields, and decades of heavy fertilizer use have built up a huge nutrient store in soil and groundwater. Judging from Swedish experiments, such fields may, in the worst case, leak nutrients at near-present rates for decades to come (Löfgren et al. 1999). Other measures may reduce the nitrogen load to the sea more quickly than reduction of agricultural fertilizer use, e.g., catch crops, buffer strips, and wetland reconstruction; however, a 50% reduction seems a long way off (Löfgren et al. 1999). The buildup of nutrients in the water of the Baltic proper lagged 5–10 years after the increase in agricultural fertilizer use (Nehring et al. 1995), and it may take decades before decreased fertilizer use will result in lower concentrations in the open sea (Wulff and Niemi 1992). This is particularly true for phosphorus, which has over twice the turnover time (13 years) of nitrogen (5–6 years) in the Baltic Sea system (Wulff and Stigebrandt 1989), and which lacks an efficient sink, like denitrification for nitrogen. In Danish waters, where runoff from land has a more direct impact, due to high nitrogen concentrations and shallow waters (Nausch et al. 1999), a somewhat more rapid improvement has been prognosticated (Iversen et al. 1998). The outlook for rapid improvement in the open Baltic proper thus seems rather bleak (Enell and Fejes 1995). The more easily removed nutrient, phosphorus, has a very long turnover time and is now present in excess. Nitrogen, which is effectively removed from the sea by denitrification, is likely to continue arriving there in great quantity since its loss from agriculture is so difficult to reduce.

No Hope for the Baltic?

There are, however, reasons for a more optimistic view. In the coastal zone, where most people meet the sea and where eutrophication is often more evident than in the open sea, local nutrient discharge from land is often its main cause. This input can be substantially reduced, particularly if derived from municipal sewage and industry. A 50% reduction of the total nitrogen load has been achieved locally in Himmerfjärden, and a major reduction also in the Stockholm archipelago. Where large reductions are achieved, local effects will probably be seen quickly, often within a few years. Furthermore, each local load reduction will contribute to reducing the overall load on the Baltic.

For the open Baltic Sea, there is some hope that current efforts at reducing the atmospheric deposition of nitrogen will lead to lower loads, directly and by way of river runoff. There is also the possibility that as slow nutrient load reductions begin to have an effect, there may be positive feedbacks. A reduction in the bottom area affected by oxygen deficiency will allow bottom fauna to recolonize, and a sediment with a normal bottom fauna has a higher capacity for removing nitrogen through denitrification (Tuominen et al. 1998, 1999). Oxidized sediment is also more effective at sequestering phosphorus and iron, elements suspected of stimulating Baltic nitrogen fixation. The result could be higher denitrification and lower nitrogen fixation. These feedbacks have yet to be quantified and included in ecological models, but offer hope that recovery of the open Baltic may be quicker than commonly predicted.

What Is the Cost, and Who Should Pay?

The political decisions to halve anthropogenic nitrogen and phosphorus loads on the Baltic Sea were taken without economic analysis. Preliminary cost analyses for the Baltic proper (Gren et al. 1997) indicated that the proposed reduction is an order of magnitude more costly for nitrogen than for phosphorus. Furthermore, as normal for such analyses, the marginal reduction cost increased steeply for large reductions, with 50% nitrogen reduction costing more than twice 45% removal. Gren et al. stressed the great uncertainty of their analyses — slight changes in assumptions can easily halve or double the cost of a given level of nutrient reduction. They also attempted a contingent valuation of the benefits from nutrient reductions and a cost-benefit analysis. The result suggested that, in aggregate, citizens in countries concerned found the predicted environmental improvement worth its cost (see also Gren et al. 2000).

A politically far from simple consideration is who should foot the bill for reduction of nutrient discharges. Gren et al. (1997) found that citizens of the transition economies were generally not willing to pay their share of the national nutrient reduction costs, whereas those in the more affluent market economies were more than willing. This was partly because the reduction, to be cost-efficient, was not evenly distributed among countries. This could be handled by an international bubble policy, but as yet there is no legal framework for such deals. In addition, at least two principles for cost distribution have been suggested. The first is the classic polluter-pays-principle, where the nation or the sector of society that causes pollution must also pay the cost of reducing that pollution. This principle is generally considered applicable to industry and municipal sewage treatment but is often set aside or modified for agriculture, where government subsidies are considered necessary as a reward for environmentally sound behavior (Bruckmeier and Teherani-Krönner 1992; Swedish National Audit Office 1999). Another possibility is to distribute cost in proportion to perceived benefit. Such calculations have been made for cost distribution between nations for nutrient load reductions to the Baltic Sea (Gren et al. 1997), suggesting that the richer market economies should pay for cleanup efforts in the poorer, former communist countries. We find it difficult to recommend this principle as a guide for international relations, since it seems to be based on an implicit "right-to-pollute principle," with dubious ethical consequences. For example, a downstream nation would have to pay to clean up the water pollution of upstream nations; a downwind nation for the air pollution reduction of windward nations. For an analysis of the complexities involved in the management of regional (transboundary) environmental quality, see Sandler (1998).

For the future, drainage basin-based planning for the coastal zone, as demanded by the draft EU Water Framework Directive, offers an opportunity to bring local stakeholders together in coastal nutrient management. Local involvement from residents, nature conservation societies, professional and sport fishermen, farmers, tourists, the tourist trade, and, of course, the local and regional governments, may help to insure that resources are efficiently used and protected, in the service of both the local and the wider regional population. In the end, the UWWTD should be rendered obsolete, or be subsumed within the Water Framework Directive. In Sweden the creation of new planning bodies is required, with new procedures, based on the drainage basin, rather than on present, historically motivated political boundaries, as well as a legal framework for national and international bubble policies. This would represent an entirely new element in Swedish coastal zone management (Ackefors and Grip 1995).

The Role of Science

In the Swedish debate, the argument heard sometimes is that "we now know what causes Baltic eutrophication, all that remains is to stop it." Stopping eutrophication is seen purely as a problem of politics and management, with no need for further input from the natural sciences, but possibly with a role for the social sciences. We disagree and list crucial questions that must be addressed for effective countermeasures to Baltic coastal eutrophication to be implemented. Some of these can be answered by natural or social science alone; however, most require transdisciplinary research:

- How can public support for reduction of coastal eutrophication be generated?
- What will society gain by reducing coastal eutrophication?
- Will there be attendant losses, for example, reduced fish stocks, or increased availability of environmental contaminants?
- How far can we restore coastal areas without restoring the open Baltic proper?
- Do we need to reduce both N and P; if so, in what proportions?
- Will positive feedbacks help reduce eutrophication? Can we model them?
- Will climate change increase or reduce the negative effects of coastal eutrophication?
- How will load reductions in noneutrophicated areas benefit adjacent areas?
- What size of reductions in load are required to reach predefined environmental goals?
- What mix of measures achieves each goal at lowest cost?
- How can suitable national and international bubble policies be developed?
- Which measures against coastal eutrophication have synergistic environmental effects?
- Will local cost-effective measures lead to cost-effectiveness for the entire Baltic Sea area?
- How can we best monitor the effects of coastal eutrophication management?

New syntheses of scientific knowledge and management experience, and also new scientific studies, will be needed to support future coastal management based on drainage basins. Bubble policies for the trading of nutrient discharge rights need to be evaluated. Investments in sewage treatment often have a planned lifetime (and depreciation time) of 30–50 years. Similar time frames may be needed for abating nutrient leakage from farms and for the open Baltic to show clear improvement. Planning must be long term and must even consider the likelihood of climate change.

ACKNOWLEDGMENTS

We thank our Dahlem colleagues, Oleg Savchuk and Karl Bruckmeier, Anders Carlberg and Ulla-Britta Fallenius of the SUCOZOMA Research Program for helpful comments. The Swedish Foundation for Strategic Environmental Research (MISTRA) funds the SUCOZOMA program, including this project. The Stockholm Marine Research Centre provided Figure 2.1 and additional funding. We also thank the many competent and dedicated colleagues, students, and technicians who have collected Himmerfjärden data over the years; without their help we would have given up long ago.

REFERENCES

Ackefors, H., and K. Grip. 1995. The Swedish Model of Coastal Zone Management. Swedish Environmental Protection Agency (SEPA) Report No. 4455. Stockholm: SEPA.

Ærtebjerg, G. 1980. Oxygen, næringssalte og primærproduktion i Kattegat. In: Forureningssituationen i Skagerrak–Kattegat, vol. 24, pp. 21–44. Report NU B1980. Copenhagen: Nordic Council and Council of Ministers.

Bianchi, T.S., et al. 2000. Cyanobacterial blooms in the Baltic Sea: Natural or human-induced? *Limnol. Oceanogr.* **4**:716–726.

Brattberg, G. 1986. Decreased phosphorus loading changes phytoplankton composition and biomass in the Stockholm archipelago. *Vatten* **42**:141–153.

Bruckmeier, K., and P. Teherani-Krönner. 1992. Farmers and environmental regulation: Experiences in the Federal Republic of Germany. *Sociologia Ruralis* **32**:66–81.

Caraco, N., J. Cole, and G.E. Likens. 1990. A comparison of phosphorus immobilisation in sediments of freshwater and marine systems. *Biogeochemistry* **9**:277–290.

Cederwall, H., and R. Elmgren. 1990. Biological effects of eutrophication of the Baltic Sea, particularly the coastal zone. *Ambio* **19**:109–112.

Cornell, S., A. Rendell, and T. Jickells. 1995. Atmospheric inputs of dissolved organic nitrogen to the oceans. *Nature* **376**:243–246.

Cronström, A. 1986. Stockholm's water supply and sewerage. A brief history. *Vatten* **42**:96–104 (in Swedish).

De Nobel, W.T., J.L. Snoep, H.V. Westerhoff, and L.R. Mur. 1997. Interaction of nitrogen fixation and phosphorus limitation in *Aphanizomenon flos-aquae* (Cyanophyceae). *J. Phycol.* **33**:794–799.

Ehlin, U. 1981. Hydrology of the Baltic Sea. In: The Baltic Sea, ed. A. Voipio, pp. 121–134. Amsterdam: Elsevier.

Elmgren, R. 1989. Man's impact on the ecosystem of the Baltic Sea: Energy flows today and at the turn of the century. *Ambio* **18**:326–332.

Elmgren, R., and C. Hill. 1997. Ecosystem function at low biodiversity — The Baltic example. In: Marine Biodiversity: Patterns and Processes, ed. R.F.G. Ormond, J. Gage, and M. Angel, pp. 319–336. Cambridge: Cambridge Univ. Press.

Elmgren, R., and U. Larsson, eds. 1997. Himmerfjärden: Changes of a Nutrient-enriched Coastal Ecosystem in the Baltic Sea. Swedish Environmental Protection Agency Report 4565 (in Swedish). Stockholm: SEPA.

Elmgren, R., et al. 1986. Eutrophication in the Marine Environment. Research Programme 1983–1989. Natl. Swedish Environmental Protection Board Report 3259, pp. 1–104 (in Swedish). Stockholm: SEPA.

Enell, M., and J. Fejes. 1995. The nitrogen load to the Baltic Sea — Present situation, acceptable future load and suggested source reduction. *Water Air Soil Poll.* **85**:877–882.

Engqvist, A. 1990. Accuracy in material budget estimates with regard to temporal and spatial resolution of monitored factors. *Estuar. Coast. Shelf Sci.* **30**:299–320.

Fisher, T.R., et al. 1999. Spatial and temporal variation of resource limitation in Chesapeake Bay. *Mar. Biol.* **133**:763–778.

Fitzmaurice, M. 1993. The new Helsinki convention on the protection of the marine environment of the Baltic Sea area. *Mar. Poll. Bull.* **26**:64–67.

Fonselius, S.H. 1969. Hydrography of the Baltic Deep Basins. III. Fishery Board of Sweden, Series on Hydrography 23, pp. 1–97.

Fonselius, S.H. 1972. On eutrophication and pollution in the Baltic Sea. In: Marine Pollution and Sea Life, ed. M. Ruivo, pp. 23–28. London: Fishing News (Books).

Granéli, E., et al. 1990. Nutrient limitation of primary production in the Baltic Sea area. *Ambio* **19**:142–151.

Gren, I.-M., T. Söderquist, and F. Wulff. 1997. Nutrient reductions in the Baltic Sea: Ecology, costs, and benefits. *J. Env. Manag.* **51**:123–143.

Gren, I.-M., R.K. Turner, and F. Wulff. 2000. Managing a Sea: The Ecological Economics of the Baltic Sea. London: Earthscan.

Gunnars, A., and S. Blomqvist. 1997. Phosphate exchange across the sediment-water interface when shifting from anoxic to oxic conditions — An experimental comparison of freshwater and brackish-marine systems. *Biogeochemistry* **37**:203–226.

Gunnarsson, J., et al. 1995. Interactions between eutrophication and contaminants — Toward a new research concept for the European aquatic environment. *Ambio* **24**:383–385.

Hallberg, R.O. 1974. Paleoredox conditions in the eastern Gotland basin during the recent centuries. *Merentutkimuslait. Julk./Havsforskningsinst. Skr.* **238**:3–16.

Häyrén, E. 1921. Studier över föroreningarnas inflytande på strändernas vegetation och flora i Helsingfors hamnområde. *Bidrag Kännedom Finland Natur och Folk* **80**:1–128.

Heiskanen, A.-S., and P. Tallberg. 1999. Sedimentation and particulate nutrient dynamics along a coastal gradient from a fjord-like bay to the open sea. *Hydrobiologia* **393**:127–140.

HELCOM (Helsinki Commission). 1993. Second Baltic pollution load compilation. Baltic Sea Environment Proceedings 45. Helsinki: HELCOM.

HELCOM. 1996. Third periodic assessment of the state of the marine environment of the Baltic Sea, 1989–1993. Baltic Sea Environment Proceedings 64B. Helsinki: HELCOM.

HELCOM. 1997. Airborne pollution load to the Baltic Sea 1991–1995. Baltic Sea Environment Proceedings 69. Helsinki: HELCOM.

Hellström, T. 1996. An empirical study of nitrogen dynamics in lakes. *Water Env. Res.* **68**:55–65.

Howarth, R.W., F. Chan, and R. Marino. 1999. Do top-down and bottom-up controls interact to exclude nitrogen-fixing cyanobacteria from the plankton of estuaries? An exploration with a simulation model. *Biogeochemistry* **46**:203–231.

Iversen, T.M., R. Grant, and K. Nielsen. 1998. Nitrogen enrichment of European inland and marine waters with special reference to Danish policy measures. In: Nitrogen, the Confer-N-s. First Intl. Nitrogen Conf., March 23–27, 1998, Noordwijkerhout, The Netherlands, pp. 771–780. Amsterdam: Elsevier.

Jansson, B.-O., and K. Dahlberg. 1999. The environmental status of the Baltic Sea in the 1940s, today and in the future. *Ambio* **28**:312–319.

Johansson, J.O.R., and R.R. Lewis, III. 1992. Recent improvements of water quality and biological indicators in Hillsborough Bay, a highly impacted subdivision of Tampa Bay, Florida, U.S.A. *Sci. Total Env.* **Suppl.**:1199–1215.

Jørgensen, B.B., and K. Richardson, eds. 1996. Eutrophication in Coastal Marine Ecosystems. Coastal and Estuarine Studies 52. Washington, D.C.: American Geophysical Union.

Karlgren, L., and K. Ljungström. 1975. Nutrient budgets for the inner archipelago of Stockholm. *J. Water Poll. Control Fed.* **47**:823–833

Larsson, U. 1986. The Baltic Sea. In: Eutrophication in Marine Waters Surrounding Sweden, ed. R. Rosenberg, pp. 16–70. Natl. Swedish Environment Protection Board Report 3054. Stockholm: SEPA.

Larsson, U. 1997. Measures against eutrophication. In: Himmerfjärden: Changes of a Nutrient-enriched Coastal Ecosystem in the Baltic Sea, ed. R. Elmgren and U. Larsson, pp. 165–175. Swedish Environmental Protection Agency Report 4565 (in Swedish). Stockholm: SEPA.

Larsson, U., and L. Andersson. 1999. Long-term changes: Nutrients in surface water. In: Östersjö '98, ed. A. Tidlund, pp. 2–5 (in Swedish). Stockholm: Stockholm Marine Research Centre.

Larsson, U., R. Elmgren, and F. Wulff. 1985. Eutrophication and the Baltic Sea — Causes and consequences. *Ambio* **14**:9–14.

Larsson, U., et al. 1999. Nitrogen fixation in the Baltic Sea proper: Direct and indirect estimates. Extended Abstract. Final Baltic Sea System Study Conf., Warnemünde, Sept. 20–22, 1999. (http://www.io-warnemuende.de/public/bio/basys/con3/ con3top.htm).

Löfgren, S., A. Gustafsson, S. Steineck, and P. Ståhlnacke. 1999. Agricultural developments and nutrient flows in the Baltic States and Sweden after 1988. *Ambio* **28**:320–327.

Nausch, G., D. Nehring, and G. Ærtebjerg. 1999. Anthropogenic nutrient load of the Baltic Sea. *Limnologica* **29**:233–241.

Nehring, D., et al. 1995. The Baltic Sea in 1995 — Beginning of a new stagnation period in its central deep waters and decreasing nutrient load in its surface layer. *Dt. Hydrogr. Z.* **47**:319–327.

Pelley, J. 1998. Lake Erie fish declines hooked to phosphorus? *Env. Sci. Technol.* **32**:253A.

Redfield, A.C., B.H. Ketchum, and F.A. Richards. 1963. The influence of organisms on the composition of seawater. In: The Sea, vol. 2, ed. M.N. Hill, pp. 26–77. New York: Wiley.

Rheinheimer, G. 1999. Pollution in the Baltic. *Naturwissenschaften* **85**:318–329.

Richardson, K. 1996. Conclusion, research and eutrophication control. In: Eutrophication in Coastal Marine Ecosystems, ed. B.B. Jørgensen, and K. Richardson, pp. 243–267. Coastal and Estuarine Studies 52. Washington, D.C.: American Geophysical Union.

Richardson, K., and A. Christoffersen. 1992. Primærproduktion. In: Planktondynamik og Stoffomsætning i Kattegat, ed. T. Fenchel, pp. 103–120. Havsforskning fra Miljøstyrelsen 10. Copenhagen: Dept. of the Environment.

Rönner, U. 1985. Nitrogen transformations in the Baltic proper: Denitrification counteracts eutrophication. *Ambio* **19**:102–108.

Rosenberg, R., et al. 1990. Marine eutrophication case studies in Sweden. A synopsis. *Ambio* **19**:102–108.

Sandler, T. 1998. Global and regional public goods: A prognosis for collective action. *Fiscal Studies* **19**:221–247.

Savchuk, O., and F. Wulff. 1999. Modelling regional and large-scale response of the Baltic Sea ecosystem to nutrient load reductions. *Hydrobiologia* **393**:35–43.

Schinke, H., and W. Matthäus. 1998. On the causes of major Baltic inflows — An analysis of long time series. *Cont. Shelf Res.* **18**:67–97.

SEPA (Swedish Environmental Protection Agency). 1997. Nitrogen from Land to Sea. Main Report. Report 4801, pp. 1–120. Stockholm: SEPA.

Skärlund, K. 1997. Loads and trends. In: Himmerfjärden: Changes of a Nutrient-enriched Coastal Ecosystem in the Baltic Sea, ed. R. Elmgren, and U. Larsson, pp. 26–31. Swedish Environmental Protection Agency Report 4565 (in Swedish). Stockholm: SEPA.

Stal, L.J., M. Staal, and M. Villbrandt. 1999. Nutrient control of cyanobacterial blooms in the Baltic Sea. *Aquat. Microb. Ecol.* **18**:165–173.

Stålnacke, P., A. Grimwall, K. Sundblad, and A. Tonderski. 1999a. Estimation of riverine loads of nitrogen and phosphorus to the Baltic Sea, 1970–1993. *Env. Monit. Assess.* **58**:173–200.

Stålnacke, P., A. Grimwall, K. Sundblad, and A. Wilander. 1999b. Trends in nitrogen transport in Swedish rivers. *Env. Monit. Assess.* **59**:47–72.

Swedish National Audit Office. 1999. Environmental subsidies for Agriculture. Swedish National Audit Office (SNAO) 1999:2 (in Swedish). Stockholm: SNAO.

Sweitzer, J., S. Langaas, and C. Folke. 1996. Land use and population density in the Baltic Sea drainage basin. *Ambio* **25**:191–198.

Tuominen, L., A. Heinänen, J. Kuparinen, and L. Nielsen. 1998. Spatial and temporal variability of denitrification in the sediments of the northern Baltic proper. *Mar. Ecol. Prog. Ser.* **172**:13–24.

Tuominen, L., et al. 1999. Nutrient fluxes, porewater profiles and denitrification in sediment influenced by algal sedimentation and bioturbation by *Monoporeia affinis*. *Estuar. Coast. Shelf Sci.* **49**:83–97.

Vallin, L., A. Nissling, and L. Westin. 1999. Potential factors influencing the reproductive success of Baltic cod, *Gadus morhua* — A review. *Ambio* **28**:92–99.

Wærn, M., and L. Hübinette. 1973. Phosphate, nitrate and ammonium in the archipelago during 1970. *Oikos* **Suppl. 15**:164–170.

Wærn, M., and S. Pekkari. 1973. Out-flow studies. *Oikos* **Suppl. 15**:155–163.

Wulff, F., and Å. Niemi. 1992. Priorities for the restoration of the Baltic Sea — A scientific perspective. *Ambio* **21**:193–195.

Wulff, F., and A. Stigebrandt. 1989. A time-dependent budget model for nutrients in the Baltic Sea. *Glob. Biogeochem. Cyc.* **3**:63–78.

3

Science and Integrated Drainage Basin Coastal Management

Chesapeake Bay and Mississippi Delta

D.F. Boesch

Center for Environmental Science, University of Maryland, P.O. Box 775, Cambridge, MD 21613, U.S.A.

ABSTRACT

Modern precepts of coastal management involve three challenging dimensions: integration, sustainability, and adaptation. The extent to which management addresses these dimensions is examined for two large coastal ecosystems heavily influenced by extensive continental drainage basins: the Chesapeake Bay and the Mississippi delta. The Chesapeake Bay, the largest estuary in the U.S.A., has been affected by eutrophication, habitat loss, and overfishing. Its biggest challenges are the control of diffuse sources of nutrient inputs from agriculture and expanding urban–suburban development and the physical restoration of once plentiful oyster habitats. The Mississippi delta is experiencing rapid loss of coastal wetlands and eutrophication of the adjacent Gulf of Mexico. River controls for navigation and flood protection and the world's most intense industrial agriculture in the upper basin affect this ecosystem greatly. Although assessments and models of nutrient dynamics in the watershed and coastal waters provide a foundation for intermedia and interdisciplinary integration, the management of both systems is not yet well integrated among sectors (e.g., fishing, transportation, and agriculture) and issues (e.g., eutrophication, overfishing, and habitat restoration). While the development of management goals is further advanced in the Chesapeake, even there a scientifically realistic vision of a sustainable future has not been developed. Management of the Chesapeake Bay is adaptive in the long term, but lacks the tight connections between models and outcomes needed for highly responsive adaptive management. Science could better serve integrated coastal management in these regions if it included: interdisciplinary and strategic research targeted to the coastal ecosystem and its watershed; more predictive approaches involving historical reconstruction, models, and experiments; more effective integration of modeling, monitoring, and research; and institutional and individual commitment to civic science.

Science and Integrated Coastal Management
Edited by B. von Bodungen and R.K. Turner

INTRODUCTION

In many regions of the world, coastal ecosystems are greatly influenced not only by anthropogenic activities in the coastal zone proper, but also by activities throughout large river drainage basins, extending hundreds or even thousands of kilometers from the coast. This is particularly true for large estuaries and river–delta ecosystems. Management of coastal environments and resources thus requires not only integration among coastal resource users and among the marine natural and social sciences, but also among resource users and sciences on continental scales (Boesch 1996). Typically, this requires integration and coordination across many political and scientific boundaries.

The large scales, socioeconomic complexity, and need to link atmospheric, landscape, aquatic, marine, and human components make the modern challenges of coastal management of such ecosystems especially steep. Three dimensions of these challenges are particularly prominent and interrelated: integration, sustainability, and adaptation.

Integration as used by the management and policy science community, as in integrated coastal management (ICM), generally implies collective consideration of the uses of products and services provided by the coastal zone to determine an "optimal mix" (Bower and Turner 1998). As Knecht and Archer (1993) point out, the integration required is itself multidimensional: intergovernmental, intermedia (land–water interface in a shoreline area from their perspective, but on larger scales from the present perspective), intersectoral (among users), and interdisciplinary.

From the natural sciences community has emerged the notion of ecosystem management (Christensen et al. 1996), which is not really different from integrated management, but provides greater emphasis on biophysical features and processes. Concepts of ecosystem management address ecosystem integrity as a critical goal; scale and boundaries; the complexity, connectedness, and dynamic nature of ecosystems that limit the predictions that can be made; and ecological principles and models. At the same time, ecosystem management acknowledges, but pays less attention to, societal needs and collaboration and consensus building among sectoral interests. Nonetheless, ecosystem management has numerous social science and governance implications (Hennessey and Soden 1999).

Sustainability has long been a concept embedded in ICM but has become a more explicit requirement in the post-UNCED era (Cicin-Sain 1993). "Sustainable" is a word that is often used, but seldom defined. In the context of sustainable development, it implies the economic development needed to sustain and improve the quality of life of human populations (i.e., sustainable economically and socially) that is environmentally sustainable, equitable among groups in society and nations, and does not foreclose options of future generations. This is a tall order. Sustainability is also listed as a goal of the natural scientists' view of ecosystem management (Christensen et al. 1996), but perhaps with a greater emphasis on environmental sustainability and less emphasis on the imperative of economic development of current populations.

Adaptation is a dimension necessitated by the inherent uncertainty in our predictions about the natural world, socioeconomic developments, and the consequences of management actions. The concept of adaptive management has been developed and applied principally in North America (Lee 1993; Hennessey and Soden 1999) but is implicit in the ICM policy cycle envisioned by the GESAMP (1996) (see also von Bodungen and Turner, this volume, and

Olsen, this volume) as a moving feedback loop through assessment, implementation, evaluation, and reassessment. Adaptive management involves implicitly learning by doing and treating management programs as experiments, with a great emphasis on accounting for outcomes.

In this chapter, I examine the status and challenges for science in advancing these three dimensions of integrated management of the two most prominent drainage basin–coastal systems in the United States: the Chesapeake Bay and its watershed, and the Mississippi delta (including the coastal environments of the delta and the nearby waters of the northern Gulf of Mexico) and its vast drainage basin. Although physiographically different, these two ecosystems share many similar issues, including eutrophication and other consequences of landscape changes within their catchments, habitat losses, fishery declines, toxic contamination, and navigation access. The management programs to address these issues and the science to support them are, in general, more advanced for the Chesapeake Bay than for the Mississippi delta, offering some instructive contrasts.

THE ECOSYSTEMS

Chesapeake Bay and Watershed

The Chesapeake Bay is the largest estuary in the United States and one of the largest in the world. Its main stem is 332 km long; the estuary has a surface area of 11,400 km^2 with 12,870 km of shoreline; and its drainage basin covers 166,000 km^2 in six states. The bay is relatively shallow (mean depth 6.5 m); therefore, the area of its drainage basin is large with respect to the estuarine volume. This, coupled with its modest tidal exchange, makes the bay very susceptible to inputs of fresh water, sediments, and dissolved materials from its catchment (Matuszeski 1996).

Approximately 15 million people live in the Chesapeake basin, with the largest concentrations at the tidal headwaters of estuarine tributaries around the Washington, D.C., Baltimore, Richmond, and Norfolk metropolitan areas. The bay includes important commercial and military ports and is an important recreational resource. Although a number of the historically important fisheries (particularly oysters) have declined, the bay supports commercial fisheries worth approximately U.S. $1 billion per year. The estuary is also heavily used for domestic and industrial waste disposal, with about 5,000 point-source discharges into the estuary or drainage basin.

The sediments laid down in the deeper parts of the Chesapeake Bay yield a chronicle of the many anthropogenic changes in the estuarine ecosystem since colonization of the region by Europeans almost 400 years ago (Boesch 1996; Boesch et al. 2001). The rate of sedimentation increased rapidly beginning in the mid-1700s as a result of erosion from land clearing to grow tobacco for export and grains to support growing populations. More plant nutrients — forms of nitrogen and phosphorus that the native forests efficiently retained — began to wash down into the bay, subtly altering its natural food web, during this agrarian period.

The industrialization that began during the late 1800s left a clear record of increased contamination by trace metals. Steam technology provided the mechanical means to overexploit the abundant oysters, effectively strip-mining the extensive reefs that gave the bay its aboriginal name, Chesapeake or "great shellfish bay." The mid-1900s brought on the petrochemical period of the bay's history. Manufactured organic chemicals, such as pesticides, and petroleum by-products appear prominently in the sediment record. More importantly, microfossil and biochemical indicators in the bay's sediments reveal a profound change in the ecosystem. Over a relatively short time during the 20th century, the estuary changed from a relatively clear-water ecosystem, characterized by abundant plant growth in the shallows, to a turbid ecosystem dominated by super-abundant microscopic plants in the water column and stressful low-oxygen conditions during the summer.

This state shift was largely due to the dramatic increase in nutrient inputs in the form of wastes from the growing population and runoff of agricultural fertilizers manufactured from petrochemicals. In addition, the burning of fossil fuels in power plants and automobiles releases nitrogen oxides into the atmosphere, which rain back down on the bay and its catchment. In all, it has been estimated that the Chesapeake Bay now receives, from land and air, about seven times more nitrogen and 16 times more phosphorus than it received when English colonists arrived (Boesch et al. 2001). In addition, the drastic depletion of the once-prodigious oyster populations has reduced by 90% or more the capacity of these and other organisms to clean the bay's waters through filter feeding. Meanwhile, humans eliminated more than half of the wetlands in the Chesapeake watershed — wetlands that served as sinks for nutrients and sediments — and built dams on rivers that prevent the upstream migration of shad and other anadromous fishes.

In the 1970s the scientific community began to understand and document the pervasive changes in the ecosystem that had taken place and to identify their causes. This led to a growing awareness by the public and political leaders, which in turn resulted in the evolution of regional management structures and restoration objectives (Hennessey 1994; Boesch et al. 2001). Through the legally established Chesapeake Bay Program, the three primary states in the region, the national capital, and the federal government have developed a series of directives and agreements related to reductions of nutrient and toxicant loadings, habitat restoration, living resource management, and landscape management. Although the bay and its drainage basin fall entirely within the United States, the commitments among the parties are quite similar to the declarations of multinational conferences and commissions for the management of European seas (Elmgren and Larsson; Mee, both this volume) because in the U.S.A., most of the responsibilities for land use and water quality fall under the jurisdiction of the states.

Because eutrophication was seen as the most pervasive and consequential human impact, the keystone agreement of the Chesapeake Bay Program called for a 40% reduction of controllable sources of nitrogen and phosphorus entering the bay by the year 2000. Large expenditures of public and private funds have already been made to reduce these inputs both from point sources (treated sewage discharges) and nonpoint sources (especially those from agriculture) or to trap the nutrients in the watershed by wetland and riparian zone restoration. As the target date for this goal approached,extensive efforts were made to assess progress and to determine the next generation of restoration goals. The assessment process relied both on the highly detailed and linked watershed and estuarine hydrodynamic models, and on an extensive water quality monitoring program. Because of the significant interannual variation in freshwater discharge and delivery of nutrients to the bay and the time lags in delivery of nutrients from nonpoint sources, it is not an easy matter to measure or predict the effects of the actual reductions. However, it appears that the 40% goal was nearly met for phosphorus;

nitrogen loadings, although reduced, did not achieved this goal. As in the Baltic Sea (Jansson and Dahlberg 1999; Elmgren and Larsson, this volume), reductions of nonpoint sources of nitrogen, particularly from agriculture, have been difficult to achieve.

On other fronts, the progress in restoring the Chesapeake Bay ecosystem has been mixed. The concentrations of a number of potentially toxic substances (some trace metals and chlorinated hydrocarbons) in sediments and organisms have declined as a result of source controls and waste treatment. Yet, the industrialized harbors in the bay remain heavily contaminated, and other subregions show elevated concentrations of toxicants or evidence of biological effects. Seagrasses have returned in some regions but cover only a small portion of the habitat occupied in the 1950s. Oyster (*Crassostrea virginica*) populations have not recovered because of the degraded reef habitat and ravages of two microbial pathogens. Populations of several anadromous fishes have increased modestly as a result of removal of barriers to upstream migration. Perhaps the most dramatic recovery has been for populations of striped bass (*Morone saxatilis*), which greatly increased as a result of a multiyear moratorium on harvest and subsequent, more conservative management of stocks.

Mississippi Delta and Basin

The coastal ecosystem associated with the Mississippi River delta is equally large as, but less well defined than, the Chesapeake Bay. It includes the Mississippi deltaic plain (Boesch et al. 1994), consisting of the mostly inactive distributaries of the river and extensive tidal wetlands, swamps, and lagoons lying between the distributaries or enclosed by fringing barrier islands. However, a large portion of the river-influenced continental shelf, which has estuary-like salinity gradients and stratified water masses, should be included as part of this coastal ecosystem. While 70% of the flow of the Mississippi River flows through its well-recognized birdsfoot delta, projecting into the Gulf of Mexico, the remaining 30% (regulated by law) flows down the only other distributary presently active, the Atchafalaya River, which enters the Gulf of Mexico 230 km to the west. This expansive wetland–estuarine–shelf ecosystem supports one of the richest fisheries in the U.S.A. and the substantial majority of the coastal and offshore oil and gas production. It is one of the most economically important coastal regions of the country — and one of the most threatened.

The catchment of the Mississippi River is vast, over 3.2 million km^2, including 41% of the conterminous United States and even a small part of Canada. It encompasses all or part of thirty of the fifty United States, from the arid west toward the Rocky Mountains to the humid forests of the Appalachian Mountains to the east. The north-central part of the basin, originally prairies, forests, and wetlands, has been extensively converted to cropland that produces most of the corn, soybeans, wheat, sorghum, and livestock grown in the U.S.A.

The average annual discharge of water through the Mississippi and Atchafalaya rivers to the coastal ecosystem is 628 km^3, essentially an order of magnitude higher than freshwater discharge into the Chesapeake Bay. The hydrology of this great river system has been greatly altered by locks, dams, reservoirs, earthwork levees, channel straightening, and spillways for purposes of flood protection, navigation, and water supply. These alterations have significantly affected the transport of water, sediments, and dissolved materials (including nutrients and toxic contaminants) in ways that have major consequences to the coastal ecosystem (Boesch 1996).

Disruption of overbank flooding in the delta, widespread hydrological modifications caused by myriad canals, and high rates of subsidence (because of the huge thickness of alluvial deposits) have conspired to result in rapid loss of tidal wetlands, particularly during the last half of the 20th century. By the late 1960s, approximately 73 km^2 of vegetated wetlands were being lost per year (Boesch et al. 1994). There is an active program to protect remaining wetland and restore degraded wetland–estuarine systems by various physical management techniques, including placement of dredged sediments, water-level controls, and diversions of river flows back into the deltaic plain. Many of these programs are conducted under the federal Coastal Wetlands, Planning, Protection, and Restoration Act.

A more recently recognized problem is the extensive seasonal hypoxia in bottom waters on the continental shelf (Rabalais et al. 1996). Hypoxic (< 2 mg l^{-1}) bottom waters have extended over 10,000 to 20,000 km^2 in the summer during the 1990s (Rabalais et al. 1998). This phenomenon and other manifestations of eutrophication have been related to the increases in nutrient loading by the Mississippi–Atchafalaya river system. In particular, flux of nitrate from the Mississippi basin to the Gulf of Mexico has averaged nearly 1 million metric tons per year since 1980, about three times higher than it was thirty years ago (Goolsby et al. 2000). The majority of the increased nitrate emanates from agricultural sources in the upper Mississippi and Ohio river basins, over 1,500 km upstream from the discharge into the Gulf of Mexico.

The popular press has provided extensive coverage of the scientific documentation of the dimensions and causes of what is frequently referred to as the Gulf of Mexico "dead zone." In response, the U.S. Congress directed the government to conduct an assessment of the causes and consequences of hypoxia in the Gulf of Mexico, including analyses of the potential for reduction of nutrient sources and associated economic costs. The resulting integrated assessment (Committee on Environment and Natural Resources 2000) presents much more comprehensive evidence concerning nutrient sources, trends, and effects on oxygen depletion for the Mississippi–Atchafalaya delta system than existed at the initiation of the Chesapeake Bay Program and its commitments for nutrient reduction, approximately fifteen years prior.

Nonetheless, agricultural interests and states upstream are fiercely contesting the assessment's findings. For example, the fertilizer industry commissioned a group of scientists to prepare an alternative analysis (Carey et al. 1999), which raises uncertainties and complications such as the role of organic carbon rather than nutrient loading and the effects of climate. The state of Illinois, which according to the integrated assessment is a major source of nutrients to the river, has vehemently criticized the assessment arguing that total nitrogen loadings have actually decreased and have no relationship with agricultural practices, questioning evidence that hypoxia has increased, observing that hypoxia in the sea is a natural phenomenon, and suggesting that efforts to reduce nutrient loading risk reducing fisheries productivity in the Gulf of Mexico.

Political jurisdictions and economic sectors responsible for the sources of nutrient pollution in the Gulf of Mexico are remote from direct relationship with the affected coastal environments. They have a natural tendency to protect the interests of agriculture, which is very economically and culturally important in these upriver regions. Furthermore, they are demanding higher levels of proof and more evidence of significant impacts on living resources in the Gulf before taking action than was the case for the Chesapeake Bay, where the responsible political jurisdictions are located on or near the bay and public concern about the environment is high. Nonetheless, in late 2000, a task force representing basin states and the federal government agreed to take steps to reduce the extent of hypoxia and recognized that this may require a 30% reduction in nitrogen loading.

In addition to wetland loss and eutrophication, integrated coastal management of the Mississippi delta region must also address significant issues in fishery management (including overfishing of some stocks, bycatch mortalities due to shrimp trawling, commercial and recreational fishery conflicts, and endangered species concerns), flood protection, navigation, oil and gas exploration and production, and migratory waterfowl management.

DIMENSIONS OF MANAGEMENT

I now provide brief perspectives on the extent to which management of these large and important coastal ecosystems addresses the three dimensions of modern management: integration, sustainability, and adaptation. I also address the contributions of science to those dimensions.

Integration

Evaluating eutrophication that results from diffuse sources of nutrients forces one to conduct assessments across environmental media and requires contributions, if not collaboration, from diverse disciplines. The Mississippi River integrated assessment (Committee on Environment and Natural Resources 2000) involves the analysis of atmospheric inputs of nutrients from data collected to monitor acid deposition; land-use characteristics; water quality data from throughout the basin and statistical models to estimate fluxes by river segment; and box models of nutrient and carbon budgets in the Gulf of Mexico. More complex, deterministic modeling is applied for the Chesapeake Bay, including atmospheric inputs as a function of meteorological variability; land-use changes; hydrologic transfer of water, nutrients, and sediments through the watershed; and three-dimensional, hydro- and ecosystem dynamics of the estuary (Boesch et al. 2001). Such integrated assessment and modeling has helped managers and scientists alike to think across environmental media, disciplines, and sectors (e.g., agriculture, environmental quality, and living resources). While there is still a long way to go in terms of improvement of models and their predictive uses, both the Mississippi River assessment and the Chesapeake Bay models are leading examples of the power of systemic science for large system management.

There is much less integration among the multiple stressors that confront these ecosystems, including eutrophication, toxic contamination, habitat modification, fishing, species introductions, and coastal land use. Surely, many of these stressors interact in important ways. For example, trace metals and organic contaminants have been shown to affect the type of algal production in enriched waters. Conversely, eutrophication-induced oxygen stresses affect immunological responses of marine animals to toxicants and pathogens.

In the Chesapeake Bay Program, different management committees exist for nutrients, toxic materials, and living resources, with relatively little interaction (except for the linkage between eutrophication and seagrasses). The scientific community as well is organized into nutrient–plankton, chemistry–toxicology, wetlands, fisheries, and social science communities, which typically sort themselves out into different lecture rooms at scientific meetings. The Chesapeake Bay Program (2000) has recently adopted a proposed comprehensive agreement (Table 3.1) that incorporates and supersedes previous agreements and directives. It establishes specific objectives or indicates actions and timetables under each goal. The renewed agreement is comprehensive in scope and many of the goals have an integrating requirement for sustaining and enhancing the living resources of the bay. Thus, a conceptual framework, if not yet a quantitative prescription, for integrated management is provided.

While the Chesapeake Bay Program provides an umbrella management structure for dealing with multiple stressors and integrated coastal management, there is no common

Table 3.1 Goals of the proposed Chesapeake 2000 Agreement (Chesapeake Bay Program 2000). Specific objectives and actions under each goal are briefly summarized.

1. Restore, enhance, and protect the finfish, shellfish, and other *living resource*s, their habitats and ecological relationships to sustain all fisheries and provide for a balanced ecosystem.
   - Oysters: tenfold increase
   - Exotic species: identify and reduce introduction
   - Fish passage: restore passage in blocked rivers
   - Multispecies management: develop and revise management plans
   - Crabs: restore health of spawning population
2. Preserve, protect, and restore those *habitats* and natural areas vital to the survival and diversity of the living resources of the bay and its rivers.
   - Submerged aquatic vegetation: recommit and raise previous restoration goal
   - Wetlands: achieve net gain through regulatory protection and restoration
   - Forests: protect and restore riparian forests
   - Stream corridors: encourage local governments to improve stream health
3. Achieve and maintain the *water quality* necessary to support the aquatic living resources of the bay and its tributaries and to protect human health.
   - Nutrients: achieve and maintain 40% goal and reduce further to protect living resources
   - Sediments: reduce loading to protect living resources
   - Chemical contaminants: no toxic or bioaccumulative impacts on living resources
   - Priority urban waters: restore urban harbors
   - Air pollution: strengthen air emission pollution prevention programs
   - Boat discharges: establish "no discharge zones"
4. Develop, promote, and achieve *sound land-use practices* that protect and restore watershed resources and water quality, maintain reduced pollutant loadings for the bay and its tributaries, and restore and preserve aquatic living resources.
   - Land conservation: protect and preserve forests and agricultural lands
   - Public access: expand public access points
   - Development, redevelopment, and revitalization: reduce rate of land development
   - Transportation: coordinate with land-use planning to reduce dependence on automobiles
5. Promote *individual stewardship* and assist individuals, community-based organizations, local governments, and schools to undertake initiatives to achieve the goals and commitments of this agreement.
   - Public outreach and education: provide information about bay to schools and public
   - Community engagement: enhance small watershed and community-based actions
   - Government by example: develop and use government properties consistent with goals

management framework in which to address the two dominant issues of the Mississippi delta: coastal wetland loss and eutrophication. Yet, river diversions intended to rebuild deteriorating deltaic wetlands may also result in removal of nitrogen delivered to the Gulf of Mexico (Lane et al. 1999), although depending on their location and efficiency of nutrient removal they could also conceivably exacerbate eutrophication in inshore or shelf waters.

A major challenge in both areas is the integration of environmental and fisheries management. Although it is generally thought that environmental degradation (hypoxia and other effects of eutrophication and wetland and other habitat modification) has diminished the capacity of these ecosystems to sustain healthy fisheries, this relationship is in fact very poorly quantified or understood in both cases. Correlations, much less cause and effect relationships, are difficult to demonstrate when other factors, including climatic variations and fishing pressures, also play such a large role. Better scientific understanding is critical both because of the recognized need to manage multispecies fishery resources in an ecosystem context and because managers are seeking performance endpoints for environmental management that are closer to what is valued by society.

In these two regions, as in everywhere else in the world, environmental management cannot be conducted detached from the pressures of socioeconomic development. The economic importance of agricultural production in the upper Mississippi basin and the role of the region in the global food supply are potent forces that constrain the options for controlling nutrient inputs into the Gulf of Mexico, just as is sustaining agriculture while accommodating population growth and development in the growing information economy in the Chesapeake region. Restoration of delta wetlands has to contend with the realities of providing flood protection and navigational access. A variety of scientific approaches help illuminate these relationships, e.g., the agricultural economic modeling conducted in the Gulf of Mexico integrated assessment or the economic land-use watershed models that predict water quality changes in the Patuxent subestuary of the Chesapeake (Voinov et al. 1999). However, we remain far from the kinds of ecological–socioeconomic models needed to guide coastal management and development toward a harmonious future.

Finally, the Chesapeake Bay and Mississippi delta case studies represent different levels of development of intergovernmental integration in management — and different degrees of difficulty. To be sure, intergovernmental agreements remain challenging in the Chesapeake Bay Program as it struggles to define post-2000 goals and approaches. There are differences in the governance structure and prevailing political philosophies among the states. For the Mississippi delta, however, even though the coastal environments in question fall essentially in one state, Louisiana, the problems cannot be addressed without the involvement of distant, noncoastal states and the federal government. In the Chesapeake, as has been the case in the Mediterranean, Baltic, and North Seas, consensus among scientists across different states (or nations) has been an empowering force for management of large marine ecosystems.

Sustainability

As the American baseball legend and bard of the obvious, Yogi Berra, is reported to have said: "It's hard to make predictions, especially about the future." Defining sustainability involves making predictions about the future, including unintended consequences, not just reconstructing the past or understanding the present. This is inherently challenging. There is always high uncertainty. Moreover, important larger-scale changes, for example, sea-level rise and other manifestations of climate change or national or global economic forces, are generally beyond the perspective, much less control, of managers.

In the Chesapeake Bay, the commitment to reduce controllable sources of nitrogen and phosphorus by 40% also meant maintaining future nutrient loadings at or below those goals once achieved. This was a step toward sustainability, albeit with an arbitrary and limited definition, with details to follow. Mississippi delta eutrophication is obviously still in an earlier stage of consensus building, but I would predict that ultimately similar "caps" that are consistent with acceptable economic impacts on agriculture will eventually be adopted.

Goals for the permanent reduction of loadings of nutrients and other pollutants such as those adopted by the Chesapeake Bay Program, the Helsinki Commission for the Baltic Sea (Jansson and Dahlberg 1999; Elmgren and Larsson, this volume), or the Paris Commission for the North Sea (Colijn and Reise, this volume) were set based on best estimates at the time of those characteristic of some prior time, such as the 1950s. They do not necessarily correspond to those conditions necessary to restore and maintain the ecosystem to a given level of health, in the sense of its vigor, organization, and resilience (Boesch 2000). The Chesapeake 2000 Agreement takes the concept of sustainability further by linking land use, individual responsibility, and community engagement to water quality, habitats and, ultimately, living resources.

One interesting approach to a broadly based vision of environmental sustainability was produced by the Chesapeake Bay Foundation, a nongovernmental conservation organization, in its annual State-of-the-Bay Report, intended to communicate the status of the bay in a way understandable to the public (Boesch 2000). The report scores 12 factors, including those related to habitats, water quality, and living resources, on a scale of 100, with 100 representing the conditions estimated at the time of arrival of European colonists. Thus the composite index (average of all 12) for 1999 was judged to be 28, up from a low of 23 in 1983, but well below the average of 70 thought to be achievable (recognizing it is not possible to return the bay to its pristine condition).

In any case, even such a progressive and ambitious effort as the Chesapeake Bay Program, with its record of commitments extending a decade or more, has not yet developed a truly long-term, intergenerational perspective. The Chesapeake Bay Program's Scientific and Technical Advisory Committee is presently undertaking an assessment of potential outcomes for the bay 30 to 50 years into the future. This requires consideration of the natural aging of the bay; the consequences of climate change on the region (higher relative sea level, warmer temperatures, and possibly more freshwater runoff); new technologies for energy production and waste treatment; and social, economic, and land-use changes. These projections require ever more efficient treatment of wastes, in order to offset population growth (estimated to be 20% within 30 years), to maintain nutrient loading caps. Even then, ground will be lost if there is not a dramatic change in the sprawling patterns of land conversion for development.

The driving forces that will determine the sustainability of the health of coastal ecosystems of the Mississippi delta are the agricultural economy in the drainage basin and river as well as river and delta management, as influenced by considerations for flood protection and navigation. A 50-year plan for coastal wetland and estuarine protection and restoration (Louisiana Coastal Wetlands Conservation and Restoration Task Force 1998) has been developed with the goals of (a) creating and sustaining marshes by accumulating sediment and organic matter, (b) maintaining habitat diversity by varying salinities and protecting key land forms, and (c) maintaining the exchange of energy and organisms. However, as discussed above, these have not yet been integrated with eutrophication reduction goals. Furthermore, even if this ambitious plan is fully implemented, it is not clear that it will produce a truly sustainable outcome over multiple generations (100 years and more) for an environment so susceptible to sea-level rise. Clearly, here as in the Chesapeake, science should contribute more to the definition of achievable futures and sustainable options; however, the choices ultimately belong to society.

Adaptation

Adaptive management embodies a simple imperative: policies are experiments; learn from doing them (Lee 1993). Practitioners must be explicit about what they expect, and they must collect and analyze information so that expectations can be compared with actuality. Finally, they must correct errors, improve their imperfect understanding, and change actions and plans.

Hennessey (1994) viewed the Chesapeake Bay Program as an excellent example of adaptive management of a large ecosystem. On the other hand, Boesch (1996) noted shortcomings in the degree of emphasis on learning and pursuit of multiple options in the face of uncertainty. The coupling among explicit expectations (from modeling), comparisons with actuality (through monitoring), and changed actions and plans is the essence of adaptive management. The program has extensive and advanced environmental and monitoring programs; however, they are relatively weakly linked in either periodic or ongoing assessments of progress (e.g., Chesapeake Bay Program 1999). In addition, the program's adaptation has been less responsive than the ideal of adaptive management would have it. This is a result of inefficiencies in the modeling–monitoring–management triad — some of which can be improved — but also because of inertia in large ecosystems (which delays the outcomes from management actions) as well as political systems (which delays policy responses).

The integrated assessment of hypoxia in the Gulf of Mexico also recommends an adaptive management framework for controlling nutrient inputs to reduce hypoxia, as does an earlier assessment of wetland protection and restoration in the delta (Boesch et al. 1994). While the emphasis on explicit expectations, collection of information for comparison with actuality, and learning provides an overall philosophical model and framework for integrating modeling, monitoring, and research, adaptive management does have its practical limits for such a large ecosystem, where many potential interventions are costly and relatively irreversible. For example, although one can certainly learn valuable lessons from monitoring the small-scale river diversions currently in place in the delta, the expense and social dislocations involved in massive river diversions require that they be considered as more than experiments, with limited degrees of control of future operations. Nonetheless, rigorous monitoring of the effects of river diversions, as they are put into place and operated at different flow rates, can provide valuable learning experiences not only for their future operations but also for planning of future diversions.

The implementation of an adaptive approach to management faces many practical challenges, including convincing the stakeholders to participate in experiments; sustaining well-supported monitoring programs in the face of waning interests and other priorities;

interpreting the ambiguous outcomes likely in complex and uncontrolled ecosystem; and resistance to changes in management approaches. Nonetheless, it is clear that our understanding, goals, and priorities do evolve over time. It stands to reason that embracing some form of formally adaptive structure would assist in the orderly and effective evolution of integrated coastal management.

TOWARD MORE EFFECTIVE SCIENCE

Among those factors limiting the effectiveness of integration, sustainability, and adaptation in coastal management are those related to the support, execution, and application of science. The following improvements in the practice of science would advance integrated coastal management:

- Development of integrated environmental and social science of large-scale problems, such as those that involve large drainage basins and coastal ecosystems. This requires the support of geographically targeted strategic research (NRC 1994), in addition to the nationally (or European Union) oriented, disciplinary research programs that now exist. Furthermore, this will also require concerted efforts within the scientific community (e.g., by scientific societies) to break down disciplinary barriers of communication and understanding of complex environmental issues. It will require similar efforts within research institutions (universities and agencies).
- Interregional comparisons that advance the general understanding of environmental phenomena and human effects and thus allow more robust extrapolation to other ecosystems. Important lessons can be learned through intercomparisons of the scientific and management experiences among the five large coastal regions treated under the topic of transboundary issues (Jickells et al., this volume). For example, what can the responses of the Danube River and Black Sea (Mee, this volume) to dramatic reduction of fertilizer application tell us with regard to nutrient management strategies in the Mississippi River basin?
- More forward-looking, predictive science based on modeling, experimentation, and reconstruction. The conservative scientific culture deters forward-looking science (with its untested hypotheses) in favor of descriptions of nature as it is or was. Support should be provided and institutional arrangements revised to foster predictive approaches to science that would enhance integrated management, which is inherently about the future.
- More effective integration of modeling, monitoring, and research (NRC 1994). Prediction, observation, and understanding are fundamental to adaptive environmental management, yet modeling, monitoring, and research are largely decoupled in program management and implementation. This leads both to doubts about model performance and unresolved discrepancies between predictions and observations.
- Advancement of civic science (*sensu* Lee 1993), in which scientists actively participate in the communication and precautionary use of science in the political process (Costanza et al. 1998; Boesch 1999). The rapidity of global and regional environmental degradation requires a new sense of urgency within the scientific community and a new social contract between science and society (Lubchenco 1998). Environmental and

social scientists and the institutions that employ them must develop mechanisms to engage in the management process on a timely basis, while preserving the standards of objective evidence required for the impartiality and long-term advancement of science.

REFERENCES

Boesch, D.F. 1996. Science and management in four U.S. coastal ecosystems dominated by land–ocean interactions. *J. Coast. Conserv.* **2**:103–114.

Boesch, D.F. 1999. The role of science in ocean governance. *J. Ecol. Econ.* **31**:189–198.

Boesch, D.F. 2000. Measuring the health of the Chesapeake Bay: Toward integration and prediction. *Env. Res.* **2**:134–142.

Boesch, D.F., R.B. Brinsfield, and R.E. Magnien. 2001. Chesapeake Bay eutrophication: Scientific understanding, ecosystem restoration, and challenges for agriculture. *J. Env. Quality* **30**: in press.

Boesch, D.F., et al. 1994. Scientific assessment of coastal wetland loss, restoration, and management in Louisiana. *J. Coast. Res. Spec. Iss.* **20**:1–103.

Bower, B.T., and R.K. Turner. 1998. Characterising and analysing benefits from integrated coastal management. *Ocean Coast. Manag.* **38**:41–66.

Carey, A.E., et al. 1999. The Role of the Mississippi River in the Gulf of Mexico Hypoxia. Environmental Institute Report No. 70. Tuscaloosa: Univ. of Alabama.

Chesapeake Bay Program. 1999. Chesapeake 2000: A Watershed Partnership. Annapolis: EPA.

Chesapeake Bay Program. 2000. The State of the Chesapeake Bay: A Report to the Citizens of the Bay Region. EPA 903-R-99-013. Annapolis: EPA (Environmental Protection Agency).

Christensen, N.L., et al. 1996. The Report of the Ecological Society of America Committee on the scientific basis for ecosystem management. *Ecol. Appl.* **6**:665–692

Cicin-Sain, B. 1993. Sustainable development and integrated coastal management. *Ocean Coast. Manag.* **21**:11–43.

Committee on Environment and Natural Resources. 2000. Integrated Assessment of Hypoxia in the Northern Gulf of Mexico. Washington, D.C.: National Science and Technology Council.

Costanza, R., et al.. 1998. Principles of sustainable governance of the oceans. *Science* **281**:198–199.

GESAMP (Joint Group of Experts on the Scientific Aspects of Marine Environmental Protection). 1996. The Contributions of Science to Integrated Coastal Management. Reports and Studies No. 61. Rome: FAO (Food and Agriculture Organisation of the United Nations).

Goolsby, D.A., et al. 2000. Nitrogen flux and sources in the Mississippi river basin. *Sci. Total Env.* **248**:75–86.

Hennessey, T.M. 1994. Governance and adaptive management for estuarine ecosystems: The case of Chesapeake Bay. *Coast. Manag.* **22**:119–145.

Hennessey, T.M., and D.L. Soden. 1999. Ecosystem management: The governance approach. In: Ecosystem Management: A Social Science Perspective, ed. D. Soden and B. Steele, pp. 29–48. New York: Brooks-Cole.

Jansson, B.-O., and K. Dahlberg. 1999. The environmental status of the Baltic Sea in the 1940s, today, and in the future. *Ambio* **28**:312–319.

Knecht, R.W., and J. Archer. 1993. "Integration" in the U.S. Coastal Zone Management Program. *Ocean Coast. Manag.* **21**:183–199.

Lane, R.R., J.W. Day, Jr., and B. Thibodeaux. 1999. Water quality analysis of a freshwater diversion at Caernavon, Louisiana. *Estuaries* **22**:327–336.

Lee, K.N. 1993. Compass and Gyroscope: Integrating Science and Politics for the Environment. Washington, D.C.: Island Press.

Louisiana Coastal Wetlands Conservation and Restoration Task Force and the Wetlands Conservation and Restoration Authority. 1998. Coast 2050: Toward a Sustainable Coastal Louisiana. Baton Rouge: Louisiana Dept. of Natural Resources.

Lubchenco, J. 1998. Entering the century of the environment: A new social contract for science. *Science* **279**:491–497.

Matuszeski, W. 1996. Case Study 1 — The Chesapeake Bay Programme, U.S.A. In: The Contributions of Science to Integrated Coastal Management, pp. 25–30. GESAMP Reports and Studies No. 61. Rome: FAO.

NRC (National Research Council). 1994. Priorities for Coastal Ecosystem Science. Washington, DC: National Academy Press.

Rabalais, N.N., et al. 1996. Nutrient changes in the Mississippi River and system responses on the adjacent continental shelf. *Estuaries* **19**:386–407.

Rabalais, N.N., R.E. Turner, W.J. Wiseman, Jr., and Q. Dortch. 1998. Consequences of the 1993 Mississippi River flood in the Gulf of Mexico. *Reg. Rivers R & M* **14**:161–177.

Voinov, A., et al. 1999. Integrated ecological economic modeling of a watershed, Maryland. *Env. Model. & Softw.* **14**:473–491.

4

Transboundary Issues

Consequences for the Wadden Sea

F. COLIJN[1] and K. REISE[2]

[1]Forschungs- und Technologiezentrum Westküste, Hafentörn,
25761 Büsum, Germany
[2]Wattenmeerstation Sylt, Alfred-Wegener-Institut für Polar- und Meeresforschung,
Hafenstrasse 43, 25992 List/Sylt, Germany

ABSTRACT

The European Wadden Sea is a narrow coastal fringe with extensive tidal mud- and sandflats, and with a conspicuous row of barrier islands. Because of its natural beauty, large flocks of migrant shorebirds, and unique intertidal habitats, the Netherlands, Germany, and Denmark have declared the Wadden Sea to be a protected nature area and are coordinating their activities and measures to implement legal instruments in environmental protection. A concerted monitoring program has been initiated and reports on the environmental quality status of the Wadden Sea have been published. However, a framework for assessing, analyzing, and evaluating natural and anthropogenic developments is still needed. The Wadden Sea is an interface between a large drainage basin of several rivers (e.g., Rhine, Elbe, and Weser) on the landward side and a large shelf sea on the seaward side. Often, riverine inputs are not direct but enter the Wadden Sea indirectly and are transformed through transports from the open sea back to the coast. Transboundary effects also become modified by an artificial shoreline architecture or by fisheries. Long-term variability in natural processes is very pronounced. These complexities are often at the root of communication problems between scientists and coastal management because they hamper the degree of certainties needed for managerial measures.

INTRODUCTION

Nutrients and pollutants (both organic compounds and heavy metals) generated in drainage basins are related to their socioeconomic (land) uses and have an impact on the coastal zones that are under the direct influence of the major rivers discharging into these areas. Therefore,

Science and Integrated Coastal Management
Edited by B. von Bodungen and R.K. Turner

eutrophication and contaminant effects in the Wadden Sea are often related to processes occurring far away. This causes problems as discharge control is dependent on different regulations and environmental standards used in different countries. In addition, different political systems (Mee, this volume), the differences in perception of pollution, and the mere distance between cause and effect (Boesch et al., this volume) are important drivers in transboundary problems and pose difficulties to solve such problems.

Due to the complexity of the ecological system of the Wadden Sea, it is difficult to ascribe observed biological changes directly to inputs of nutrients, organic pollutants, and heavy metals arriving from the drainage basins into the coastal waters and its food webs, and there are multiple effects of coastal fisheries. Coastal engineering and infrastructural works, like reclamation and harbor areas, might change currents and sedimentation/erosion balances in the Wadden Sea. These may add to or interact with imports from the drainage basins. The effects may be intensified, balanced, or even counteracted.

The distance to the sources is a special scientific problem because during transport, compounds may be transformed chemically, deposited in sediments, retained, and resuspended. Moreover, dissolved compounds absorb to particles in estuarine regions of the rivers and, therefore, behave differently. The final destination of these compounds is often uncertain depending on, for example, mixing processes, salinity gradients, sediment composition, and bioavailability. Thus a direct linking of matter fluxes from the drainage basins to coastal ecosystems is often problematic.

THE WADDEN SEA ECOSYSTEM

The Wadden Sea is an international protected nature reserve in the Netherlands, Germany, and Denmark. It is a shallow lagoonal sea of almost 500 km in length and is protected from the sea by a row of barrier islands (Figure 4.1). A series of rivers discharge into the Wadden Sea such as the Ems, the Weser, the Elbe, and the Eider. Part of the Rhine outflow runs directly into the western Wadden Sea through the IJsselmeer sluices. Together with the Rhine, a total drainage area of about 500,000 km^2 affects the Wadden Sea. The adjacent mainland consists of low-lying agricultural land, diked to prevent flooding by storm tides of up to 4 m above normal high tides. The Wadden Sea proper covers 8000 km^2 and has an average width of only 16 km. Salt marshes comprise 4%, intertidal sand- and mudflats 54%, permanently submerged shallows and deep channels 30%, and 12% of the area goes to 58 sandy barrier islands, high sands, and some marsh islands.

The Wadden Sea came into existence some 7000 years ago when postglacial sea-level rise established approximately the present coastline of the North Sea basin. In the 8th century, the Frisian people immigrated into the Wadden Sea region. They began to embank and irrigate marshlands and tidal flats. This gradually halved the size of the Wadden Sea. The dynamic islands became artificially stabilized. From a geomorphological perspective, the seaward boundary may follow the row of barrier islands, while hydrographically and ecologically a seaward body of coastal water of about 10 km in width should be included, roughly delineated by the 15 m depth contour. However, this zone is not included in the areal figures given above (cf. Table 6.1 in Jickells et al., this volume). Outside the four major estuaries, salinity does not drop below half the marine North Sea waters. Residual currents run parallel to the coast from

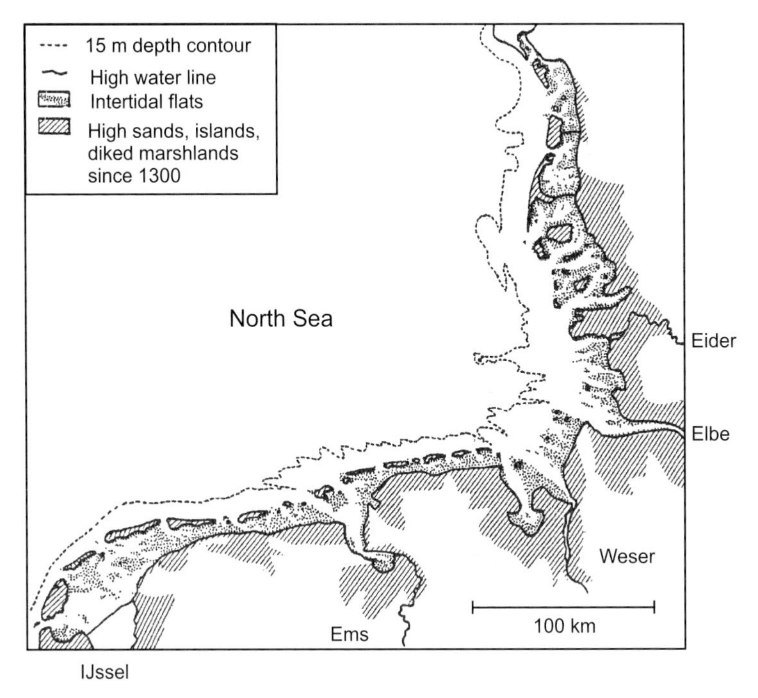

Figure 4.1 Map of the European Wadden Sea which stretches along the southeastern coast of the North Sea from the Netherlands over Germany to Denmark. The seaward boundary is either seen in the row of barrier islands or along the 15 m depth contour.

southwest to northeast. About half of the water volume is exchanged by each tide. The mean tidal amplitude ranges from 1.5 m at Den Helder in the southwest to 3.5 m in the mouth of the Elbe and diminishes again to 1.5 m at Esbjerg in the northeast. Average water temperatures are 8°C in winter and 15°C in summer, with extreme values from around zero to summer highs above 20°C.

Productivity in the Wadden Sea is high because of a net inflow of organic matter and because of its shallowness. The Wadden Sea serves as a nursery for fish and shrimp of the North Sea. The region is a central feeding ground for coastal birds of the East Atlantic flyway, which use the Wadden Sea with its rich macrobenthic fauna as a recovery area during the biannual migrations between Africa and their Arctic breeding grounds. The macrobenthos in the Wadden Sea is mainly formed by bivalves (e.g., *Cerastoderma edule*, *Mytilus edulis*, *Mya arenaria*, and *Macoma balthica*), snails (*Hydrobia ulvae*), crustaceans (*Corophium volutator*, *Crangon crangon*, and *Carcinus meanas*), and many polychaetes (e.g., *Arenicola marina*, *Heteromastus filiformis*, *Nereis spp.*). Food sources for all these animals are derived from the autochthonic primary production by phytoplankton and microphytobenthos and from an input of allochthonous material from the North Sea and rivers.

Economically, tourism dominates the islands while agriculture dominates the mainland. Main ports along the Wadden Sea are Den Helder, Delfzijl, Emden, Wilhelmshaven, Bremerhaven, Bremen, and Hamburg. Within the Wadden Sea proper, shrimps and mussels are the main fishing catch.

TOWARDS A COMMON MANAGEMENT

To improve scientific knowledge on natural properties and their threads in the Wadden Sea, a comprehensive monograph on the ecology of the Wadden Sea was published (Wolff 1983) and augmented by a detailed habitat inventory (Dijkema et al. 1989). During the last decades, starting with the Ems–Dollard estuary (Baretta and Ruardij 1988), ecosystem studies have been performed in the European Wadden Sea in pursuit of the following goals: to understand the natural variability in the system, to study its resilience to natural and anthropogenic perturbations (Dittmann 1999; Gätje and Reise 1998), and to establish the degree of human impacts on the system. Every three years, scientific Wadden Sea symposia are held. Scientists exchange research findings, discuss management issues, and formulate recommendations for politicians and policy makers.

Since 1978, the protective status has been agreed upon at several international conferences as part of a trilateral agreement between the Netherlands, Germany, and Denmark. Guiding management principles and a set of common objectives were adopted in 1991 (Wadden Sea 1991) to achieve, as far as possible, a natural and sustainable ecosystem in which natural processes can proceed in an undisturbed way. For coordination, a Common Wadden Sea Secretariat (CWSS) was established in 1987. Seaward of the dikes, the Wadden Sea has been designated a *biosphere reserve* by all three states, and most of the German Wadden Sea attained the status of a National Park in 1985–1986. The chemical and biological quality of the Wadden Sea has been described in a "quality status report" as part of the North Sea Quality Status Report in 1993 (de Jong et al. 1993). One of the latest developments is the formulation of a trilateral assessment and monitoring program (TMAP), where all monitoring data are compiled and evaluated.

To be able to assess marine and coastal ecosystems, a reliable framework is needed but is still lacking, as are the targets and goals to test the state of the system. Therefore, within the European Environmental Agency, a project is underway to develop indicators of environmental quality (van Buuren, pers. comm.). In a recent proposal to the 5th Framework Programme of the European Union (EU) entitled "Socioeconomic Forcing of Fluxes from Catchments to Coastal Seas," it has been suggested that the effects of major European catchments on the coastal zone be studied. One aspect of this proposal is the development of suitable indicators for coastal areas under stress.

Very recently, Costanza and Mageau (1999) presented a contribution to the discussion on the health of ecosystems. They concluded with a series of indicators which, in combination, could be used as endpoints, from which a performance of the system (health) could be derived (Figure 4.2). Their indicators encompass system resilience against perturbations (resilience is the property of systems to return to the former state after a perturbation), life expectancy of the system (how long does the system keep its original properties), organization (biodiversity, the number of species, functional groups or links between organisms), and vigor (metabolism, productivity, scope for growth). Some of these parameters can be measured, whereas others need complicated analysis before reliable and useful data are available. In an earlier paper, Costanza et al. (1992) developed a definition of ecosystem health which encompassed the idea of sustainability as "a comprehensive, multiscale, dynamic measure of system resilience, organization, and vigor." The problem with this kind of valuation is that most of the indicators needed for the estimation of ecosystem health are difficult to assess and/or need

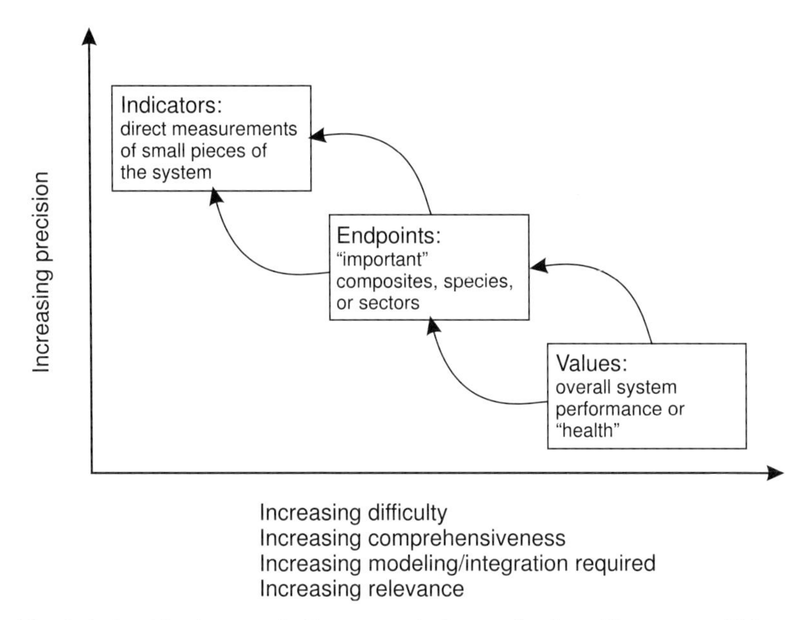

Figure 4.2 Relationships between indicators, endpoints, and values (Costanza and Mageau 1999). Reprinted by permission of Kluwer Academic Publishers.

network (food web) analysis or simulation modeling. This again means that all major relations in the organization of the system should be known and quantified, which is usually not the case.

Other attempts to cope with the problem of quantifying the state of an ecosystem were published by Radach (1998), who defined an ecological development index (EDI) that has been used to describe long-term changes in the German Bight based on data on nutrients, salinity, temperature, and phytoplankton composition. As an alternative to the more or less statistical and complex approaches, a rather simple method based on a comparison between present state and a reference state has been developed by Ten Brink et al. (1991) for the Dutch national water management plan (Colijn et al. 1996). In principle, this method is based on the estimation of numbers or population densities of a selection of up to 32 species of plants and animals or habitats (salt marshes) during a period of low human impact (reference) and high impacted periods. Many assumptions were needed to derive the numbers for these two periods, which can be taken to represent approximately the 1930s and the mid 1980s. The data were presented in a radar plot which showed some resemblance to an amoeba (a mobile protist), hence the procedure is called the AMOEBA method (Ten Brink and Colijn 1990) (Figure 4.3). The differences between reference state and present state were interpreted in terms of increased eutrophication, changes due to infrastructural works, overfishing in the North Sea, and protective measures regarding several species. The method has limitations; it is difficult to derive reliable reference data as well as data concerning unknown intrinsic developments and variabilities of natural ecosystems. Within the policy-making framework, the method worked well because it provided an easy to understand visualization of the data. Moreover, the recommendations strongly influenced the policy of the Ministry of Transport and Public Works and many of the recommendations were subsequently put into practice:

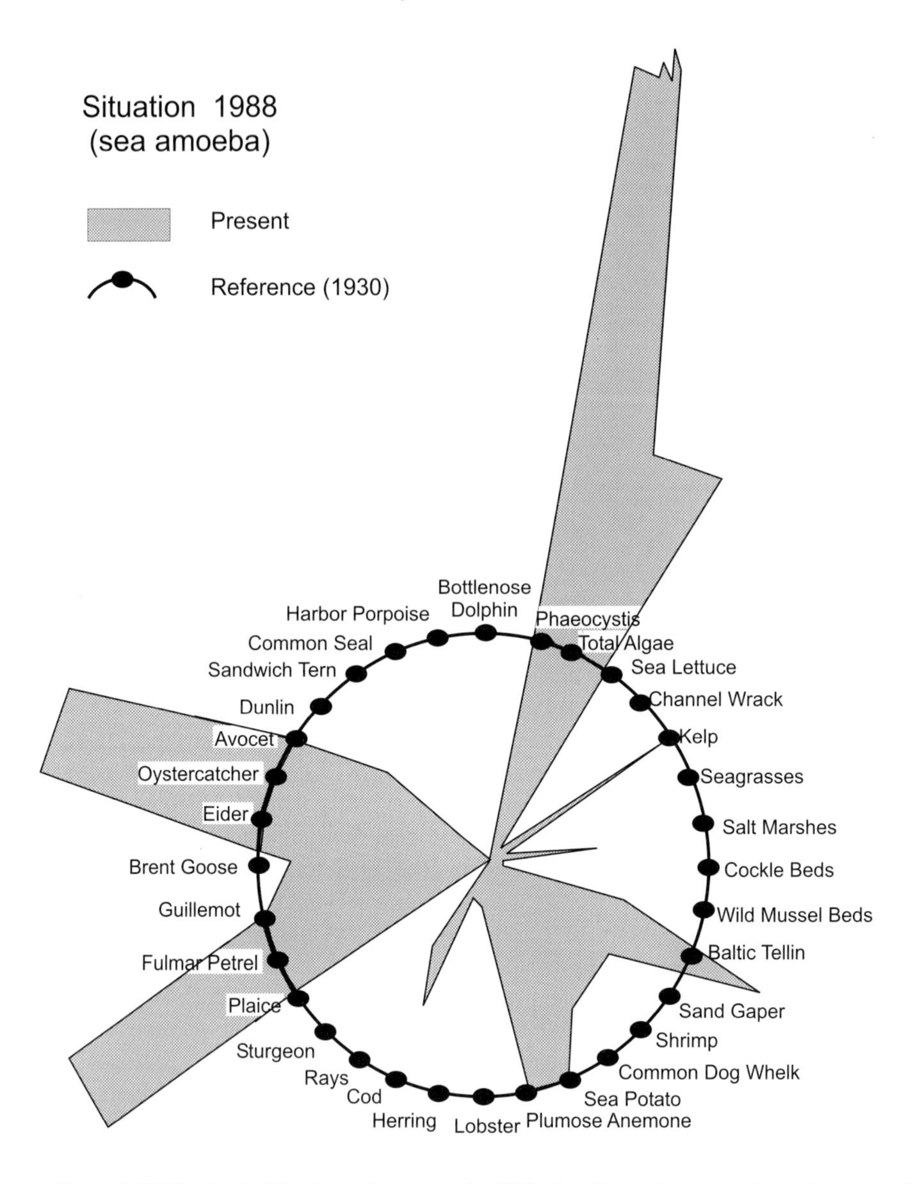

Figure 4.3 AMOEBA for the Dutch marine waters in 1988. For all species or entities, reference values are represented on the circle (100%). The present values are given by connecting the values to the gray area, an amoeba-like figure. (All values are given in Ten Brink and Colijn 1990).

nutrient reduction programs were intensified, a complete ban on organic pollutants like PCBs was accepted, and major reduction targets for several heavy metals were agreed. Restoration programs (e.g., to reintroduce seagrasses) were decided upon and implemented (de Jonge et al. 1996b). Because all measures concerning fisheries were under the responsibility of the Ministry of Agriculture, Nature Conservation (sic!), and Fisheries, the respective suggestions made in the policy plan for their sector had to be abandoned.

Keeping the caveats of these examples in mind, a framework for analyzing natural and anthropogenic effects is needed to evaluate the state of ecosystems. Such a framework is also urgently needed for integrated coastal management (ICM). The valuation problem linked to setting targets is treated both by Turner et al. (1996) and by von Bodungen and Turner (this volume).

THE INPUT FROM DRAINAGE BASINS

The amount of information available about discharges from European rivers into coastal zones is considerable. This is partly due to international agreements, like the International Rhine Commission, or to international agreements on the North Sea (International Ministerial Conferences about the North Sea , the Oslo and Paris Commissions), and the Baltic Sea (Helsinki Commission), which demand mandatory measurement and calculation of annual loads. Thus, regular updates on amounts of nutrients and pollutants are published and the goals are controlled in quality status reports (e.g., de Jong et al. 1993; NSQSR 1993), which are published regularly. A consequence of the political agreements is that for many compounds, especially heavy metals, clear effects of reductions can be observed (Laane et al. 1999) (Figure 4.4). For dissolved phosphate, concentration in the Rhine dropped to values of the early 1960s (de Vries et al. 1998). For phosphate, but not for nitrate, values have also dropped in the German Bight (Radach 1998; Hickel 1998) (Figure 4.5). In the northern German Wadden Sea, winter concentrations of phosphate and nitrogen are directly linked to the Elbe outflow, whereas in summer large amounts of phosphate are remobilized from the sediments and partly exported to the adjacent North Sea. This P-export can exceed the respective discharges of the river Elbe at this time of the year by a factor of 8 (Dick et al. 1999) and thus heavily counteract the efforts in P-reduction. In Table 4.1, a compilation of annual P-loads in the Rhine is presented, showing the long-term decline in loads.

Table 4.1 Annual loads of phosphate for the Rhine outflow from 1977–1998 in $10^3 t y^{-1}$ (after Lenhart et al. 1996; Lenhart, pers. comm.).

year	10^3 t y^{-1}	year	10^3 t y^{-1}	year	10^3 t y^{-1}	year	10^3 t y^{-1}
1977	14.9	1983	17.6	1989	12.6	1995	7.2
1978	17.1	1984	17.1	1990	8.9	1996	6.0
1979	17.9	1985	15.9	1991	6.7	1997	5.7
1980	21.0	1986	15.7	1992	6.8	1998	4.6
1981	19.5	1987	12.5	1993	7.7		
1982	16.0	1988	16.8	1994	7.7		

RELATIONS BETWEEN LAND USE AND DISCHARGES

Billen et al. (1991) were among the first to discuss the relation between land use and river discharges. Many other studies have attempted to relate agricultural uses to riverine concentrations of nutrients, especially nitrogen compounds (Boesch, this volume; Hickel 1998; de

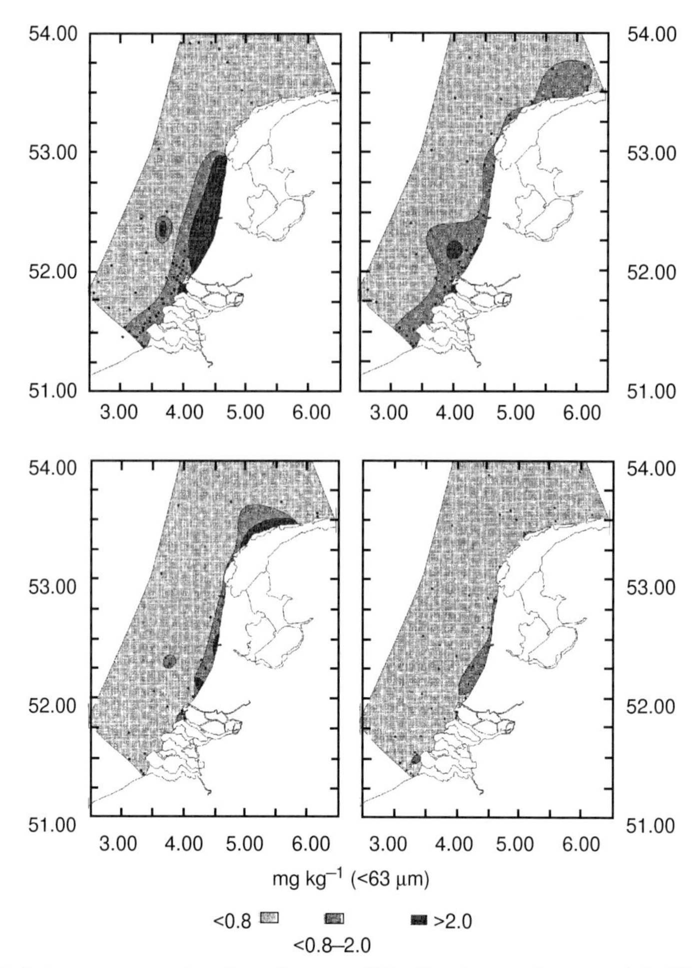

Figure 4.4 Cadmium concentrations in sediments off the Dutch coastal zone and the Wadden Sea in 1981, 1986, 1991, and 1996 (Laane et al. 1999). Reprinted by permission of Elsevier Science.

Jonge et al. 1995, 1996a). Present European legislation tries to reduce the amount of nutrient discharges towards rivers by prescribing maximal levels of nitrogen fertilizers per hectare and year. This should ultimately lead to a reduction of the diffuse sources of nitrogen compounds that have proved much more difficult to combat compared to the phosphate point sources from purification plants, where household waters are treated and dephosphatized in the so-called third purification step. In addition, the ban on phosphate-containing detergents has made a major contribution toward combating the high phosphate concentrations.

Differences can be observed among European rivers. The Rhine has a typical transboundary nature, a well-documented and implemented recovery strategy, and a clean-up plan at the catchment level. The Elbe catchment and associated estuary and coastal sea are examples of a river system subject to rapid changes in economic conditions in a major part of the catchment area, from a central to a market economy with a concomitant change in nutrients and other fluxes to the coast. In many other European estuaries and coastal areas, the legacy of

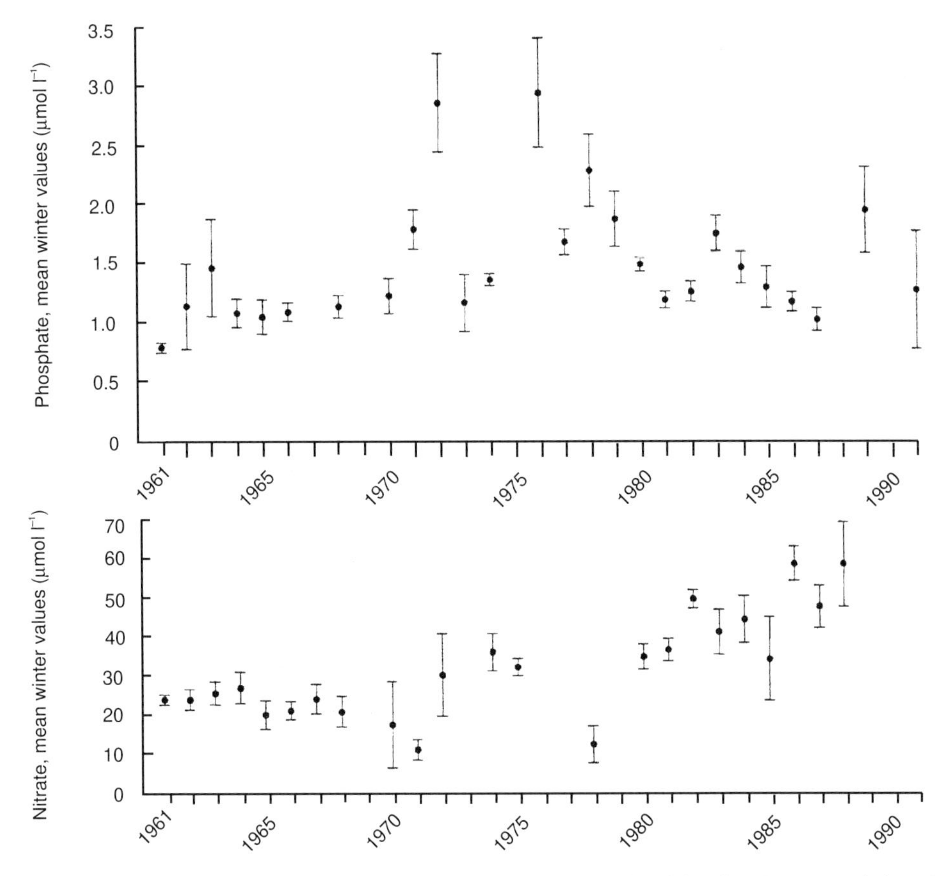

Figure 4.5 Dissolved nutrient concentrations near Helgoland in the German Bight after normalization to a salinity of 30 to show the long-term trend in the winter values (December–February) (from NSQSR 1993). Reprinted by permission of ICES Journal of Marine Science.

past industrialization and absence of regulations on emissions is still present in the sediments. A full overview of the daily nutrient loads for the European continental rivers during 1977 to 1993 is available (Lenhart et al. 1996), whereas discharges of organic contaminants and heavy metals are presented for the North Sea in the North Sea Quality Status Report (NSQSR 1993).

Together with clay and silt, transported from the North Sea onto the tidal flats and salt marshes of the Wadden Sea, heavy metals accumulate in the sediments. At present, concentrations of Zn, Pb, and Cu in the tidal sediments are only half of those found seaward of the Wadden Sea at a muddy basin 20 to 30 m in depth (Irion 1994). This is the primary site of most riverine depositions. From there, via secondary transport, heavy metal concentrations are expected to increase in the tidal sediments over the next decades, although the loads from the rivers have been decreasing since the 1980s. Some of the contaminants also arrive at the Wadden Sea directly from the rivers, which is indicated by higher concentrations of Hg and Cd in mussels from the Elbe estuary compared to other locations within the Wadden Sea

(NSQSR 1993). Industrial heavy metals accumulating since about 1880 may be expected to transcend gradually from the biosphere to the geosphere in salt marsh deposits. Prevalent erosion at salt marsh edges — a process largely caused by modern coastal architecture — brings these contaminants back to the biologically most active tidal flats.

A particular problem concerning the effects of inputs from drainage basins is the contribution that comes in from the seaward side in coastal seas. For many compounds, the natural (background) concentration is lower than in the river waters. However, the contribution of nitrate from the Atlantic to the North Sea is at least ten times larger than the input from land (Radach and Gekeler 1996; NSQSR 1993). Considerations about coastal processes must thus include offshore mixing and transport processes (Dick et al. 1999; Sündermann et al. 1999).

MIGRANT FAUNA

The relation between a drainage basin and the coastal sea is strongly asymmetric. The land exports nutrients and pollutants to the sea, but this is a unidirectional process. A number of fish, however, actively migrate from the sea upstream, far into the drainage basins. Anadromous fish invariably have declined or have become regionally extinct because of the additive effects of overfishing in the estuaries, various obstructions in the rivers such as dams, weirs and artificial river banks, and pollution. Notable examples of fish that used to spawn in the Rhine, Ems, Weser, Elbe, and Eider during the summer and which stayed in the Wadden Sea the rest of the year are lampern and lamprey, sturgeon, salmon and trout, houting, twaite shad, and Allis shad (Lozán et al. 1994). Today, only smelt is still present in significant numbers.

Thus, an entire ecosystem component has been decimated. Only recently have some populations begun to recover, and attempts are being made to reintroduce a few of the lost species. Other migrant fish, particularly slowly growing and reproducing species such as rays, have been completely lost from the Wadden Sea because these became a frequent by-catch of the commercial fishery system exploiting the entire North Sea.

A reverse development occurred with black-headed gulls (Exo 1995). This once common breeding bird in the drainage basins moved to the coast because of a loss of suitable inland breeding sites. This gull increased exponentially from zero in 1950 to the most common breeding gull in the 1990s, with 43,000 pairs reported around the Wadden Sea. In addition, other coastal birds such as herring gull, lesser black-backed gull, common gull, eider duck, cormorant, and oystercatcher increased likewise because of less hunting and collecting of eggs, protection of breeding sites, and discards of the fisheries offered as an additional food supply (Meltofte et al. 1994). As a result, a general switch from fish to birds occurred in the upper trophic levels of the coastal food web. This was essentially caused by differential, sectoral management. While fishery management failed largely to maintain the fish stocks of the Wadden Sea, habitat and resource shifts together with very effective bird protection measures caused a recovery in coastal bird numbers in the Wadden Sea.

INTRODUCED SPECIES

About 80 nonindigenous species are assumed to have been introduced into the coastal North Sea (Reise et al. 1999). Half of these are well established in the Wadden Sea. Most of the exotic invertebrates originate from the Western Atlantic and were introduced by transoceanic shipping. Most algae stem from the Pacific. They arrived accidentally with introduced oysters, mainly first at the French Atlantic coast and then by secondary dispersal immigrating into the Wadden Sea. Most of the arrivals became established in brackish environments. Here the share of macrobenthic species is up to 20%, while at the open marine coast the share is about 6%. Only a few species had the capacity to transform the ecosystem locally. An example is the deliberately introduced cordgrass, *Spartina anglica*. It completely changed the vegetation at seaward edges of salt marshes. Another example is the razor clam, *Ensis americanus,* which added a large amount of biomass to mobile nearshore sediments where there was almost none before its introduction in 1978 with ballast water. A more elaborate review of introduced species is given by Nehring and Leusch (1999).

Management must seek to enforce effectively the already adopted quarantine procedures for aquaculture organisms. For ballast water, effective treatments to reduce introductions still need to be developed.

CHANGES IN THE WADDEN SEA

The previous and present biological situation, in regard to nutrients, primary production, phytoplankton, seagrasses, and macrobenthos, has been described in numerous papers, which show that contradictory statements on the causes and effects of eutrophication have been made, e.g., the role of phosphorous transport through the IJsselmeer, effects of the Rhine discharge, and import of nutrients from the North Sea (Cadée and Hegeman 1993; de Jonge et al. 1995, 1996a, b; de Vries et al. 1998; Reise 1994). Possible effects of other contaminants are even more difficult to assess: only a few good studies of direct effects of pollutants exist, for example, on the impact of DDT on birds (mainly terns), on PCBs affecting the fertility of seals, and on tributyltin compounds affecting gastropods. Despite these studies, a clear framework is still lacking to discriminate the causes and to assess the state of the ecosystem.

One of the reasons why the causal link between inputs of nutrients and the eutrophication effects is so difficult to unravel is that, in contrast to freshwater lakes or enclosed seas, open marine systems have a wide variety of transport, diffusion, and flushing mechanisms. Whereas the hydrographic situation is relatively simple in front of the Dutch coast, where the Rhine river flows directly into a coastal water mass that is subsequently transported by residual currents to the north, the hydrography of the German Bight is rather complicated and strongly dependent on meteorological (wind) conditions. Depending on the strength and direction of the wind, the Elbe plume mixes with water masses in the Wadden Sea or flows to the west and gets incorporated into a large eddy in the German Bight. Thus, the retention time of these Elbe water masses in front of the German coast is at times much longer than that of the Rhine waters off the Dutch coast (de Vries et al. 1998). Therefore, it is clear that predictions on nutrient concentrations and phytoplankton biomass seldom show good correlations (de Vries et al. 1998; Hickel 1998; de Jonge et al. 1996a; Hesse et al. 1995; Bauerfeind et al. 1990). However, Hickel (1998) observed a close relation between nitrate winter concentrations and the nanophytoplankton biomass (Figure 4.6). These relations are also strongly affected by variability in underwater irradiance levels (Colijn 1982; Cloern 1999) which do play a substantial role as limiting factor for phytoplankton growth in turbid coastal zones. Model calculations by Lenhart (1999) support the conclusion that a 50% nutrient load

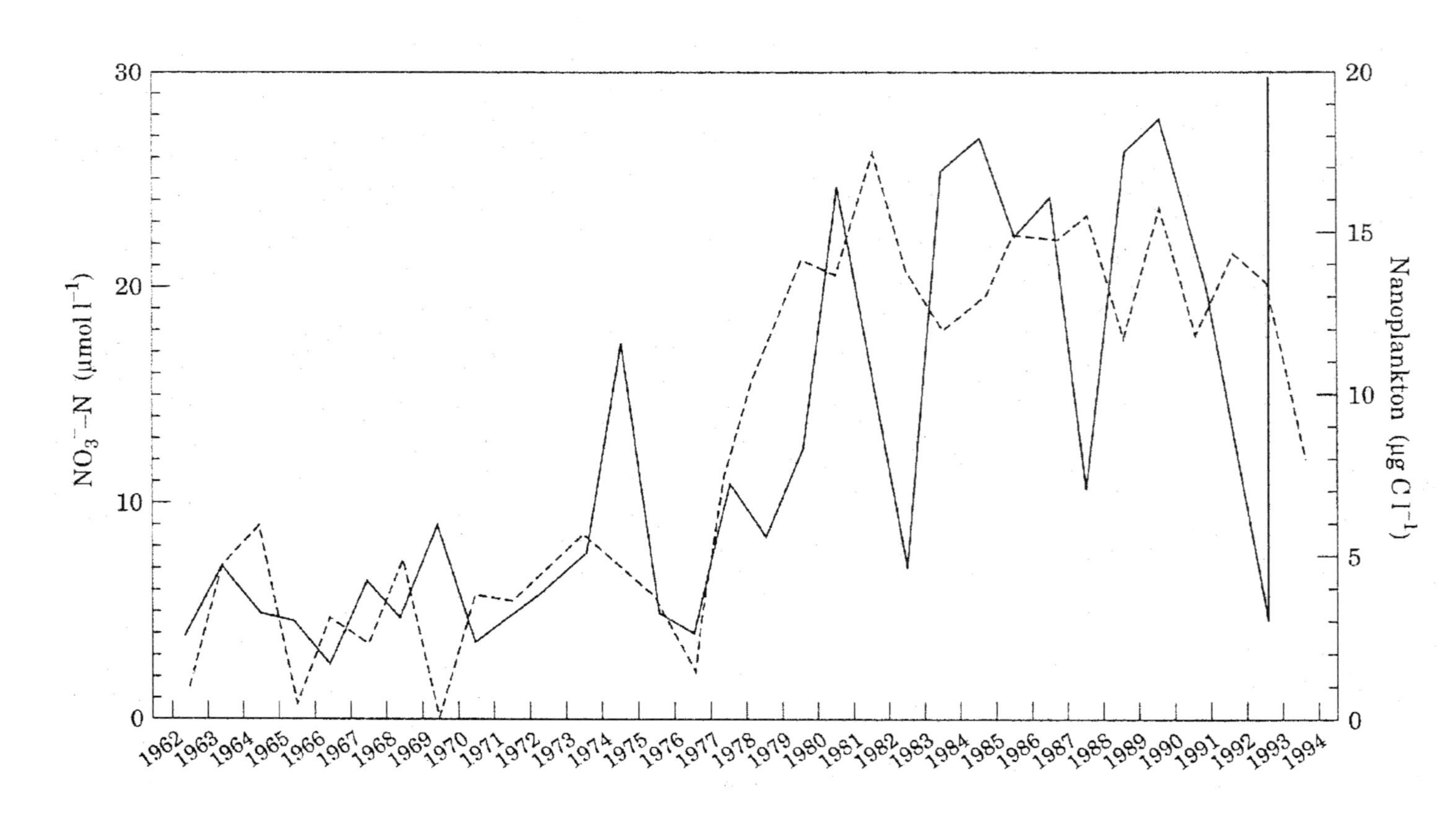

Figure 4.6 Long-term trend in nitrate concentration and nanoplankton biomass at Helgoland from 1962 to 1994 (after Hickel 1998). Reprinted by permission of ICES Journal of Marine Science.

reduction in N and P cannot be linearly transferred to a 50% reduction in primary production (e.g., Cadée and Hegeman 1993).

In a recent study, van Beusekom (pers. comm.) compared time series of nutrients and chlorophyll from different parts of the Wadden Sea. During years of high nitrogen input into the coastal zone, he observed that the Dutch Wadden Sea showed a more intense remineralization of organic matter (ammonium and nitrite values in autumn enhanced) than during years with a low nitrogen input. Van Beusekom explains this correlation with a higher production in the coastal zone and, consequently, a higher import of organic matter from the coastal zone into the Wadden Sea during years with a high nitrogen input. This view is in line with observations by Schaub and Gieskes (1991) and Cadée and Hegeman (1993), who found that more phytoplankton is present in years with a high river discharge than in dry years. It also reaffirms the notion that nitrogen is the single limiting nutrient for primary production in the North Sea (van Beusekom and Diel-Christiansen 1994; Hydes et al. 1999).

On a worldwide scale, massive occurrences of green algae covering intertidal sediments in sheltered lagoons and estuaries have been recorded over recent decades and have generally been attributed to anthropogenic eutrophication of coastal waters (Fletcher 1996). Several species of green algae are involved with filamentous as well as foliose thalli, seasonally generating coherent mats. Underneath such algal mats, anoxic, hydrogen-sulphidic conditions often occur, and most of the infauna emigrates or dies (Raffaelli 1999). Historically, in the Wadden Sea, green algae occurred only marginally. The first sign of an increase, as observed in the late 1950s, was followed by strong, rather sudden massive development in the 1980s. It reached its maximum level in 1991 and gradually declined thereafter, but it has remained well above the assumed baseline abundance (Kolbe et al. 1995; Reise and Siebert 1994).

Neither this temporal development nor the spatial distribution on the intertidal mud- and sandflats within the Wadden Sea is closely correlated with nutrient supply. In addition to nitrogen, green algae need proper light and temperature in spring as well as an abundance of biogenic substrates for germination and attachment on the otherwise mobile sediments (Schories and Reise 1993). Storms were a crucial factor during further development. Waves dislocate the benthic green algae and hence interrupt or eventually terminate further growth into coherent mats. Thus, the coincidence of several independent factors is required for massive green algal growth.

In contrast to the proliferation of benthic green algae, seagrass beds (*Zostera noltii* and *Z. marina*) have declined dramatically in the southern part of the Wadden Sea, where the riverine loads are highest (Reise 1994). The decline over the last three decades is assumed to be ultimately caused by anthropogenic eutrophication (Philippart 1994, 1995). The causal chains involve competition with bioturbating lugworms (*Arenicola marina*), heavy overgrowth with epiphytes, and insufficient abundances of epiphytic grazers.

Algal blooms, duration of blooms, chlorophyll concentrations, primary production, and species composition of phytoplankton have all been directly linked to eutrophication. The occurrence of green algae in large masses, the decline of seagrasses, and low oxygen concentrations in deep water of the German Bight, primarily in the 1980s (von Westerhagen et al. 1986), have been linked to eutrophication. Whereas relations between nutrients and plants are generally more direct, for higher trophic levels these relations become increasingly more difficult to detect. In the western Wadden Sea, the total zoobenthic biomass roughly doubled after the stock of planktonic algae was doubled in the late 1970s (Beukema and Cadée 1997).

For higher trophic levels like fishes, birds, and marine mammals, such relations have not yet been studied in enough detail to allow for clear conclusions (e.g., Boddeke 1996; Prins et al. 1999; Zwarts and Wanik 1993).

MANAGEMENT EFFECTS

The coastal biota of the Wadden Sea have undergone a history of changes due to a combination of various factors. The entire coastal architecture has changed over time with important ecological consequences (Reise 1998). Successive embankments of brackish reed marshes, salt marshes, and mudflats decreased the share of these habitats in the Wadden Sea. These served as sinks for organic matter transported with the tides. The tidal area has become gradually narrower, by about one-third of its natural width, and a landward dissipation of hydrodynamic energy is being blocked by seawalls (Reise et al. 1998; Pethick, this volume). This decreases sediment stability on the remaining tidal flats, as it is reflected in grain size distributions, deficient of smaller particles (Flemming and Nyandwi 1994). This and a concomitant increase in turbidity is expected to counter the effects of anthropogenic nutrient enrichment on benthic and pelagic primary production. In part of the northern Wadden Sea, embankments plus causeways from the mainland to the islands seem to have generated a situation where altered hydrodynamics dominate over eutrophication, while in other parts of the Wadden Sea eutrophication effects overrule those of increased hydrodynamics (Reise et al. 1998). In principle, it would be possible to mitigate the situation by converting previously embanked areas into retention sites for sediments and organic substances. This can be achieved by allowing tides to overflow a lowered dikeline at seaward polders while enforcing the landward dikeline which then has to guarantee the safety of populated marshlands. These retention sites could be managed to regain salt marshes and brackish reed marshes and thus may also serve as a biological filter for runoffs from the agricultural hinterland. In addition, retention sites may attract more tourists and related revenues, which can counterbalance financial losses due to precluded agricultural use. Public discussion for such alternatives in coastal architecture is, however, still at a very early stage.

Because of the transboundary character of eutrophication, it gets more and more difficult to tackle the effects through legislation in one country alone. Moreover, other changes in the system, like the reduction of intertidal flats opposed to natural situations, might reduce the capacity to remineralize the organic matter introduced. This would then reduce eutrophication effects because less organic nitrogen is converted to ammonium and nitrite.

Toxic pollutants may potentially suppress effects of eutrophication. This is discussed by Herman et al. (1991) for the Westerschelde estuary. As measured in mussels, contamination of the estuary with heavy metals was high, but concentrations of cadmium and copper have fallen again since 1982. Phytoplankton abundance was low and began to increase in the 1980s. A surplus of nutrients such as phosphorous and nitrogen was already present and concentrations remained at a high level. It is assumed that algal growth was formerly suppressed by toxic metals and that eutrophication effects could not proliferate. A reverse case was presented by Jak (1997), who observed interactions between contaminants and nutrients in mesocosms. He argued that the expression of eutrophication was more obvious when contaminant levels were high and thus zooplankton grazing was reduced.

In some small areas of the Wadden Sea, mussel beds increased in size and number, while elsewhere mussel beds decreased in comparison to the first half of the 20th century (Reise 1994). An expansion of mussel beds may be attributed to an improved food supply mediated by eutrophication. However, the mussel fishery evidently reversed this trend in most areas, resulting in a food shortage for oystercatchers, which feed on mussels (Beukema 1992). Seal reproduction in the Wadden Sea was shown to be affected by PCB contamination (Reijnders 1986). Nevertheless, seal population size increased exponentially because hunting was phased out and because resting sites were protected from disturbing visitors.

These counteracting processes may generate spurious similarity to a reference state of the system. Thus, in addition to the obvious advantage of having indicator species, we still need to know about the underlying proximate and ultimate causes to advise coastal management properly. Monitoring needs to be augmented by research on the ever-changing interplay of coastal factors.

NATURAL VARIABILITY

The lack of insight into the long-term development of marine ecosystems is probably due to other complications: the lack of information on natural variability and the lack of discriminatory techniques for distinguishing between anthropogenic and natural causes. The effects of climate change have been the subject of many long-term studies, but long-term observations are rarely available, with the exception of the Helgoland plankton time series since 1962 (Hickel 1998) and the Continuous Plankton Recorder series, which now spans 40 to 60 years (SAHFOS 1999). These data show that long-term changes in biodiversity in the North Sea occur (Nehring 1998), that exceptional inflow events with appearance of specific organisms occur, and that long-term changes in the phytoplankton biomass were correlated with the North Atlantic Oscillation (NAO) index, an index that indicates large-scale weather patterns over the North Atlantic Ocean (Sündermann et al. 1996). Speculations on North Sea regime shifts, based on phytoplankton and zooplankton observations and large changes in fish catches, have also been related to positive NAO indices. This correlates with the hypotheses presented by Lindeboom et al. (1994), who stressed the importance of natural factors for regulation of ecosystem dynamics in the North Sea as part of their explanation of sudden changes in the Wadden Sea observed in the late 1970s.

Occasionally, severe winters occur in the Wadden Sea, when the prevailing Atlantic deep pressure systems are interrupted by a long-lasting continental high pressure system. A severe winter constitutes a cutting event in the benthic ecology of the Wadden Sea (Beukema 1985, 1990). First, ice and low temperatures cause massive losses in bivalves and other invertebrates. Second, various species respond with a particularly high recruitment during the following summer, leading to elevated benthic biomass some years later as the animals grow. Conversely, an extended period of mild winters causes the average biomass to decrease (Beukema 1992). This phenomenon is mainly caused by epibenthic predators becoming even more sensitive to severe winters than their prey.

The Wadden Sea ecosystem is not a closed system but represents a fringe of the larger North Sea, which is connected with the Northern Atlantic. Therefore, it is subject to changes and events originating in these adjacent, much larger water bodies. These may often overrule the impacts originating in the drainage basin of the Wadden Sea.

SCIENCE AND AGENCIES

In the continental coastal areas of the North Sea (Dutch coast, German Bight) and the Wadden Sea, comprehensive knowledge is available on inputs and local processes as well as on the biological structure and function of major parts of the ecosystem (de Jong et al. 1993; Dittmann 1999; Gätje and Reise 1998). However, even under favorable conditions, predictions on future development are difficult, and some events cannot be predicted at all (e.g., introduction of a new species; Nehring 1998; Reise et al. 1999). Only a few ecological models exist that are able to describe the main bulk processes, such as primary production, respiration, and consumption, for at least part of the most important functional groups of organisms (phytoplankton, microphytobenthos, seagrasses, macrobenthos, zooplankton, bacteria, etc.). The coastal system is in a permanent state of change. This limits the value of established knowledge, and experience needs to be supplemented by ongoing surveys and process analyses.

The linkage of processes occurring in the river basins and nearshore coastal areas like the Wadden Sea is also hampered because the information needed for managerial decisions is distributed over many services and institutions, which are themselves governed by different agencies or ministries. These agencies and ministries have different tasks to complete and different visions as to the development of terrestrial and marine systems. Therefore, no mechanism is available for a constructive discussion on these visions. Moreover, the ministries and lower governmental institutions have different levels of expertise (information), political power, and are often not willing to share their power with others, thus keeping their own position as strong as possible. Consequently, they are not a priori open for cooperation.

This is not in line with the current philosophy of ICM, which envisages a dynamic process in which a coordinated strategy is developed and implemented for the allocation of environmental, social, cultural, and institutional resources, to achieve the conservation and the sustainable multiple use of the coastal zone. Main drawbacks of current management efforts are the lack of consistency, the inability of governments to adapt their strategies towards an integrated policy, and several gaps in knowledge, including problems with setting criteria for assessments and evaluation.

A link between science and management is the availability and development of tools and criteria for assessment. If these are not available, how can we discuss environmental problems? Is a link to economic and monetary issues needed or can this discussion be held without taking socioeconomic factors into account? Another unsolved problem involves communicating the research results and managerial consequences to potential users, like fishermen, and how to reach a final consensus after discussion of these often negatively appreciated consequences. Several mechanisms will help to improve the quality of the Wadden Sea; its international status is an important prerequisite for ongoing protection. The wealth of natural processes occurring is an impetus for scientists to study these processes. An interesting question is to what extent results and knowledge can be transferred to other areas (transboundary science). The outcome of actual restoration measures, like the planting of seagrass in the Netherlands, is uncertain because of complex interactions between nutrients and salinity (van Katwijk et al. 1999). Habitat restoration of rivers will probably have a more direct effect on migrant fish stocks mentioned above. Together with an effective management of fisheries, improvements may be expected.

ACKNOWLEDGMENTS

We would like to thank Justus van Beusekom for the use of data from his unpublished report on Wadden Sea specific eutrophication criteria.

REFERENCES

Baretta, J., and P. Ruardij. 1988. Tidal Flat Estuaries, Simulation, and Analysis of the Ems Estuary. Ecological Studies 71. Berlin: Springer.

Bauerfeind, E., W. Hickel, U. Niermann, and H. von Westernhagen. 1990. Phytoplankton biomass and potential nutrient limitation of phytoplankton development in the southeastern North Sea in spring 1985 and 1986. *Neth. J. Sea Res.* **25**:131–142.

Beukema, J.J. 1985. Zoobenthos survival during severe winters on high and low tidal flats in the Dutch Wadden Sea. In: Marine Biology of Polar Regions and Effects of Stress on Marine Organisms, ed. J.S. Gray and M.E. Christiansen, pp. 351–361. Chichester: Wiley.

Beukema, J.J. 1990. Expected effects of changes in winter temperatures on benthic animals living in soft-sediments in coastal North Sea areas. In: Expected Effects of Climate Change on Marine Coastal Ecosystems, ed. J.J. Beukema, W.J. Wolff, and J.J.W.M. Brouns, pp. 83–92. Dordrecht: Kluwer.

Beukema, J.J. 1992. Expected changes in the Wadden Sea benthos in a warmer world: Lessons from periods with mild winters. *Neth. J. Sea Res.* **30**:73–79.

Beukema, J.J., and G.C. Cadée. 1997. Local differences in macrozoobenthic response to enhanced food supply caused by mild eutrophication in a Wadden Sea area: Food is only locally a limiting factor. *Limnol. Oceanogr.* **42**:1424–1435.

Billen, G., C. Lancelot, and M. Meybeck. 1991. N, P, and Si retention along the aquatic continuum from land to ocean. In: Ocean Margin Processes in Global Change, ed. R.F.C. Mantoura, J.M. Martin, and R. Wollast, pp. 19–44. Dahlem Workshop Report. Chichester: Wiley.

Boddeke, R. 1996. Changes in the brown shrimp (*Crangon crangon L.*) population off the Dutch coast in relation to fisheries and phosphate discharge. *ICES J. Mar. Sci.* **53**:995–1002.

Cadée, G.C., and J. Hegeman. 1993. Persisting high levels of primary production at declining phosphate concentrations in the Dutch coastal area (Marsdiep). *Neth. J. Sea Res.* **31**:147–152.

Cloern, J.E. 1999. The relative importance of light and nutrient limitation of phytoplankton growth: A simple index of coastal ecosystem sensitivity to nutrient enrichment. *Aquat. Ecol.* **33**:3–16.

Colijn, F. 1982. Light absorption in the waters of the Ems–Dollard estuary and its consequences for the growth of phytoplankton and microphytobenthos. *Neth. J. Sea Res.* **15**:196–216.

Colijn, F., R. Laane, H.R. Skjoldal, and J. Asjes. 1996. Ecological Quality Objectives in Perspective. Proc. Sci. Symp. on the 1993 North Sea Quality Status Report, pp. 249–254. Copenhagen: Ministry of Environment and Energy, Danish Environmental Protection Agency.

Costanza, R., and M. Mageau. 1999. What is a healthy ecosystem? *Aquat. Ecol.* **33**:105–115.

Costanza, R., B. Norton, and B.J. Haskell. 1992. Ecosystem Health: New Goals for Environmental Management. Washington D.C.: Island Press.

De Jong, F., et al., eds. 1993. Quality Status Report of the North Sea, Subregion 10, The Wadden Sea. Wilhelmshaven: Common Wadden Sea Secretariat.

De Jonge, V.N., J.F. Bakker, and M. van Stralen. 1996a. Recent changes in the contributions of river Rhine and North Sea to the eutrophication of the western Dutch Wadden Sea. *Neth. J. Aquat. Ecol.* **30**:27–39.

De Jonge, V.N., D.J. de Jong, and J. van den Bergs. 1996b. Reintroduction of eelgrass (*Zostera marina*) in the Dutch Wadden sea: Review of research and suggestions for management measures. *J. Coast. Conserv.* **2**:149–158.

De Jonge, V.N., et al. 1995. Responses to developments in eutrophication in four different North Atlantic estuarine systems. In: Changes in Fluxes in Estuaries, ed. K.R. Dyer and R.J. Orth, pp. 179–196. Fredensborg: Olsen and Olsen.

De Vries, I., et al. 1998. Patterns and trends in nutrients and phytoplankton in Dutch coastal waters: Comparison of time series analysis, ecological model simulation, and mesocosms experiments. *ICES J. Mar. Sci.* **55**:620–634.

Dick, S., et al. 1999. Exchange of matter and energy between the Wadden Sea and the coastal waters of the German Bight. Estimations based on numerical simulations and field measurements. *Dt. Hydrogr. Z.* **51(2/3)**:181–220.

Dijkema, K.S., G. van Tienen, and J.G. van Beek. 1989. Habitats of the Netherlands, German, and Danish Wadden Sea 1:100,000. Research Institute for Nature Management, Texel. Leiden: Veth Foundation.

Dittmann S., ed. 1999. The Wadden Sea Ecosystem: Stability, Properties, and Mechanisms. Berlin: Springer.

Exo, K.-M. 1995. Das Wattenmeer — Unverzichtbarer Lebensraum für Millionen Küstenvögel. Schriftenreihe Schutzgemeinschaft Deutsche Nordseeküste (SDN) No. 8, pp. 8–46. Wilhelmshaven: SDN.

Flemming, B.W., and N. Nyandwi. 1994. Land reclamation as a cause of fine-grained sediment depletion in backbarrier tidal flats/southern North Sea. *Neth. J. Aquat. Ecol.* **28**:299–307.

Fletcher, R.L. 1996. The occurrence of "green tides" — A review. In: Marine Benthic Vegetation, ed. W. Schramm and P.H. Nienhuis, pp. 7–43. Ecological Studies 123. Berlin: Springer.

Gätje, C., and K. Reise, eds. 1998. Ökosystem Wattenmeer, Austausch-, Transport- und Stoffumwandlungsprozesse. Berlin: Springer.

Herman, P.M.J., H. Hummel, M. Bokhorst, and A.G.A. Merks. 1991. The Westerschelde: Interaction between eutrophication and chemical pollution? In: Estuaries and Coasts: Spatial and Temporal Intercomparisons, ed. M. Elliot and J.-P. Ducrotoy, pp. 359–364. Fredensborg: Olsen and Olsen.

Hesse, K.-J., U. Tillmann, and U. Brockmann. 1995. Nutrient–phytoplankton relations in the German Wadden Sea. ICES Annual Science Conference Papers No. T:8. ICES: Copenhagen.

Hickel, W. 1998. Temporal variability of micro- and nanoplankton in the German Bight in relation to hydrographic structure and nutrient changes. *ICES J. Mar. Sci.* **55**:600–609.

Hydes, D.J., B.A. Kelly-Gerreyn, A.C. Le Gall, and R. Proctor. 1999. The balance of supply of nutrients and demands of biological production and denitrification in a temperate latitude shelf sea — A treatment of the southern North Sea as an extended estuary. *Mar. Chem.* **1–2**:117–131.

Irion, G. 1994. Morphological, sedimentological, and historical evolution of Jade Bay, southern North Sea. *Senckenbergiana marit.* **24**:171–186.

Jak, R.G. 1997. Toxicant-induced changes in zooplankton communities and consequences for phytoplankton development. Ph.D. diss., Univ. of Amsterdam.

Kolbe, K., et al. 1995. Macroalgal mass development in the Wadden Sea: First experiences with a monitoring system. *Helgoländer Meeresunters.* **49**:519–528.

Laane, R.W.P.M., et al. 1999. Trends in the spatial and temporal distribution of metals (Cd, Cu, Zn, and Pb) and organic compounds (PCBs and PAHs) in the Dutch coastal zone sediments from 1981–1996: Possible sources and causes for Cd and PCBs. *J. Sea Res.* **41**:1–17.

Lenhart, H.-J. 1999. Eutrophierung im kontinentalen Küstenbereich der Nordsee. Reduktionsszenarien der Flußeinträge von Nährstoffen mit dem Ökosystem-Modell ERSEM. Ph.D. diss., Univ. of Hamburg.

Lenhart, H.-J., J. Pätsch, and G. Radach. 1996. Daily nutrient loads for the European continental rivers during 1977–1993. Technical report. Berichte aus dem Zentrum für Meeres-und Klimaforschung (ZMK) Hamburg. Hamburg: ZMK.

Lindeboom, H., et al. 1994. (Sudden) changes in the North Sea and Wadden Sea: Oceanic influences underestimated? *Dt. Hydrogr. Z.* **Suppl. 2**:87–100.

Lozán, J.L., et al. 1994. Warnsignale aus dem Wattenmeer. Berlin: Blackwell.

Meltofte, H., et al. 1994. Numbers and distribution of waterbirds in the Wadden Sea. International Waterfowl and Wetlands Research Bureau (IWRB) Publication 34/Wader Study Group Bull. No. 74, pp. 1–192. Wageningen: Wetlands International.

Nehring, S. 1998. Establishment of thermophilic phytoplankton species in the North Sea: Biological indicators of climatic changes? *ICES J. Mar. Sci.* **55**:818–823.

Nehring, S., and H. Leusch. 1999. Neozoa (Makrobenthos) an der deutschen Nordseeküste — Eine Übersicht. Bericht BfG-1200. Koblenz: Bundesanstalt für Gewässerkunde.

NSQSR (North Sea Quality Status Report). 1993. Oslo and Paris Commissions, London. Fredensborg: Olsen and Olsen.

Philippart, C.J.M. 1994. Interactions between *Arenicola marina* and *Zostera noltii* on a tidal flat in the Wadden Sea. *Mar. Ecol. Prog. Ser.* **111**:251–257.

Philippart, C.J.M. 1995. Effect of periphyton grazing by *Hydrobia ulvae* on the growth of *Zostera noltii* on a tidal flat in the Dutch Wadden Sea. *Mar. Biol.* **122**:431–437.

Prins, T.C., et al. 1999. Effects of different N- and P-loading on primary and secondary production in an experimental marine ecosystem. *Aquat. Ecol.* **33**:65–81.

Radach, G. 1998. Quantification of long-term changes in the German Bight using an ecological development index. *ICES J. Mar. Sci.* **55**:587–599.

Radach, G., and J. Gekeler. 1996. Annual cycles of horizontal distributions of temperature and salinity, and of concentrations of nutrients, suspended particulate matter, and chlorophyll on the north-west European shelf. *Dt. Hydrogr. Z.* **48 (3/4)**:261–298.

Raffaelli, D. 1999. Impact of catchment land-use on an estuarine benthic food web. In: Biogeochemical Cycling and Sediment Ecology, ed. J.S. Gray, W. Ambrose, Jr., and A. Szaniawska, pp. 161–171. Dordrecht: Kluwer.

Reijnders, P.J.H. 1986. Reproductive failure in common seals feeding on fish from polluted coastal waters. *Nature* **324**:456–457.

Reise, K. 1994. Changing life under the tides of the Wadden Sea during the 20th century. *Ophelia* **Suppl. 6**:117–125.

Reise, K. 1998. Coastal change in a tidal backbarrier basin of the Northern Wadden Sea: Are tidal flats fading away? *Senckenbergiana marit.* **29**:121–127.

Reise, K., S. Gollasch, and W.J. Wolff. 1999. Introduced marine species of the North Sea coasts. *Helgoländer Meeresunters.* **52**:219–234.

Reise, K., and I. Siebert. 1994. Mass occurrence of green algae in the German Wadden Sea. *Dt. Hydrogr. Z.* **Suppl. 1**:171–180.

Reise, K., et al. 1998. Austauschprozesse im Sylt-Rømø Wattenmeer: Zusammenschau und Ausblick. In: Ökosystem Wattenmeer, Austausch-, Transport- und Stoffumwandlungsprozesse, ed. C. Gätje and K. Reise, pp. 530–557. Berlin: Springer.

SAHFOS (Sir Alister Hardy Foundation for Ocean Science). 1999. Annual report of the Sir Alister Hardy Foundation for Ocean Science. Plymouth: SAHFOS.

Schaub, B.E.M., and W.W.C. Gieskes. 1991. Eutrophication of the North Sea: The relation between Rhine river discharge and chlorophyll-*a* concentration in Dutch coastal waters. In: Estuaries and Coasts: Spatial and Temporal Intercomparisons, ed. M. Elliot and J.-P. Ducrotoy, pp. 85–90. Fredensborg: Olsen and Olsen.

Schories, D., and K. Reise. 1993. Germination and anchorage of *Enteromorpha spp.* in sediments of the Wadden Sea. *Helgoländer Meeresunters.* **47**:275–285.

Sündermann, J., et al. 1996. Decadal variability on the Northwest European shelf. *Dt. Hydrogr. Z.* **48**:365–400.

Sündermann, J., K.J. Hesse, and S. Beddig. 1999. Coastal mass and energy fluxes in the southeastern North Sea. *Dt. Hydrogr. Z.* **51(2/3)**:113–132.

Ten Brink, B.J.E., and F. Colijn, eds.. 1990. Ecologische ontwikkelingsrichtingen Zoute wateren. Ministerie van Verkeer en Waterstaat, Rijkswaterstaat, Dienst Getijdewateren, Notanummer GWWS–90.009. Den Haag: Ministry of Transport, Public Works, and Water Management.

Ten Brink, B.J.E., S.H. Hosper, and F. Colijn. 1991. A quantitative method for the description and assessment of ecosystems: The AMOEBA approach. *Mar. Poll. Bull.* **23**:265–270.

Turner, R.K., S.E. Subak, and W.N. Adger. 1996. Pressures, trends, and impacts in the coastal zones: Interactions between socioeconomic and natural systems. *Env. Manag.* **20**:159–173.

Van Beusekom, J.E.E., and S. Diel-Christiansen. 1994. A Synthesis of Phyto- and Zooplankton Dynamics in the North Sea Environment. Godalming: WWF (World Wildlife Fund).

Van Katwijk, M.M., G.H.W. Schmitz, A.P. Gasseling, and P.H. van Avesaath. 1999. Effects of salinity and nutrient load and their interaction on *Zostera marina*. *Mar. Ecol. Prog. Ser.* **190**:155–165.

Von Westernhagen, H., et al. 1986. Sources and effects of oxygen deficiencies in the southeastern North Sea. *Ophelia* **26**:457–473.

Wadden Sea. 1991. The Wadden Sea: Status and Developments in an International Perspective. Report to the Sixth Trilateral Governmental Conf. on the Protection of the Wadden Sea, Esbjerg, Nov. 13, 1991. Wilhemshaven: National Forest and Nature Agency, Danish Ministry of the Environment, and the Common Wadden Sea Secretariat.

Wolff, W.J., ed. 1983. Ecology of the Wadden Sea. 3 vols. Rotterdam: Balkema.

Zwarts, L., and J.H. Wanik. 1993. How the food supply harvestable by waders in the Wadden Sea depends on the variation in energy density, body weight, biomass, burying depth, and behaviour of tidal-flat invertebrates. *Neth. J. Sea Res.* **31**:441–476.

5

Eutrophication in the Black Sea and a Basin-wide Approach to Its Control

L.D. MEE

Plymouth Environment Research Centre, Environmental Policy Unit,
University of Plymouth, Drake Circus, Plymouth PL4 8AA, Devon, U.K.

ABSTRACT

The Black Sea ecosystem has been severely altered at every trophic level as a result of eutrophication. The loss of the benthic ecosystem of the northwestern shelf of the Black Sea was caused by the formation of a vast hypoxic "dead zone," present seasonally since 1973. The phenomenon of eutrophication itself results from human activities in all 17 countries of the Black Sea basin. The most significant source of nutrients is agriculture, mostly through the excessive application of chemical fertilizers and their subsequent loss to streams and the atmosphere. Large intensive animal farms are also a major cause of nutrient discharge. The economic collapse of most central and eastern European countries has resulted in decreased nutrient loadings, providing a temporary reprieve for the system. The countries of the Black Sea basin are endeavoring to develop a legal and policy framework for capping nutrient discharges at their 1997 levels to allow the Black Sea to recover. This chapter discusses the scientific basis for this action and practical options for limiting nutrient emissions in the future.

INTRODUCTION

There is ample evidence of serious degradation of the Black Sea environment. This enclosed sea (Figure 5.1), with an area of 420,000 km^2, is shared by six countries, each of which uses its coastline for recreation, major trading ports, fishing, urban development, and industry. This complex and poorly regulated situation leads to a huge stress on nature and landscapes in the coastal zone and on the sea itself. Some 16 million people live in the coastal zone, but as

Science and Integrated Coastal Management
Edited by B. von Bodungen and R.K. Turner

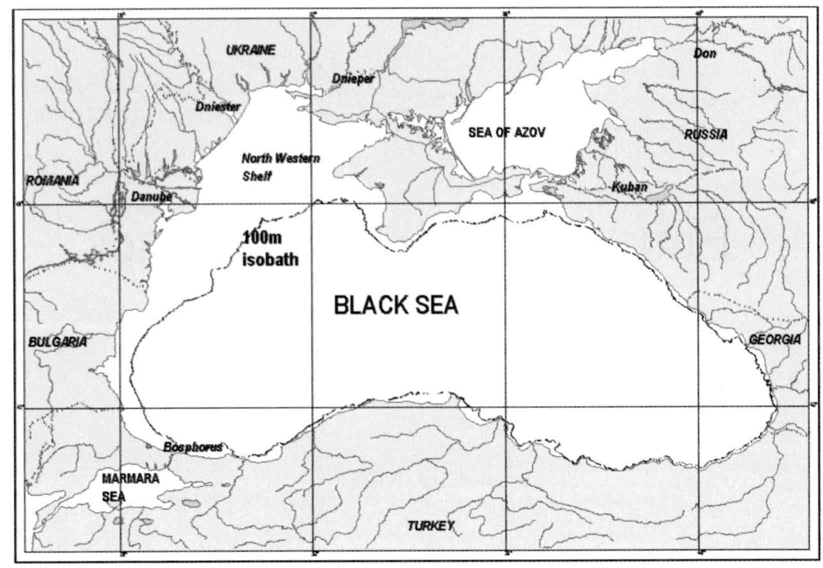

Figure 5.1 Map of the Black Sea showing the major rivers, national borders, the northwestern shelf, and the 100 m isobath.

many as 160 million people living in 17 countries[1] inhabit the catchment area of the Black Sea. The catchment area covers about one-third of the European continent and includes three of Europe's most significant riverine systems: the Danube, Dnieper, and Don rivers. With such a large and populous hinterland, the state of the Black Sea ecosystem is obviously heavily influenced by human activities throughout its entire basin. In this chapter I examine the impact of the Black Sea catchment on the marine and coastal environment and identify the measures that are being taken to reverse previous damage and conserve the Black Sea ecosystem. Particular attention is given to the issue of eutrophication, widely considered to be the main cause of the collapse of much of the marine ecosystem of the Black Sea. Finally, I present a conceptual model which illustrates the time scales for degradation and recuperation of the system as a result of eutrophication.

THE BLACK SEA ECOSYSTEM IN DECLINE

Context

In a working group formed between the International Commission for the Protection of the River Danube (ICPDR) and the Istanbul Commission for the Protection of the Black Sea (ICPBS) in 1998, the challenge was given to Black Sea scientists to demonstrate the linkage between Danube discharge and eutrophication. Marine scientists have been conducting

[1] Coastal countries are: Bulgaria, Georgia, Romania, Russia, Turkey, and Ukraine. Noncoastal countries are: Austria, Belarus, Bosnia and Herzegovina, Croatia, Czech Republic, Germany, Hungary, Moldova, Slovakia, Slovenia, and Yugoslavia.

research on the Black Sea for more than 130 years; a recent bibliography (Mamaev et al. 1996) cites 4,256 publications in the period 1974–1994 alone! Until recently, however, there has been little effort to integrate the information and there are still many gaps in knowledge regarding recent changes. Additionally, the economic collapse in many Black Sea countries has resulted in the suspension of many monitoring programs. Notwithstanding, the National Reports to the Working Group (ICPBS/ICPDR 1999), combined with several recent publications, provided much insight on the changes affecting the Black Sea that are attributable to eutrophication.

The Black Sea Prior to the 1960s

The Black Sea is Europe's newest sea. It was formed a mere seven or eight thousand years ago when sea-level rise caused Mediterranean water to break through the Bosphorus valley refilling a vast freshwater lake tens of meters below the prevailing sea level (Ryan et al. 1997). The salty water sank to the bottom of the lake, filling it from below and forming a strong pycnocline between the Mediterranean water on the bottom and the fresh water mixed with some seawater near the surface. The depth of this natural density barrier depended (and still depends) upon the supply of fresh water from rivers and rain, and the energy available from the wind and the sun for mixing it with the underlying seawater. The oxygen in the incoming water was quickly exhausted by the demands of bacteria associated with decaying biota and terrestrial organic material falling through the density gradient into the bottom water. Within a few hundred years, the sea, below some 100–200 m depth, became depleted of oxygen. The bacterial population changed to anaerobic forms releasing toxic hydrogen sulfide and the resulting water body became the largest volume of anoxic water on our planet.

For several thousand years, therefore, only the surface waters, down to the "liquid bottom" pycnocline, were capable of supporting higher life forms. Though not very biologically diverse compared with open seas at similar latitudes, the Black Sea developed remarkable and unique ecosystems, particularly in its expansive northwestern shelf where the sea is relatively shallow. The seabed in this part of the Black Sea was well oxygenated since it is well above the pycnocline. This area, and the adjacent shallow Sea of Azov, also receives the inflow of the major rivers mentioned earlier. The areas adjacent to the river discharges (including the entire Sea of Azov) were comparatively productive. On the northwestern shelf a particularly unique ecosystem developed, based on the "keystone" benthic red algae, *Phyllophora sp.*, which formed a vast bed with a total area equivalent to that of Belgium and the Netherlands. This particular keystone was also a place of great beauty as vast underwater fields of red algae, home to a myriad of dependent animals, linked together a complex web of life.

Evidence for the Onset of Decline

Since the 1970s, it has been evident that the Black Sea's benthic ecosystems are declining. First evidence came from the northwestern shelf, where a succession of "die-offs" of benthic macrophytes were observed, starting with the sensitive 150 m fringe of *Cytoseira barbata* in the 1970s. In the Romanian part of the Black Sea, for example, total *Cytoseira* declined from 5,400 tons (fresh weight) in 1971 to 755 tons in 1973 and 123 tons in 1979 (Cociasu et al. 1999). It was replaced by opportunistic species with a short life cycle. Even more dramatic was the decline of *Phyllophora*. Zernov's phyllophora field on the northwestern shelf (named after its discoverer) occupied 11,800 km^2 in the 1960s and had a total biomass of 9 million tons (Zaitsev 1993). By the end of the 1970s, the *Phyllophora* biomass was 1.4 million tons; one decade later it did not exceed 0.3 million tons and occupied only 500 km^2. During cruises in 1998, Ukrainian scientists (Alexandrov et al. 1999) found no trace of the field.

At the same time, a lens of suboxic or anoxic water formed every summer in the area previously occupied by *Phyllophora*. After 1973, hypoxic bottom waters were often present on the northwestern shelf for periods in excess of two weeks in summer and covered up to 15,000 km^2 from 8 to 40 m depth in the 1980s (Zaitsev 1993; Alexandrov et al. 1999). They have recently receded to areas of less than 1,000 km^2 (Mikhailov 1999). There is also evidence to suggest that the suboxic zone of the open Black Sea has enlarged towards the sea surface by about 10 m since 1970 (Konovalov et al. 1999).

The consequence of summer anoxia on the shelf was the mass mortality of benthic animals. Thousands of tons of dead animals littered the beaches of Romania and Ukraine every summer. In the Ukrainian sector alone, between 1973 and 1990, losses were estimated at 100–200 t km^{-2}, amounting to a total of 60 million tons of bottom animals including 5 million tons of fish (Zaitsev 1993). In the Romanian sector, catastrophic collapses were seen in benthic communities in muddy bottom habitats: 15 species of crustacean were present in 1977, 2 in 1980; 20 species of mollusk were found in 1977, 4 in 1980 (Cociasu et al. 1999). The biomass of macrobenthic animals was reduced in the proportion: 30 (1977), 10 (1978), to 1 (1980). In the sandy sublittoral environment, species diversity declined dramatically with 14 species of polychaete in 1965 and only 2 in 1982. Similarly, there were 17 species of amphipod in 1965 and 2 in 1982; the density of zoobenthos declined from some 100,000 individuals per m^2 in 1965 to 4,000–60,000 in 1982. In areas away from the northwestern shelf, a decrease in species density and depth range was observed during the 1970s and 1980s (Zaika 1990), but in living *Cytoseira* beds[2], there is relative stability in zoobenthos (SCEP 1999).

The change in the Black Sea ecosystem extended far beyond that of the benthic environment. Profound alterations were recorded in plankton communities. For example, an increase of about 40-fold was recorded in phytoplankton biomass between the period 1961–1963 and 1983–1990 along the coast of Bulgaria (Boyanovsky et al. 1999). A general increase in phytoplankton density was recorded throughout the entire Black Sea, prior to 1992, particularly in the rim current (Yilmaz et al. 1999). The species changes in the Black Sea also resulted in the sudden replacement of diatoms (consistently dominant before 1972) by dinoflagellates (dominant in Bulgarian waters from 1972 to 1989; Boyanovsky et al. 1999). Extensive studies from Romania, Bulgaria, and Ukraine have revealed that as eutrophication has progressed, blooms of nanoplankton and picoplankton have become increasingly important. Species of cyanobacteria (blue-green algae) and coccolithophorads, which were rare before 1970, have become frequent or even dominant (Cociasu et al. 1999). In 1991, a bloom of the coccolithophore, *Emiliania huxleyi,* dominated the entire Black Sea flora, reducing the mean transparency (as measured by Secchi Disk) to 6 m from the historical average of about 20 m (Vladimirov et al. 1999).

---

[2] The coast of Russia has intact beds of *Cytoseira*. These have been reduced in size, possibly through oil pollution, but are living remnants of the earlier Black Sea coastal benthic ecosystem.

In zooplankton communities, the initial effect of eutrophication was an increase in the populations of copepods. These are food species for pelagic species such as sprats, whose population also increased sharply. On the Romanian sector of the northwestern shelf, for example, the summer biomass of copepods increased sixfold through the decades of the 1970s and 1980s (Cociasu et al. 1999). This reflects the shift in the system from dominance by benthic production to pelagic production as eutrophication advanced. The favorable situation for pelagic consumers did not continue, however. As eutrophication advanced still further, the production of gelatinous zooplankton such as *Noctiluca scintillans* was increasingly favored. These species do not serve as fodder for organisms higher in the trophic chain and represent an ecological "dead end." In the 1980s, the entire northwestern shelf saw the development of huge blooms of *N. scintillans*, sometimes contributing to as much as 99.8% of summer biomass (Cociasu et al. 1999). As a result, the summer biomass of nongelatinous "fodder" species declined to about 20% of late 1960s values (Boyanovsky et al. 1999; Yilmaz et al. 1999). Interestingly, this situation was also observed in Novorossiisk Bay in the eastern Black Sea where copepods declined from being 44% of the zooplankton biomass in the 1950s and 1960s to 9% in the early 1990s (SCEP 1999).

Jellyfish, another group of nonfodder species, also increased spectacularly during the onset of eutrophication. The total average biomass of the jellyfish *Aurelia aurita* increased from 670 thousand tons in 1949–1962 up to 222 million tons in 1976–1981 and 300–500 million tons by the end of the 1980s (Alexandrov et al. 1999).

To complicate the situation further, the Black Sea has been subjected to a series of invasions by opportunistic species over the past 50 years or so. With its high primary productivity (both natural and by eutrophication) and low salinity (average about 22), the Black Sea presents ecological conditions which in many ways resemble those of the lower reaches of a large temperate estuary. It provides a ready niche for species from similar environments, transported on ship's hulls or in ballast or bilge waters. The disturbance of the system through the loss of many benthic species and the presence of trophic "dead ends" may have made it less robust and more vulnerable to invasions. The most spectacular invasion, however, came when the gelatinous ctenophore *Mnemiopsis leidyi* was accidentally introduced in about 1982, probably transported in ballast water from the eastern seaboard of North America. It developed in an extraordinary explosion, attaining a biomass estimated at 1 billion tons (2 kg m^{-2}) by 1990 (Shushkina and Vinogradov 1991; GESAMP 1997). *Mnemiopsis* is an efficient and wide-ranging predator of zooplankton and the presence of such a vast bloom appears to have disrupted the entire Black Sea ecosystem, at least on a temporary basis.

In the case of commercial fisheries, it is difficult to discern the impact of eutrophication from possible overfishing. After the mid-1970s, benthic fish populations (e.g., turbot) collapsed and pelagic fish populations (small pelagic fish, such as anchovy and sprat) started to increase. This may have resulted from habitat loss as the benthic algal beds were lost and the pelagic ecosystem dominated the food chain (Zaitsev 1993). The commercial fisheries diversity declined from some 25 fished species to about five in twenty years (1960s to 1980s). Certainly anchovy stocks and fisheries increased rapidly from the late 1960s to 1988, attaining over 500,000 tons annual catch. With the arrival of *Mnemiopsis*, the catch plunged to less than 100,000 tons in one year. Since then, it has gradually recovered and is currently over 400,000 tons. The recovery is entirely on the southern side of the Black Sea (mostly along the coast of Turkey) and there is evidence that spawning grounds switched from the north to the south.

Most fish stocks in the northwestern Black Sea are still depleted (Yilmaz et al. 1999; McLennan et al. 1997; Prodanov et al. 1996).

First Signs of Recovery

The decade of the 1990s has witnessed a decreasing input of nutrients to the Black Sea. There were some signs of recovery in certain aspects of the Black Sea's pelagic and benthic ecosystems. One aspect, the reduction in summer anoxia on the Black Sea shelf, has already been described. No associated recovery of benthic macroalgae has been reported, however. In the Danube prodeltaic area, Romanian scientists have reported an apparent recovery of zoobenthos species diversity: 14 species reported in 1994, 23 in 1995, 25 in 1996, and 30 in 1997. On the Bulgarian shelf, phytoplankton populations have receded by about 30% from their 1990 maxima (Boyanovsky et al. 1999). Work in Bulgaria and Ukraine has shown some recovery of diatoms from 1990–1995, and the flora are more diverse than previously observed. Furthermore, the incidence of unusually intense blooms seems to have diminished since 1992. A small recovery in fodder species of zooplankton was recorded in Romania in 1996–1997 (Cociasu et al. 1999), but this has not yet been confirmed from other areas.

The ctenophore *Mnemiopsis* is still present in large quantities in both eastern and western parts of the Black Sea. Data from 1997 cruises show biomasses of some 600 g m^{-2} in the east and 300 g m^{-2} in the western Black Sea (Yilmaz et al. 1999). Similar biomasses of the jellyfish *Aurelia* were also recorded. Interestingly, a newcoming invader, the ctenophore *Beroe ovata*, which preys on *Mnemiopsis*, was first identified in the Black Sea in October 1997 (Boyanovsky et al. 1999). Bulgarian scientists are concerned that this will herald a new invasion.

As mentioned previously, some populations of pelagic fish have started to recover. This may be a consequence of shifts in spawning grounds and the timing of spawning (earlier than the development of summer populations of *Mnemiopsis*). According to Kolarov (pers. comm.), 1999 was an exceptional year for Bulgarian fisheries, marked by the return of species such as bluefish which have been missing from the commercial capture for more than a decade.

LINKAGE BETWEEN HUMAN ACTIVITIES, EUTROPHICATION, AND THE DEMISE OF ECOSYSTEMS

The Causal Chain

Protection of the Black Sea ecosystem requires actions to mitigate eutrophication. An understanding of the causal chain that led to this phenomenon is essential if it is to be controlled in the future. The linkages in the chain and the response time between the links will be explored in this section. They are described in order, beginning with the phenomenon itself, by investigating the cause of each link, immediate, secondary, tertiary, etc.

The Immediate Cause: Increased Nutrient Discharges

There is substantial evidence that eutrophication in the Black Sea is the result of very large increases in the discharge of nitrogen (N) and phosphorus (P) compounds from the 1960s onwards. This triggered heavy phytoplankton blooms, a decrease in seawater transparency, and an increase in the load of organic detritus reaching the seafloor (Mee 1992; Zaitsev 1999). On the northwestern shelf, benthic macroalgae were deprived of light and the replenishment rate of oxygen in bottom waters was insufficient to satisfy the demand for aerobic bacteria oxidizing the decaying detritus. Unfortunately, there is no single set of monitoring data covering nutrient inputs to the Black Sea from the time at which eutrophication appears to have begun. It is necessary to assemble the available information as pieces in an environmental jigsaw puzzle. Additionally, there is continued controversy as to whether nitrate, phosphate, or in some cases silicate (Si), limits the growth of phytoplankton. Field studies suggest that phosphate may now be the limiting nutrient close to the Danube discharge (Cociasu et al. 1999); however, nitrate may well limit, more generally, production over the open Black Sea itself (Yilmaz et al. 1999).

In 1996, a first attempt was made to assess the significance of all land-based sources of nutrients to the Black Sea. This involved the careful application of a rapid assessment technique in each coastal country in order to evaluate inputs from potential point sources, whether industrial or domestic. River inputs were included using current monitoring data considering each river as a point source, and atmospheric sources were calculated from modeling data. The results of the survey (Topping et al. 1999) are shown in Table 5.1. Over 50% of the dissolved nutrients appear to be discharged through the Danube River.

The existence of a very well-developed program for the protection of the Danube has provided much useful information for the characterization of nutrient sources and fluxes through the river and its tributaries. Although I focus on this system in this chapter, it should be remembered that this only represents some 50–60% of the land-based nutrient discharge to the Black Sea, the remainder coming from the other major international rivers (Dnieper, Dniester, Don), national rivers, and coastal direct discharges.

Even in the case of the Danube, historical data on discharge loads is fragmentary with many contradictions due to inadequate intercomparison of techniques or insufficient sampling frequency. A comparison of data for the period 1959–1960 and 1980–1997 suggests the following changes in annual discharges from the Danube to the Black Sea: inorganic phosphate load increased from 12×10^3 t P y^{-1} to 18×10^3 t P y^{-1}; inorganic nitrogen increased from 140×10^3 t N y^{-1} to 600×10^3 t N y^{-1}; silicate decreased from 790×10^3 t Si y^{-1} to 330×10^3 t Si y^{-1} (Cociasu et al. 1999). This information should be treated with some caution as the measurement techniques for some of these variables have changed considerably over the 40-year period. Humborg et al. (1997) have suggested that the decrease in silicates is the result of the construction of the Iron Gates dam on the Danube almost 1000 km from the sea, in the early 1970s. Silicate limitation may be one of the main reasons for the disappearance of diatoms as dominant species on the northwestern shelf (Humborg et al. 1997).

Table 5.1 Estimates of nutrient loads, thousands of tons per year.

	Danube load	All other rivers and point sources	Atmospheric load	Total
Total nitrogen	346	301	400	1047
Total phosphorus	25	25	–	50

The longest consistent data set for Danube discharge is that of Cociasu et al. (1999), who report the following trends from the mid-1980s to 1997:

- Increase in inorganic phosphate to 27×10^3 t y^{-1} average 1987–1991, then almost 50% decrease to 15×10^3 t y^{-1} average 1994–1997.
- Gradual decrease in total nitrogen discharge from 800×10^3 t y^{-1} in 1988 to 300×10^3 t y^{-1} in 1997.
- Irregular decrease of silicate from 500×10^3 t y^{-1} in 1980 to 330×10^3 t y^{-1} in 1997.
- Molar N:P ratios varied from 50–70 in 1988–1992, rising to over 100 in 1993–1994 and gradually falling to current values of about 40.
- Molar Si:P about 40 from 1988–1994, then steadily rising to over 110 in 1997.

Clearly, there has been a decrease in nutrient loads in recent years, a matter that I will discuss later in this chapter.

Monitoring data from the Black Sea itself broadly reflects the situation in the Danube. Romanian scientists have been monitoring nutrient concentrations in their coastal waters since the 1960s (Cociasu et al. 1999). For offshore stations, away from the direct influence of coastal sources, the situation can be summarized as follows:

- Pre-1969 surface PO_4 below 0.2 μmol^{-1} (annual average) then rapid increase to an average of 0.8 μmol^{-1} in the period from 1969 to 1980 (max. almost 3 μmol^{-1} in 1974). Levels then receded to below 0.2 μmol^{-1}, except for 1985–1989 when values as high as 0.5 μmol^{-1} were recorded.
- Gradual decline in total inorganic nitrogen from 24 μmol^{-1} in 1976 (no prior data) to 4–6 μmol^{-1} from 1989–1997.
- N:P ratios have decreased from 50–200 (1980–1990) to below 50 (1990–1997).

This closely mirrors the reported variations in Danube nutrient loads.

Recently, new data have become available on temporal (1968–1997) variations in nutrient concentrations in the intermediate layers (down to about 650 m) of the open Black Sea (Konovalov et al. 1999). These support the notion of eutrophication as a phenomenon affecting the entire Black Sea. A significant increase in nitrogen and phosphorus inventories was observed over this period though there is some evidence that this increase lessened in the past three to four years.

Secondary Causes: Land-based Sources of Pollution

A study by the Environmental Programme for the Danube River Basin (EPDRB 1997) estimated that for 1988–1992, 23–35% of the total P and 42–45% of the total N discharged into the Danube or its tributaries is eventually discharged to the Black Sea (see Table 5.2).

Furthermore, the recently published Danube Water Quality Model (DPRP 1999) has examined the human activities that result in the increased nutrient discharges in the Danube basin. Work has also been conducted on the Black Sea itself (Topping et al. 1999). For the Danube, it is estimated that some 74% of total N emissions are from diffuse sources while 26% are from point sources (Haskoning Institute 1994). The figures for total P are 58% and 42%, respectively. It was estimated that 48% of the emissions of total N and 47% of total P are from agriculture. For industry, the figures are 5% and 8%, respectively, domestic discharges

Table 5.2 Emissions of nutrients to the Danube and its tributaries and discharges to the Black Sea during two time periods.

	Emissions to surface water (10^3 t y^{-1})	Discharge to the Black Sea (10^3 t y^{-1})	Discharge as % of emissions
Total P, 1988–1989	130	46	35
Total P, 1992	105	25	24
Total N, 1988–1989	990	447	45
Total N, 1992	820	345	42

are 27% and 39%, respectively, and unclassified sources constitute 20% and 6% of each of the totals. Thus, for both N and P, agricultural activity comprises about half of the total liquid emissions.

The transport pathways of compounds emitted to and through the rivers differ widely depending on their source. Nitrogen compounds, for example, tend to be rather soluble and a significant proportion of the nitrogen load finds its way into groundwater or is washed into streams and rivers. Alternatively, it may enter the lower atmosphere as ammonia or nitrous oxide, some of which is oxidized to nitrate and purged by rain to enter the hydrological cycle. A considerable part may also be mobilized by farm animals as major consumers of fertilized fodder crops and released as manure or ammonia. When present in rivers as nitrate, dissolved nitrogen is highly mobile and does not readily adsorb onto suspended particles (other than living phytoplankton). Phosphate, on the other hand, is readily adsorbed on soil or sediment particles. It tends to be washed into streams with soil erosion but may also be introduced in dissolved form from sewage, industry, or through the use of polyphosphate detergents. Reservoirs that trap particulate matter tend to retain a higher proportion of phosphorus compounds than those of nitrogen, unless they are subject to intense phytoplankton blooms. The Dnieper River, for example, currently consists of a chain of eutrophic reservoirs that retain a large proportion of the nutrient emissions to the river (GEF 1997). In the case of the Danube, the Iron Gates reservoir, almost 1000 km upstream from the Black Sea, retains some 28% of the P discharged into it whereas N retention appears to be negligible (DPRP 1999).

Tertiary Causes: Inappropriate Practices

A close examination of available data regarding time trends for recognized sources of N and P contamination gives important insights into the development of eutrophication itself. Information on the changes in demography, fertilizer use, and agricultural production is available from the FAO data base (FAO 1999). I have reaggregated the FAO country data using statistics for the amount of arable land within each Danube basin country (ICPBS/ICPDR 1999) to provide annual basin-wide estimates from 1961–1997 for a variety of indicators.

Figure 5.2 illustrates the data for total N and P fertilizer applications (hereafter referred to as Nf and Pf, respectively) since 1961 in the Danube basin. It demonstrates the fourfold increase in Pf and fivefold increase in Nf during the period 1961–1982, the period known as the "green revolution." The decade of the 1980s was one of fairly constant use of Nf and an

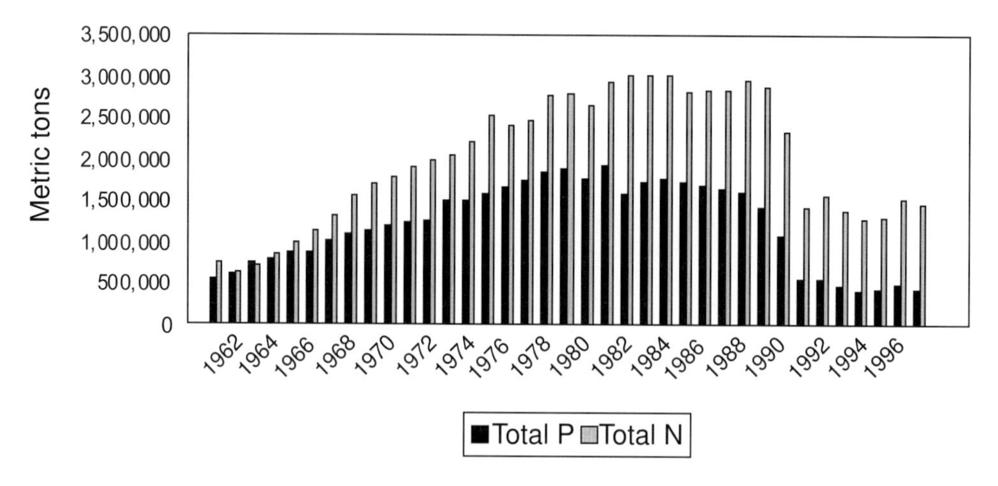

Figure 5.2 Total N and P fertilizer application, Danube basin.

approximately 10% decline in Pf (largely attributable to reduced application in Germany and Austria). However, in the three-year period from 1989 to 1991, Pf fell to about 28% of levels in the previous decade and Nf to some 50%. By 1997, Pf applications were 20% below those of 1961 while Nf had fallen to the 1967 level. Nf:Pf ratios had changed from 1:1 to 3:1.

It is interesting to observe how agricultural yields responded to these changes. Figure 5.3 illustrates key indicators for Romania. The two decades from 1963–1983 saw an increase in the yields of "roots and tubers" by as much as 2.2-fold and vegetables by sixfold. Cereal crop yields approximately doubled. Except for vegetables (where pesticide applications may have also improved yields), the increases in yields were proportionally lower than the increases in fertilizer application. Following the recent decline in fertilizer use, yields of all three crop classes have remained well above the 1961 values. These results suggest that in the 1980s, fertilizers were applied in excess of requirements and with considerable wastage.

A comparison of Pf and Nf data with information available on discharges of total nitrogen and phosphorus to the Black Sea (Figure 5.4) for the period since 1986, suggests that variations in N and P discharges correspond to decreases in fertilizer application. A lag time is observed between decreasing application and discharge of about seven years for N and four years for P. Of course, there are year-to-year variations due to changes in rainfall and river transport as well as the presence of other variable sources. Nevertheless, a strong correlation was recorded between measurements of annual discharge to the Black Sea and average values of Nf for the seven preceding years in the case of N ($r^2 = 0.78$), or the Pf for the four preceding years in the case of P ($r^2 = 0.69$). This suggests that the primary source of the increase in nutrient fluxes was the failure to retain nitrogen and phosphorus fertilizers on land.

This, however, is not the only practice leading to increased discharges of N and P. Many human settlements downstream from Austria lack sewage systems. Sewage treatment, in general, and nutrient removal, in particular, are absent from many major cities. Figure 5.5 shows the proportion of Danube basin inhabitants connected to sewage systems in 1996. As

Figure 5.3 Average values for agricultural yield and fertilizer use, Romania.

sanitation gradually improves in the region, increasing amounts of N and P from human sewage will be discharged to the river, unless treatment practices are improved.

Another problem has been the discharge from intensive livestock farms, particularly pig farms. By 1983, there were 45 million pigs in the basin and some of the intensive pig breeding units in Romania housed as many as one million pigs. The number has declined drastically in recent years due to the economic failure of intensive animal farming; however, many of the pigs have now been transferred to the "informal sector" (backyard smallholdings) and are no longer reported in official statistics. Figure 5.6 shows the annual nitrogen emission of the human and pig population of the entire basin since 1961. Phosphorus follows a similar pattern. Of course, not all of these emissions are discharged to rivers and the sea although most large intensive pig farms discharge their liquid waste into rivers. Certainly, between 1974 and 1991, emissions from pigs were greater than those from the entire human population (which is not currently growing in the region). Incidentally, the decrease in N and P emissions, which resulted from the reduction of livestock between 1990 and 1997, was equivalent to the removal of a city of 22.5 million people from the basin!

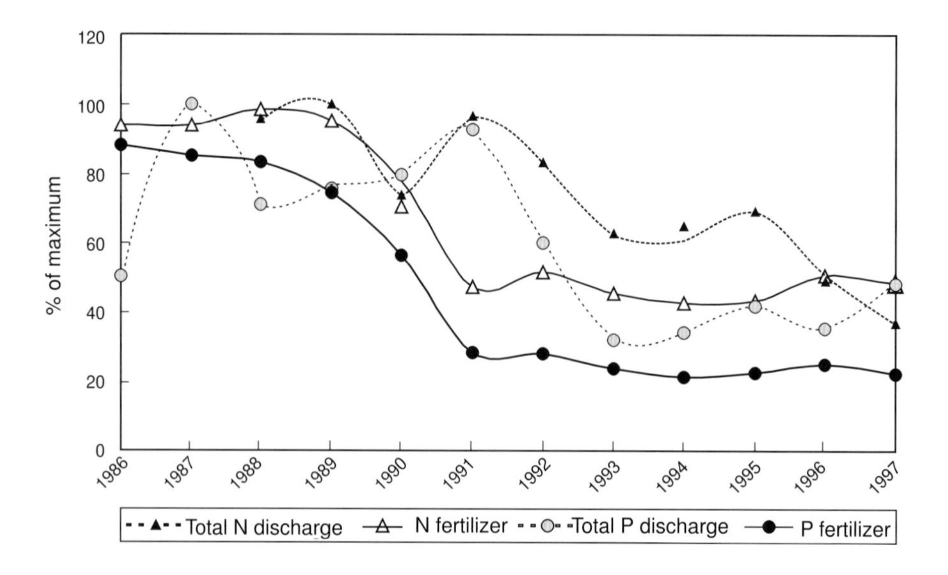

Figure 5.4 Normalized basin-wide fertilizer usage and N and P discharge to the Black Sea.

Many of the industrial sources of N and P are also related to the agricultural industry: fertilizer and food processing industries. A notable exception is the use of polyphosphate detergents, already prohibited in most Danubian countries but still widely employed around the Black Sea itself.

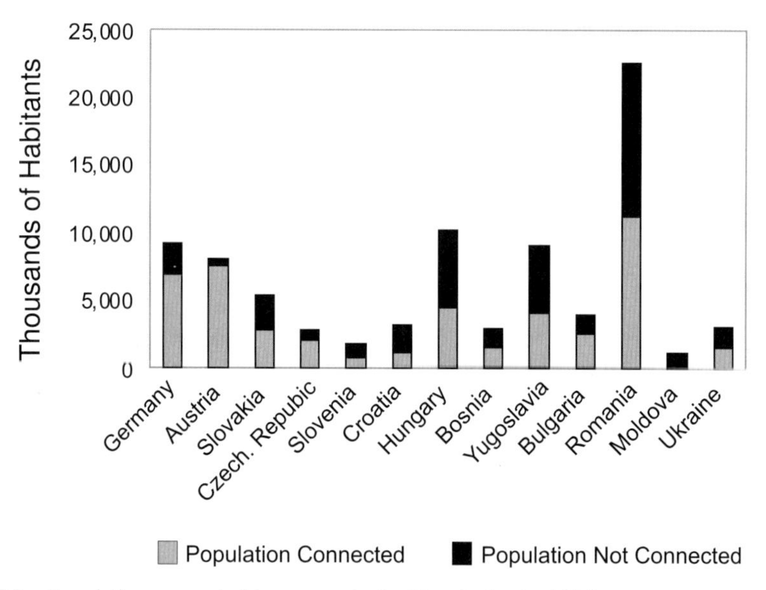

Figure 5.5 Population connected to sewage in the Danube basin, 1996.

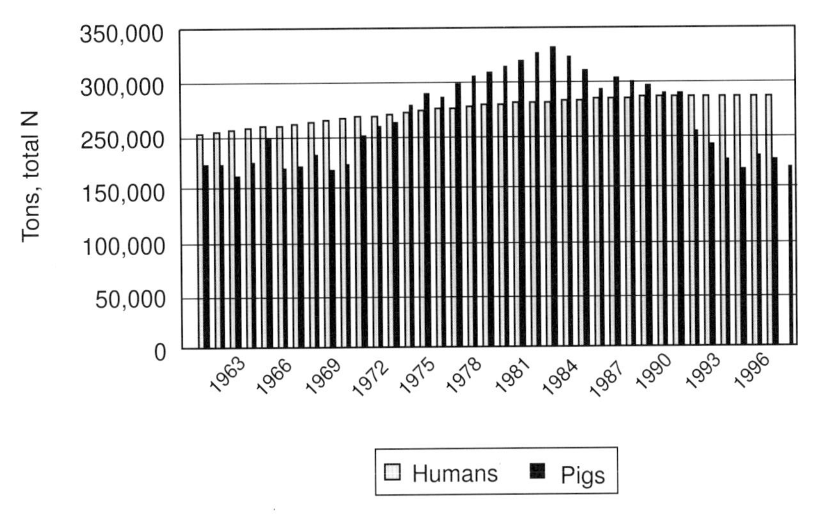

Figure 5.6 Estimated maximum environmental releases of N by pigs and humans, Danube basin.

Quaternary Causes: Economic Subsidies and the Limited Value Assigned to Environmental Protection

Economic subsidies have been a major reason for the excessive application of agrochemicals worldwide (Smil 1997). In countries with centrally planned economies, fertilizer factories were encouraged to meet or exceed production quotas. Farm managers were supplied with predetermined amounts of these substances and had little or no incentive to economize in their application to crops. This situation suddenly changed after 1989–1990 when countries embarked on the transition to market economies, but without the credit lines and guarantees normally open to farmers in the West. Within a year or two, many farmers were suddenly unable to apply any fertilizers to their land, resulting in the massive reduction in N and P application described earlier (and ultimately benefiting the Black Sea environment).

This situation may well be a temporary one, however, since many farmers are keen to return to practices and yields enjoyed earlier. The gradual increase in pesticide importation (Figure 5.7, based on the FAO database) suggests that some investment in agrochemicals continues to increase.

If new economic constraints are to influence agrochemical usage, market mechanisms should be combined with a common basis for regulation. The directives of the European Commission have been effective in regulating phosphate fertilizers and may eventually reduce nitrate emissions. Phosphate regulation has been relatively easy owing to the cost-effectiveness of control measures. The application of these directives is a major factor distinguishing those who are within the community (Austria and Germany), those who aspire to membership, and those who are likely to remain outside.

Undervaluing the environment is a more complex problem, extending beyond economics to societal values and ethics. The control of eutrophication requires a reexamination of lifestyles. There is a demand for cheap food, but part of its production cost is subsidized by the free use of the natural environment for the disposal of agricultural waste. Currently, there is a trade-off between the use of water bodies for the disposal of waste, on one hand, and the

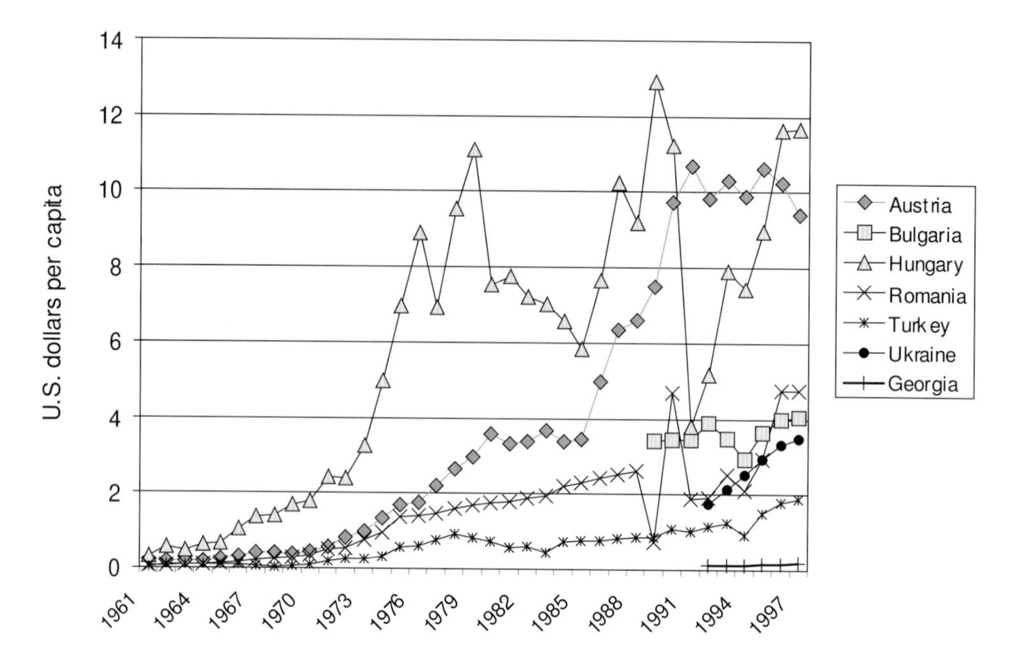

Figure 5.7 Pesticide imports in U.S. dollars per inhabitant per year.

multiple opportunities offered by a clean and biologically diverse natural environment, on the other. The current value base is a very utilitarian one. In the case of many Black Sea countries, poverty, ignorance, disempowerment, and poor governance has led to a pathway of unsustainable development and environmental decline.

ACTIONS TO RESOLVE THE PROBLEM

Reversing the Causal Chain

The Black Sea provides a remarkable example of a system that appeared to be chronically ill but has now entered a phase of "remission." The challenge is to understand the relationship between eutrophication and the human activities that cause it and to propose cost-effective actions to reverse the causal chain. The necessary actions require international cooperation and a harmonized policy approach throughout the basin, followed by a program of investments to ensure that the policies can be implemented. The present section explores the policy developments, the scientific basis of proposed actions, and the nature of the remedial and preventative actions themselves.

Legal and Policy Developments

As a result of the new cooperative spirit between countries in the region that followed "perestroika" in the mid-1980s, a range of legal and policy tools is now in place in the Danube basin and the Black Sea (for a summary, see Table 5.3 and Mee 1999). In addition, there are

Table 5.3 International legal and policy instruments in place in the Black Sea basin

Instrument	Secretariat	Observations
The Bucharest Convention for the Protection of the Black Sea Against Pollution: signed 1992; ratified 1994.	The Istanbul Commission with a Secretariat in Istanbul (from October 2000).	The Convention is in force since 1994.
The Danube River Protection Convention: signed 1994; ratified 1998.	The International Commission for the Protection of the Danube River, Vienna (operative).	Since the Convention has gone into force, ICPDR's role has rapidly developed.
The Odessa Declaration for the Protection of the Black Sea: signed April, 1993	Black Sea Environmental Programme, Programme Coordination Unit (became PIU in 1998, see below).	A pragmatic 3-year policy agreement largely implemented with financial and technical support from the GEF and CEC.
The Black Sea Strategic Action Plan: signed and adopted October, 1996.	Currently managed by a Project Implementation Unit (PIU) in Istanbul, Turkey.	A wide-ranging decadal plan covering many aspects of environmental protection in the Black Sea; sets goals and milestones.
Strategic Action Plan for the Danube River basin: adopted 1995.	Managed by the Environmental Programme for the Danube River basin with a coordinating unit in Vienna, Austria.	A wide-ranging basin-wide plan developed with the support of the European Commission.
Danube Pollution Prevention Programme: agreed 1999.	Its implementation is the responsibility of the ICPDR.	Developed through a project funded by the GEF, it includes detailed studies of actions to reduce transboundary pollution in the Danube River.

initiatives currently under development in the Dnieper and Don river basins. Furthermore, a large number of environmentally focused projects financed by the Global Environment Facility (GEF), World Bank, European Commission, and other donors are operative or planned within the Black Sea basin. Recently, the GEF has decided to integrate its initiatives into a "basin-wide initiative" with the main objective of addressing transboundary problems such as eutrophication.

There are relatively few motives or incentives for individual countries to take action to control nutrient emissions, especially where the primary challenge is to restore economic growth, provide employment, and alleviate poverty. For the lower Danubian countries, the main incentive is the aspiration to European Union (EU) membership and the grants and loans available to assist them in this process. Recent positive overtures towards Turkey by EU member states have also brought it into the group of pre-accession countries. The process of membership imposes a prerequisite of approximation to European Commission (EC) environmental legislation. Compliance with the EC Nitrate and Phosphate Directives would certainly provide better perspectives for controlling discharges of dissolved N and P to rivers and the sea. The main problem in the next decade, however, will be how to encourage Russia, Ukraine, and Georgia to achieve environmental standards similar to their Central European neighbors.

Cooperation between the Istanbul Commission for the Bucharest Convention (Black Sea) and the International Commission for the Protection of the Danube River (ICPDR) has led to the recommendation (ICPBS/ICPDR 1999) that in the long term, all states in the Black Sea basin should "take measures to reduce the loads of nutrients and hazardous substances to such levels necessary to permit Black Sea ecosystems to recover to conditions similar to those observed in the 1960s." It was agreed, however, that "as an intermediate goal, urgent control measures should be taken by all states in the Black Sea basin in order to avoid that the discharges of nutrients and hazardous substances into the Sea exceed those which existed in 1997." The group recognized the shortages of data that made this recommendation somewhat empirical.

Testing the Policy Recommendation

Would a cap on nutrient discharges at their 1997 level really work? In this chapter I have presented some evidence that phosphorus discharges from agriculture and intensive pig farms are currently at a level similar to the 1960s. Unfortunately, the data set for total nitrogen is much weaker, but levels clearly exceed those of 1961. Some additional insight may be obtained based upon the strong correlation previously described between nitrogen fertilizer application and total nitrogen discharge. Figure 5.8a presents a plot of Nf, averaged for the seven years prior to each data point, and the extent of the summer hypoxia on the Black Sea shelf as reported by Zaitsev (1993) and Mikhailov (1999). No attempt has been made to plot a correlation curve between the two sets of observations, as the data on hypoxia are the result of occasional cruises rather than fully systematic monitoring. Notwithstanding this limitation, the results are remarkable. No hypoxia was observed before 1973, and cruise data from 1996 suggest that the phenomenon had virtually disappeared for the first time in 23 years. Most importantly, this had occurred at similar levels of Nf (about 1.5×10^6 t N), suggesting that a lower threshold for hypoxia had been reached. This corresponds to a value of 393 $\times 10^3$ t of annual total N discharge to the Black Sea to which a standard error of 84 $\times 10^3$ t can be ascribed (derived from the correlation between Nt and total N discharge from 1986–1996). This implies that a discharge load of about 310 $\times 10^3$ t total N per year is the level below which hypoxia on the northwestern shelf should be unlikely. This is close to the reported load for 1997 (300.4 $\times 10^3$ t; Cociasu et al. 1999).

A similar plot for Pf (Figure 5.8b) shows a different situation. Hypoxia first appeared on the shelf at values of Pf of just over 1.2×10^6 t y^{-1}. Such values were restored in 1991, a bad year for hypoxia. Indeed, even 1994 levels of 0.5×10^6 t y^{-1} had still not resulted in the disappearance of the phenomenon. This suggests that on an annual scale, riverborne phosphate is not the limiting nutrient for phytoplankton production on the shelf (away from the immediate discharge of the river). This is a logical consequence of temporarily anoxic conditions; these would rapidly remobilize phosphate from surface sediments and make it available for new

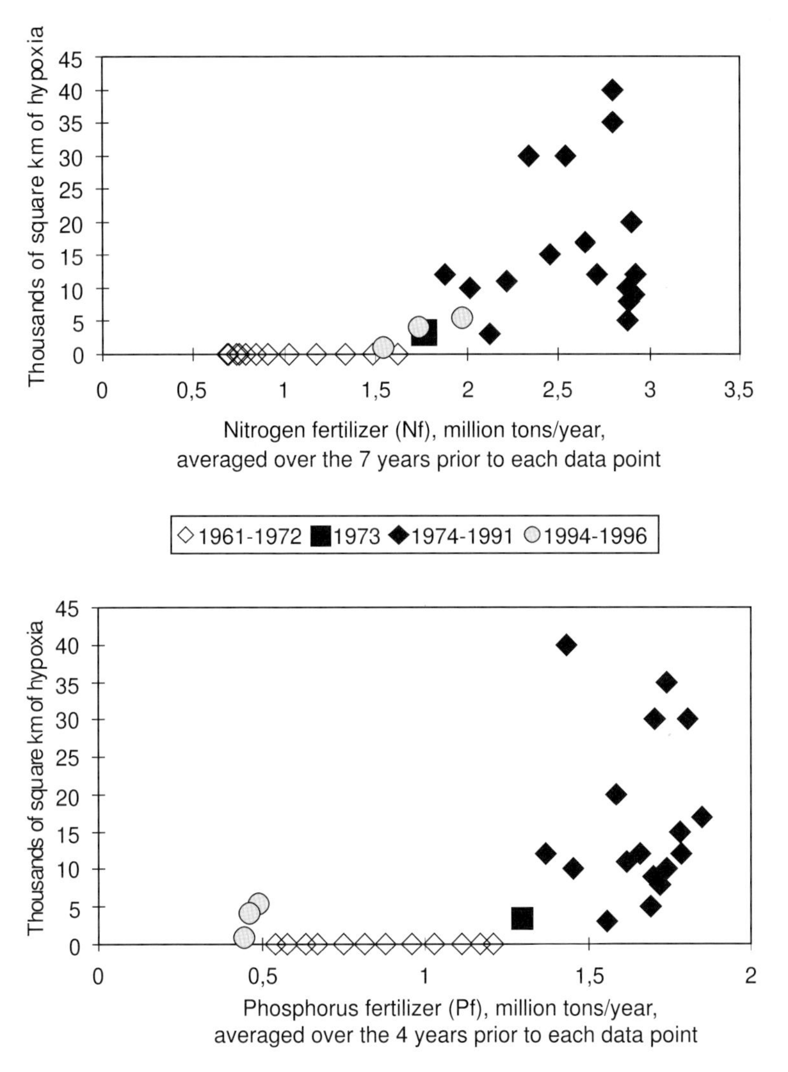

Figure 5.8 In the Danube basin, variation in observed area of northwestern shelf summer hypoxia with increasing loading of nitrogen fertilizer and increasing loading of phosphorus fertilizer.

plant production. Indeed, phosphorus supply to the entire shelf system may be dominated by remobilization from the sediment pool. The presence of huge macrophyte beds and oxidized surface sediments before the 1970s would have helped to keep the system strongly buffered for maintaining low dissolved phosphate, this capacity has since been lost.

It remains to be seen whether the recent observation of the disappearance of hypoxia will be maintained over the next few years and whether or not the benthic ecosystems will be able to gradually recover. This process may be slow, and any recovery of *Phyllophora* has yet to be observed. If conditions of oxygen and transparency are adequate, some experimental replanting of *Phyllophora* might be attempted.

Potential Low-cost Approaches to Controlling Future Nutrient Emissions

As a result of activities of the ad hoc Danube–Black Sea Working Group, the Black Sea Environmental Programme Steering Committee was presented with suggestions regarding low-cost measures which could be taken to prevent increases in nutrient discharge to the Black Sea and enable the "cap" to be implemented. These fall into four general categories (Mee 1998; ICPBS/ICPDR 1999):

1. *Reform of agricultural policies.* The agricultural sector is currently being restructured in many countries to improve productivity. If a return to large increases in nutrient runoff is to be avoided, it is important to include relatively simple policy provisions in the restructuring process. These include such things as leaving strips of unploughed land near streams, rivers, and lakes; provision of storage clamps for overwinter storage of manure; erosion control through practical demonstration projects, and incentives for "biofarming"[3]. Regulations concerning buffer zones for streams and rivers are already in place in some countries (e.g., Ukraine) but enforcement is still rather poor. Another area requiring attention is freshwater fish farming, and extensive (low-feeding) aquaculture should be encouraged rather than intensive rearing, which results in very large nutrient discharges. Intensive farms should be subjected to discharge permits and levies as an incentive for proper treatment of waste. Effective levies should also be imposed on intensive animal rearing facilities which do not treat or recycle their waste.
2. *Improved wastewater treatment, particularly through the use of alternative technologies.* Conventional primary and secondary domestic wastewater treatment does not prevent large nutrient discharges. Tertiary treatment (including nutrient removal) implies very high operation and maintenance charges that may be unaffordable under current economic conditions. For small communities, a good example of low-cost alternative technology is the use of reed bed techniques for sewage treatment following screening. This is increasingly employed for small towns in western countries and has a low capital and operation cost. This technique, however, has not been successfully applied for larger towns or cities. One option that should be properly evaluated for towns in Russia, Georgia, and Turkey is the use of deep discharge diffusers that carry waste to depths well below the pycnocline (the density gradient at about 100 m depth in the Black Sea). With careful design, these can be extremely efficient but the continued risk of microbiological contamination makes the technique controversial. It also compounds the notion of the ocean as a dumpsite. Industrial waste, particularly from the food processing industry, is often untreated and nutrient removal should be a statutory requirement.
3. *Rehabilitation of key basin ecosystems.* The creation of protected areas, particularly in the case of wetlands, encourages the natural assimilation of plant nutrients. The reflooding of wetlands results in nutrient removal in two stages: a fast initial removal as aquatic plants grow and then a slower continuous removal as phosphorus is bound into sediments and nitrogen returned to the atmosphere by denitrification. The

[3] The term "organic farming" is commonly employed in some countries. In the U.K., for example, standards for this practice are set by the Soil Association.

protected or reflooded wetlands serve as biodiversity reserves and productive areas for fisheries. The creation of terrestrial protected areas is also very important as it allows buffer zones to enhance carbon and nitrogen removal. An urgent priority is to afford protection to the remaining areas of marine macroalgae such as the *Cytoseira* beds in Russia or the *Phyllophora* beds in Ukraine to seed recovery of the Black Sea's ecosystems. These beds are currently under threat as a result of development of the oil industry (Russia), tourism development (all areas), and trawling (all areas).
4. *Changes in consumer practices (including use of phosphate-free detergents)*. The prohibition of polyphosphate-based detergents leads to a major reduction in phosphate discharge to aquatic systems. These detergents are already banned in most Danubian countries and the ban should be extended to all Black Sea countries as soon as possible. (Such a ban should be part of a new ministerial agreement.) Public awareness of the eutrophication issue should be raised and clear information provided on the consumer practices that lead to higher nutrient discharges. Awareness should also be raised of the need for protected areas and the consequences of their loss to developers.

Recent Developments in the Danube

The recently announced recommendations of the Danube Pollution Reduction Programme (DPRP) present proposals to reduce pollution emissions from over 700 point sources, together with the restoration of a large number of wetlands. The cost of implementing this ambitious program has yet to be announced. Simulations from the Danube Water Quality Model (DPRP 1999) suggest that the implementation of the program would result in a 14% reduction in total nitrogen loads to the Black Sea and a 27% reduction in total phosphorus. This illustrates the relative difficulty in reducing total nitrogen fluxes.

CONCLUSIONS

The current reprieve in eutrophication in the Black Sea will be temporary unless there is basin-wide agreement on measures to prevent nitrogen and phosphorus discharges from exceeding loads reported in 1997. With the 1997 loads there is clear evidence of rapid recovery in the pelagic ecosystem but little widespread recovery in benthic ecosystems. Further research and monitoring will be necessary to review this situation as part of a program of adaptive management. It is recognized, however, that even if suitable water conditions are maintained, such a recovery may take years or decades. Actions to maintain the current loads should allow for additional economic growth in those basin countries with depressed economies. This will require innovative and cost-effective approaches to nutrient control by ensuring that developments are conducted in a manner that limits emissions or that the assimilative capacity of river and coastal wetlands is enhanced to absorb any additional loads.

ACKNOWLEDGMENTS

The cooperation of the GEF Black Sea Environmental Programme for the preparation of this report is appreciated and acknowledged. The detailed peer review by I. de Vries was an important contribution to the final draft.

REFERENCES

Alexandrov, B.G., et al. 1999. Report on the Ecological Indicators of Pollution in the Black Sea: Ukraine. Danube River Pollution Reduction Programme and Black Sea Environmental Programme Special Report. Vienna: UN Development Programme (UNDP) Danube Programme Coordination Unit.

Boyanovsky, B., et al. 1999. Report on the Ecological Indicators of Pollution in the Black Sea: Bulgaria. Danube River Pollution Reduction Programme and Black Sea Environmental Programme Special Report. Vienna: UNDP Danube Programme Coordination Unit.

Cociasu, A., P.E. Mihnea, and A. Petranu. 1999. Ecological indicators of the Romanian coastal waters in the Black Sea. In: Black Sea Pollution Assessment, ed. L.D. Mee and G. Topping, pp. 131–170. Black Sea Environmental Series, vol. 10. New York: United Nations Publications.

DPRP (Danube Pollution Reduction Programme). 1999. Danube Water Quality Model Simulations, in support of the transboundary analysis and the pollution reduction programme. Danube Pollution Reduction Programme. Vienna: UNDP Danube Programme Coordination Unit.

EPDRB (Environmental Programme for the Danube River Basin). 1997. Nutrient Balances for Danube Countries — Project EU/AR/102A/91, Final Report. Vienna: EPDRB.

FAO (Food and Agriculture Organisation of the United Nations). 1999. FAOstat statistics database. Rome: FAO.

GEF (Global Environment Facility). 1997. Transboundary Diagnostic Analysis for the Dnieper River Basin: Synthesis Report. New York: UNDP and UNEP (UN Environment Programme).

GESAMP (Joint Group of Experts on the Scientific Aspects of Marine Environmental Protection). 1997. Opportunistic settlers and the problem of the ctenophore *Mnemiopsis leiydi* invasion in the Black Sea. Reports and Studies 58. London: IMO (International Maritime Organisation).

Haskoning Institute. 1994. Environmental Programme for the Danube River Basin: Danube Integrated Environmental Study Final Report DG1A. Brussels: European Commission.

Humborg, C., V. Ittekot, A. Cociasu, and B. von Bodungen. 1997. Effect of Danube River dam on Black Sea biogeochemistry and ecosystem structure. *Nature* **386**:385–388.

ICPBS/ICPDR. 1999. Eutrophication in the Black Sea: Causes and Effects. Summary report by the Joint ad hoc Technical Working Group established in January 1988 between the International Commission for the Protection of the Black Sea (ICPBS) and the International Commission for the Protection of the Danube River (ICPDR) based on the studies by Black Sea scientists and the discussion of the "Group." Vienna: UNDP Danube Programme Coordination Unit.

Konovalov, S.K., L.I. Ivanov, J.W. Murray, and L.V. Eremeeva. 1999. Eutrophication: A plausible cause for changes in hydrochemical structure of the Black Sea anoxic layer. In: Environmental Degradation of the Black Sea: Challenges and Remedies, ed. S. Besiktepe et al., pp. 61–74. NATO Science Series. Dordrecht: Kluwer.

Mamaev, V.O., D. Aubrey, and V.N. Eremeev, eds. 1996. Black Sea Bibliography. Black Sea Environmental Series, vol. 1. New York: UN Publications.

McLennan, D.N., T. Yasuda, and L.D. Mee. 1997. Analysis of the Black Sea fishery fleet and landings. Technical paper. Istanbul: Black Sea Environmental Programme.

Mee, L.D. 1992. The Black Sea in crisis: A need for concerted international action. *Ambio* **21**:278–286.

Mee, L.D. 1998. Concept paper. Black Sea Environmental Programme: Implementation Phase. Paper presented to the Steering Committee of the Black Sea Environmental Programme, Dec. 1998. Istanbul: Black Sea Environmental Programme.

Mee, L.D. 1999. Pollution control and prevention in the Black Sea. In: Black Sea Pollution Assessment, ed. L.D. Mee and G. Topping, pp. 303–342. Black Sea Environmental Series, vol. 10. New York: UN Publications.

Mikhailov, V. 1999. Assessment of the state of impact zones and shelf areas of Ukraine. In: Black Sea Pollution Assessment, ed. L.D. Mee and G. Topping, pp. 263–277. Black Sea Environmental Series, vol. 10. New York: UN Publications.

Prodanov, K., et al. 1996. Environmental Management of Fish Resources in the Black Sea and their Rational Exploitation. Studies and Reviews of the General Fisheries Council for the Mediterranean No. 68. Rome: FAO.

Ryan W.B.F., et al. 1997. An abrupt drowning of the Black Sea shelf at 7.5 kyr BP. *Mar. Geol.* **138**:119–126.

SCEP (State Committee for Environmental Protection of the Russian Federation). 1999. Report on the Ecological Indicators of Pollution in the Black Sea: Russia. Special Report. Vienna: DPRP and Black Sea Environmental Programme.

Shushkina, E.A., and M.Y. Vinogradov. 1991. Long-term changes in the biomass of plankton in open areas of the Black Sea. *Oceanology* **31**:716–721.

Smil, V. 1997. Global population and the nitrogen cycle. *Sci. Am.* **7**:76–81.

Topping, G., H. Sarikaya, and L.D. Mee. 1999. Land-based sources of contamination to the Black Sea. In: Black Sea Pollution Assessment, ed. L.D. Mee and G. Topping, pp. 33–56. Black Sea Environmental Series, vol. 10. New York: UN Publications.

Vladimirov, V.L., et al. 1999. Hydro-optical studies of the Black Sea: History and status. In: Environmental Degradation of the Black Sea: Challenges and Remedies, ed. S.T. Besiktepe et al., pp. 245–256. NATO Science Series. Dordrecht: Kluwer.

Yilmaz, A., A.E. Kideys, and Z. Uysal. 1999. Report on the ecological indicators of pollution in the Black Sea: Turkey. In: Danube River Pollution Reduction Programme and Black Sea Environmental Programme Special Report, ed. Basturk et al., pp. 33–69. Vienna: UNDP Danube Programme Coordination Unit.

Zaika, B.E. 1990. Change in macrobenthic populations in the Black Sea with depth (50–200 m). *Doc. Acad. Sci. Ukrainian SSR Series B* **11**:68–71 (in Russian).

Zaitsev, Y.P. 1993. Fisheries and environment studies in the Black Sea system. 2. Impact of eutrophication on the Black Sea fauna. Studies and Reviews. General Fisheries Council for the Mediterranean No. 64, pp. 59–86. Rome: FAO.

Zaitsev, Y.P. 1999. Eutrophication of the Black Sea and its major consequences. In: Black Sea Pollution Assessment, ed. L.D. Mee and G. Topping, pp. 57–68. Black Sea Environmental Series, vol. 10. New York: UN Publications.

Standing, left to right: Franciscus Colijn, Peter Frykblom, Laurence Mee, Fred Wulff, Tim Jickells
Seated, left to right: Maren Voss, Josef Pacyna, Ragnar Elmgren, Donald Boesch

6

Group Report: Transboundary Issues

T.D. JICKELLS, Rapporteur

D.F. BOESCH, F. COLIJN, R. ELMGREN, P. FRYKBLOM, L.D. MEE, J.M. PACYNA, M. VOSS, and F.V. WULFF

INTRODUCTION

Effective integrated coastal management (ICM) requires that the problem in question be constrained with appropriate boundaries that fully contain the cause and effect. Many coastal environmental management issues arise from activities within the coastal zone itself; others arise from outside the zone. Outside activities taking place in the adjacent shoreline are considered elsewhere in this book (Schwarzer et al., this volume). Here we focus on processes and activities that take place farther afield but still impact the coastal zone. The most obvious situation is where processes within a drainage basin affect the downstream coastal waters, but, as rapidly became clear in our discussions, there are many other situations in which coastal management issues arise from activities outside the political or physical boundaries of the coastal area in question.

In this chapter we consider how the effective use and communication of scientific knowledge can be strengthened in the ICM policy process for coastal regions heavily influenced by activities taking place outside the physical boundaries of the coastal system itself. For such regions, ranging from small lagoons and estuaries to large regional seas, effective coastal management must integrate both scientific understanding and policies concerning the entire drainage basin, and beyond in the case of atmospheric inputs, as well as the marine environment and coastal zone itself. Many boundaries must be crossed to achieve successful ICM in the face of transboundary issues, including those between environmental media, political entities, socioeconomic groups, scientific disciplines and between scientists and policy makers.

AGENTS OF ENVIRONMENTAL CHANGE

In addressing the specific workshop goal, it is useful to first make clear what the dominant transboundary physical and socioeconomic processes operating in coastal systems are. These

Science and Integrated Coastal Management
Edited by B. von Bodungen and R.K. Turner

processes can be categorized in a number of ways. Here we start by considering natural and anthropogenic agents of change as drivers.

Natural Change

Coastal ecosystems are regulated in the natural state by inputs of water, sediments, energy, and nutrients from the land, air, and offshore, and these interact to control primary, secondary, and tertiary productivity and the associated cycling of carbon, nutrients, and trace elements. These are not static systems, but rather undergo large natural changes on time scales ranging from short (minutes/hours/days) to long (years/decades/centuries) up to geological time scales (Mann and Lazier 1991). Some changes can take the form of natural extreme events (e.g., invasion of the crown of thorns starfish on the Great Barrier Reef, floods, earthquakes) that dramatically perturb coastal systems, while others are more insidious (Hsü 1991; Dobson et al. 1997). These natural changes provide the context within which anthropogenically induced change must be assessed. In the central North Sea, for example, it is clear that large-scale changes in phytoplankton abundance have taken place related primarily to natural climatic forcing, not to increased anthropogenic inputs such as nutrient inputs (Jickells 1998).

Anthropogenic Change

The ultimate drivers of anthropogenic change in coastal systems are similar to those in most other parts of the globe — population growth and increasing, but spatially very unequal, per capita human consumption of resources. Related to these pressures are the effects of physical changes in the coastal region arising from recreation, commerce, or hazard protection. These issues interact with cultural values and issues of governance to create different pressures in different environments. Furthermore, it is clear that population growth in coastal areas is greater and will probably continue to be greater than overall population growth (Jickells 1998). Hence pressures due to human migrations for political or economic reasons to the coastal area will continue to increase and exacerbate problems of coastal zone management. These pressures on the coastal zone include processes at the shoreline itself (Schwarzer et al., this volume) and overfishing, which is a worldwide problem (Botsford et al. 1997; McGlade, this volume). Other pressures derive from activities remote from the immediate coastal zone and are the focus of this chapter. Some of these pressures are directly connected to the coastal zone via the drainage basin or offshore exchange processes, others are less directly connected. As with natural change, anthropogenic change can be abrupt or gradual.

Mechanisms of Interaction

Drivers operating through the drainage basin include changes in freshwater flows as a result of land-use changes and water abstraction, both of which can alter the physical and chemical characteristics of the coastal aquatic system. Changes in the fluxes of nutrients, sediments, and contaminants arise due to agricultural, industrial, and waste disposal activities, land-use change (particularly deforestation), and damming in the catchment. The serious problems that deforestation in catchments can pose for mangrove forests and coral reefs are considered by Lacerda et al. (this volume). Climate change is also likely to play an important role in the future (Planque and Taylor 1998; Conversi and Hameed 1998; Schellnhuber and Sterr 1993). In general, nutrient and contaminant fluxes rise as a result of increased human activity (Jickells 1998), although sediment and Si fluxes can fall drastically as a result of damming (Humborg et al. 2000; Milliman 1991). Land-use change influences fluxes through changes in sediment storage capacity within catchments and additionally for nitrogen via changes in denitrification (Cornwell et al. 1999; Jickells et al. 2000; Howarth et al. 1996). Resulting changes in nutrient (N:P:Si) ratios can have profound effects on coastal phytoplankton ecosystems (Meybeck, this volume; Humborg et al. 2000; Justic et al. 1995; Philippart et al. 2000). These catchment processes operate on a wide range of spatial and temporal scales. The longer time and space scales present particularly difficult challenges to ICM, as shown, for example, by the effects of nitrate inputs from agriculture in the upper reaches of large river systems on the coastal zone (Figure 6.1) and by the effects of long-term changes in sediment fluxes due to damming (Milliman 1991).

Drivers operating through offshore exchange include energy fluxes that regulate water column stratification, currents that influence biological migration or nutrient transport, and upwelling. These drivers are subject to large-scale physical forcing, which provides a context for understanding the functioning of the ecosystem (Mantoura et al. 1991), but they are in general beyond the control of ICM, at the local scale, except where coastal engineering plays a role (Nienhuis and Smaal 1994). In some systems where offshore exchange dominates, ICM of inputs such as nutrients may not even be necessary. However, these offshore

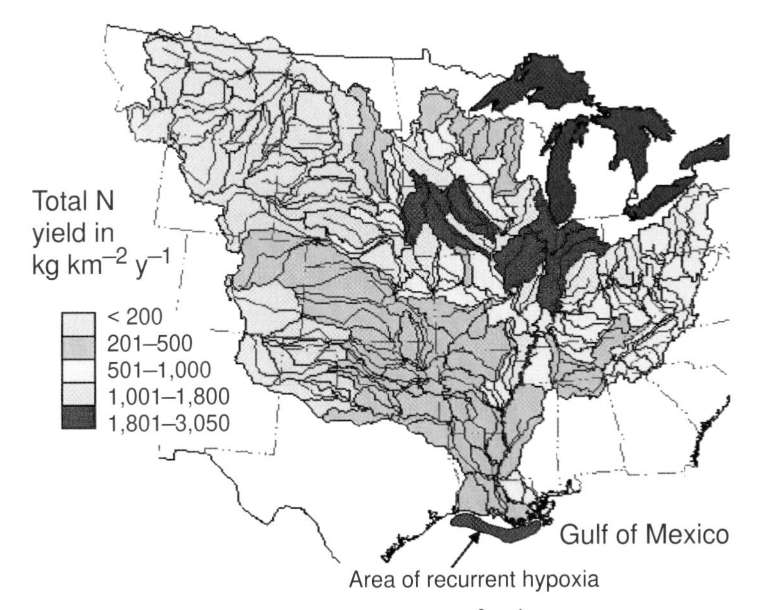

Figure 6.1 Estimated yield of total nitrogen (kg km^{-2} y^{-1}) to the Gulf of Mexico by the major sub-basins of the Mississippi River system (U.S.A.) and the location of seasonal hypoxia in bottom waters of the continental shelf of the Gulf of Mexico (after Goolsby 2000; Rabalais et al. 1996). The predominant source of nutrients linked with eutrophication in the Gulf are the areas of intensive agriculture in the U.S. Corn Belt located 1,500 km upstream.

exchange drivers are all potentially sensitive to climate change, representing a particularly complex, long-range interaction between coastal systems and human activity.

Another driver of change in coastal zones that is not physically directly connected to the coastal zone is atmospheric inputs arising from industrial and agricultural activities. These can be important sources of contaminants and fixed nitrogen and can involve rapid, highly episodic, long-range transport and delivery to the coastal zone (Nixon 1995; Nixon et al. 1996; Jickells 1998). The management of these inputs requires a separate set of regulatory instruments to those for catchment-derived inputs (e.g., see EMEP 1999 and Appendix 6.1).

External trading pressures also impact on the coastal zone in a major way. In the case of fisheries, the patterns of international trade, particularly for tropical shrimp (Kay and Alder 1999; McGlade, this volume), represent a major external pressure on coastal zones of one region by demand in another part of the world. Similar pressures arise from the external demand for wood and the mineral/fossil-fuel extraction industries, which again manifest themselves as direct local effects in the coastal zone. Shipping is an important issue for many coastal zones (e.g., Table 6.1). Much of the pressure for this transportation through the coastal zone arises from external trading pressures, but the environmental consequence (e.g., the problem of handling spoil from dredging to maintain navigation, introduction of alien species, or the threat of oil spills) is felt locally (Gollasch et al. 2000). Tourism represents another external source of pressure and generally involves seasonal movement of people into and out of the coastal zone, bringing economic benefits and environmental pressures. In all of these cases, the economic pressures are generally driven from outside the coastal zone but create environmental impact at local levels, for example, large-scale mangrove destruction in many countries as a result of the creation of shrimp ponds, or loss of coastal ecological habitats to resort or port development.

Physical and Socioeconomic Setting

The forces described above impact on coastal zones that are themselves extremely heterogeneous geographically, ranging from exposed coasts with direct vigorous connections with the open sea, to restricted exchange systems with long water residence times and sometimes very large catchment areas. The physical structure of each particular coastal zone represents the dominant control on its biogeochemical behavior and its management (Mantoura et al. 1991). In Figure 6.2, we present a schematic overview of the different types of systems. Simple uniform regulation of inputs to, and activities in, a coastal zone regardless of its physical structure will seldom be an efficient method of protecting the ecosystem despite its obvious simplistic appeal on the basis of equity. Thus the physical and biogeochemical plus socioeconomic nature (state) of the coastal zone must be central in any ICM process. The same drivers will have very different effects in different systems, as seen by contrasting the North Sea and the Baltic (Jickells 1998) or large coastal ecosystems in the United States (Boesch 1996). In addition, management by setting simple concentration thresholds is inappropriate since they imply that there is a threshold for no effect, which is not the case. Hence, a "critical load" approach is difficult to apply to coastal zones unless quantifiable criteria are set. The use of water quality objectives based on biological criteria is also problematic because of statistical variability and problems of definition (Colijn and Reise, this volume; Ten Brink et al. 1991). Therefore, ICM requires a case by case evaluation rather than a uniform application of a specific formula (Elmgren and Larsson, this volume).

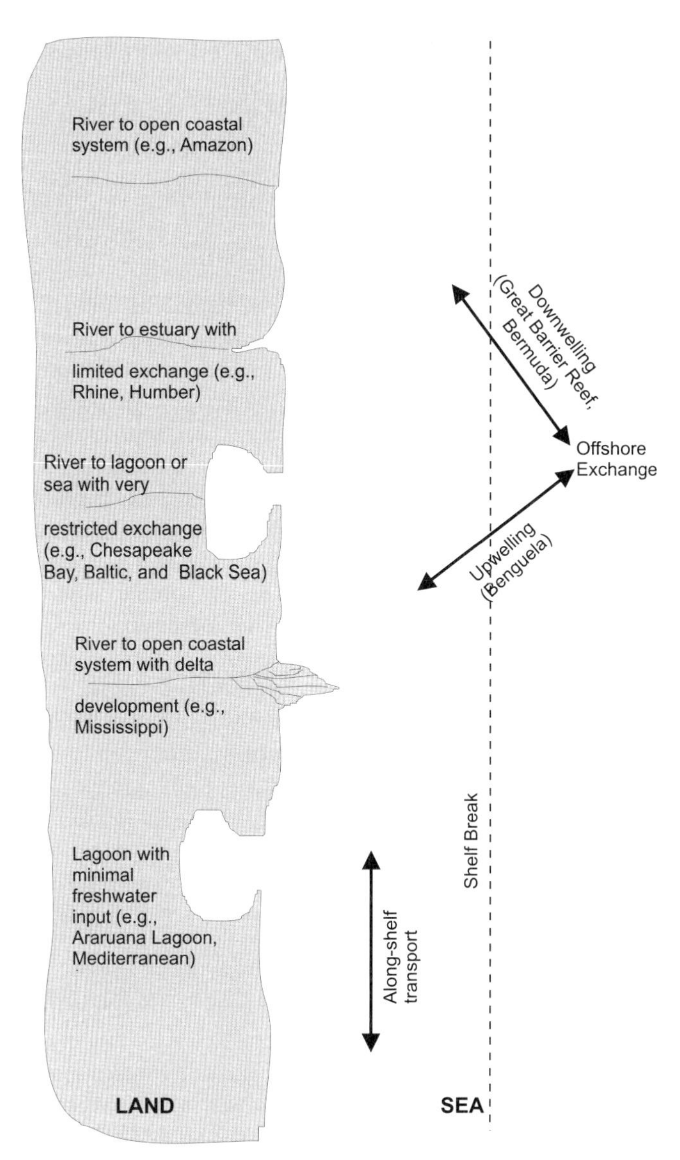

Figure 6.2 Examples of physical types of coastal systems (see also Mantoura et al. 1991). Key physical variables in describing the system include:

(a) catchment size and population vs. area/volume of receiving waters.
(b) length, flow, and variability of flow of rivers.
(c) residence time of water in the coastal zone and the relative importance of flushing by rivers (and groundwater) and offshore exchange.
(d) nature of water exchange with offshore (e.g., upwelling or lack of it, seasonality).
(e) sediment balance and turbidity both via its direct effect on particle water interactions and via its impact on light and phytoplankton growth.
(f) extent and seasonality of vertical stratification which restricts mixing and can promote oxygen depletion.

LESSONS FROM SUCCESSFUL AND UNSUCCESSFUL MANAGEMENT

The four background papers on the Baltic Sea, Chesapeake Bay and the Mississippi Delta, Wadden Sea, and Black Sea provide fascinating insights into the problems and success of coastal management in some areas (see Elmgren and Larsson, Boesch, Colijn and Reise, as well as Mee, all this volume). Some of the essential features of these four systems are summarized in Table 6.1.

The role of science in the development of ICM varies in these different examples. It is clear in many cases that at various stages of the ICM process, major sociopolitical events provided catalysts for activity. In general, scientific research prepared the ground for these changes in advance, but the catalytic event created the political climate that initiated progress. Such a situation has been noted in other ICM developments (Healey and Hennessey 1994). In other cases, such as Chesapeake Bay, Wadden Sea, Mississippi, and Baltic, scientific research was a major contributor to alerting public opinion to the problem. In part, this reflects the fact that the problems in the Mississippi involving deep-water anoxia were initially only detectable by oceanographic research. This would also have been the case in the Baltic, the Wadden Sea, and the Black Sea, but the anoxia became sufficiently shallow to create large-scale "die-offs" (Baden et al. 1990; von Westernhagen et al. 1986; Mee, this volume). This highlights the importance of scientific research and survey activities in the coastal zone in identifying environmental issues before they reach a crisis point. The process by which sociopolitical events set the agenda is not an ideal one in terms of having scientific drivers of ICM but, in general, the resulting impetus toward improved environmental quality is to be welcomed. This analysis of the processes creating the climate for the development of ICM requires scientists to recognize that they are only part of the constituency bidding to develop government policy, and that they need to interact positively with nongovernmental organizations (NGOs) and the public at large to achieve optimum results (Boesch 1999). Without such engagement, ad hoc development triggered by political concerns can lead to inappropriate management, for example, the regulation of nitrates in fresh waters only in terms of human health issues rather than in terms of eutrophication (Addiscott et al. 1991).

An important lesson from studies of the development of public concern in a region reflects the heterogeneity within a coastal system. Thus the perception of people living on the northern European mainland in the riparian states of the North Sea is profoundly influenced by the evidence of beach foams, "black spots," dead seals, and other indicators of environmental change along the Wadden Sea coast of Holland, Germany, and Denmark (de Jong et al. 1999). Similar problems are not evident on beaches in England, and thus political pressure there is different, while scientific assessments based on sampling within the open sea provide a different perspective again on the North Sea (North Sea Task Force 1993). The reality is that the North Sea is heterogeneous. The Wadden Sea region is within the plume of the Rhine and Elbe and heavily impacted by nutrients, while some other parts of the North Sea are much less impacted (North Sea Task Force 1993). Similar heterogeneities are seen in the Mediterranean, where the Adriatic subregion has a similar history of different response to eutrophication pressures (Justic et al. 1987). It is necessary to recognize both the heterogeneity and its relation to public/managerial perspectives in ICM.

Another interesting lesson that can be drawn from the various case studies is that geographically large-scale interactions can be tackled within large drainage basins (Baltic, Black

Table 6.1 Physical characteristics and management practices in the Baltic Sea, Black Sea, Wadden Sea, Chesapeake Bay, and Mississippi Delta.

	Baltic Sea	Black Sea	Wadden Sea	Chesapeake Bay	Mississippi Delta
Physiographic type	Shallow enclosed sea	Deep enclosed sea	Coastal lagoon and adjacent coastal sea	Drowned river valley estuary	River delta and inner continental shelf
Catchment area (km^2)	1,700,000	1,874,904	455,359	166,000	3,200,000
Land:water area	4	4.5	21	15.8	95
Land area:water volume (km^{-1})	80		1,337	2,410	5,000
Residence time	20–25 y	~1000 y (surface waters decades)	1–7 days (Wadden Sea); 33 days (German Bight)	7 months	1 y
Freshwater discharge ($km^3\ y^{-1}$)	440	350	121	66	590
Nitrogen inputs (10^6 t N y^{-1})	1.25	1.05	0.60	0.15	1.60
Phosphorus inputs (10^3 t P y^{-1})	56	50	28	11	136
Key environmental drivers	• Freshwater inflow • Saltwater inflow • Stratification	• Freshwater inflow • Mixing at boundary currents • Flows to/from Mediterranean • Land-based pollutants	• Exchanges with sea • River inputs to coastal waters • Atmospheric deposition	• Freshwater inflows	• Freshwater inflows • Sea-level rise and subsidence • Wind-driven shelf circulation, mixing • Hurricanes
Key management issues	• Eutrophication, hypoxia • Habitat degradation • Coastal conservation	• Eutrophication • Sewage pathogens • Habitat loss • Oil transport, operational discharges • Opportunistic predators • Loss of fish stocks	• Use of resources • Habitat restoration • Nature reserves • Coastal defense	• Eutrophication, hypoxia • Habitat degradation	• Eutrophication, hypoxia • Wetland loss • Physical

Table 6.1 continued

	Baltic Sea	Black Sea	Wadden Sea	Chesapeake Bay	Mississippi Delta
Socioeconomic drivers	• Agriculture • Population (sewage) • Industry (including transport and tourism • Fishing	• Agriculture • Fishing • Tourism • Transport of oil and cargo • Industrial development	• Agriculture • Fisheries • Tourism • Industrial activities • Harbors	• Agriculture • Land development • Fishing (overfishing, multispecies)	• Agriculture • River management, flood protection • Oil and gas development • Fishing (trawling effects)
Transboundary governance	Helsinki Convention (HELCOM)	• Bucharest Convention • Odessa Ministerial Declaration • Black Sea Strategic Action Plan • Istanbul Commission	• International Wadden Sea Ministerial Conferences, Treaty • North Sea Agreement	Chesapeake Bay Program, a multistate-federal program involving agreements	Only existing program involves only 5 coastal states; commitments for nutrient control by states in basin being discussed
Models used for integrated management	Under development (MARE research program)	Models developed at the Black Sea Activity Center for Coastal Zone Management (Russia)	• Ecosystem models • Operational storm and tidal predictions • Transport models for parts of Wadden Sea	Complex, linked watershed-estuary models used for projecting future conditions, setting targets	Crude, black-box model used to estimate effects of nutrient loading reduction on hypoxia
Status of environmental monitoring	• Monitoring since ~1970 • State of the Baltic Report (every 5 years)	• Institutions mostly inactive (except for Romania/Ukraine), basin-wide monitoring discontinued since 1989 • Comprehensive Black Sea Assessment reports (1996–1999)	• Internationally agreed monitoring program (TMAP) • National monitoring programs • Quality Status Reports 1992–2000	• Extensive coordinated monitoring program (14 years) • Periodic State of the Bay Report	• River discharges well monitored; wetland changes periodically mapped; shelf observations only associated with research programs • CENR Integrated Assessment of Hypoxia

Table 6.1 continued

	Baltic Sea	**Black Sea**	**Wadden Sea**	**Chesapeake Bay**	**Mississippi Delta**
Successes in use of science in ICM	• Determination of the role of N and P in ecosystem degradation • Determination of N and P loads and establishment of reduction targets • Elimination of pollution hot spots • Banning use of PCBs and DDT	• Black Sea Diagnostic Analysis (1996) • Black Sea Strategic Action Plan (1996) • New environmental quality standards developed (1998/99)	• Identification of effects of eutrophication • Regulations on fisheries to guarantee food for birds • Shellfish cultivation agreement	• Determination of the role of N in ecosystem degradation • Linking eutrophication to loss of seagrasses	• Identification of scope and causes of hypoxia • Determination of causes of wetland loss and assessment of options
Challenges in use of science in ICM	• Integrating modeling where cost-benefit analyses are coupled to environmental targets on different scales	• Setting objectives for adaptive management • Better linkage between monitoring and ICM • Improved knowledge of ecosystem service functions • Economic scenarios for management	• Natural habitats • No-take refuges • Marine mammal health • Sustainable fisheries • Mismanagement in past	• Integrating modeling and monitoring • Provide basis for multispecies management • Managing land use in the context of small catchment dynamics	• Science to support simultaneous delta restoration and reduction of hypoxia • Linking wetland loss and eutrophication to fisheries production
Important mechanisms to communicate scientific knowledge	• Public media • Reports by agencies, universities, and NGOs	• No established permanent mechanisms • TV talk shows • Environmental education programs	• *Wadden Sea Newsletter* • Biannual Wadden Sea Conferences	• Scientific & Technical Advisory Committee • *Bay Journal* newspaper • News media	• CENR Integrated Assessment • National press attention

Sea, Chesapeake Bay, Wadden Sea/Rhine) and even in terms of atmospheric transport (EMEP 1999: Appendix 6.1). In the case of the European Monitoring and Evaluation Programme (EMEP), the Rhine and the Baltic, regulatory systems across international boundaries have been established with some success, even now trading pollution abatement and debt in the Baltic (Appendix 6.1). In the case of the Black Sea, western European models have been adopted for coastal zone management, driven largely by the interest of the relevant countries in joining the European Union. Such external pressure for environmental management is an important driver in other transboundary pollution problems. Interestingly, similar catchment basin management has not yet been achieved in the Mississippi, although this may reflect the relatively recent recognition of the scale of the problem (Boesch, this volume; Table 6.1), though an agreement to this effect has been signed as this book goes to press (Showstack 2000). This slow progress may also reflect the fact that the dominant issue here is nitrate from agriculture, an issue that in general has also not been successfully handled by some of the other drainage basin management systems (Baltic, Black Sea, Rhine, Chesapeake Bay) despite their success in managing other contaminants, such as phosphorus (Boesch, this volume). This reflects the particular difficulty of managing a very mobile contaminant (nitrate) released from diffuse sources by the activity of an economically and politically powerful lobby (agriculture).

SCIENTIFIC CHALLENGES TO EFFECTIVE TRANSBOUNDARY ICM

Uncertainties/Ignorance

Scientific research is often viewed, particularly from outside the scientific community, as dealing in issues of fact and certainty. In practice, it deals with probabilities in a world of incomplete information. Uncertainty is an inherent problem, both in science and in management. In natural sciences we try to accept as little uncertainty as possible, while we accept larger ranges of uncertainty in social and economic sciences. Ecological modeling has high levels of uncertainty simply because of a lack of known causal links between organisms. In the face of this uncertainty, scientists, like everybody else, can have different interpretations and opinions of a situation. For managers it is difficult to cope with uncertainties because taking the wrong decisions might be expensive, a situation which can lead to no decisions being taken.

In transboundary processes, particularly large uncertainties may arise as a result of the distance between source and sink effects, for instance, making it difficult to predict with confidence the outcome of activities upstream in the catchment area on the coastal zone. There is a perception that sociopolitical and socioeconomic decisions are only taken in a situation of certainty, but we found numerous examples of effective ICM operating within uncertain and incomplete information (Table 6.1). However, it is very important to distinguish between uncertainty and ignorance (McGlade, this volume) and to convey this effectively to managers and the public. Uncertainty can be specifically factored into the ICM process; ignorance, by contrast, means that the effect of an action cannot be predicted at all. An interactive ICM process, or adaptive management, in which there is a continuous cycle of scientific evaluation of the effectiveness of ICM does allow progress to be made in the face of uncertainties without abandoning the scientific principle of estimating outcomes with associated uncertainties. It is important that scientists make every effort to convey the scale of uncertainties/ignorance, and also for stakeholders and managers to accept the importance of both the scientific advice and its associated uncertainty in policy making. Despite these caveats about uncertainties, it is clear that scientists most effectively influence the management process when they can offer a consensus view to the stakeholder community (Boesch 1999), albeit one constrained by the uncertainties.

Scales of Interaction

A first step in any ICM should involve a scaling exercise to define the physical boundaries and the magnitude and relative importance of the different physical and biogeochemical drivers on the ecosystem (i.e., riverine, atmospheric, or offshore) for the issue in question (e.g., nutrients, sediments) (see also Liss et al. 1991). A similar scaling of socio/economic/political drivers will then focus attention on the key issues of concern in this area. This assessment will, in turn, inform the debate about future scientific research requirements, aimed at describing the system with sufficient certainty to allow politicians to consider options for management of the system and to set appropriate milestones and goals for the ICM process. In our deliberations, eutrophication issues were frequently a major focus of concern because (a) contaminant threats are less generally pervasive and are also more amenable to technological control, (b) nutrient fluxes have increased with increasing human activity globally and seem in many areas likely to be subject to further substantial increases in the near future, and (c) eutrophication control requires scientific management to achieve good results and the expertise to achieve this is available. Effective reductions in nutrient inputs have been shown to reverse at least some symptoms of eutrophication (Mee, Elmgren and Larsson, and Boesch, all this volume).

Socioeconomic and physical drivers operate on a wide range of space scales. It is relatively straightforward to consider effects, make predictions, and describe uncertainties at local and short time scales. Issues become much more difficult at larger space scales (e.g., Boesch and Mee, both this volume) and particularly on longer time scales.

As an example of long-term issues, damming in catchments has a variety of short-term effects, ranging from reduced seasonality on an annual time scale, through wetland habitat loss on interannual time scales, to erosion of deltas and changes in phytoplankton species on decadal time scales (e.g., Mee, this volume; Phillippart et al. 2000). On even longer intergenerational time scales, issues such as long-term changes in water flows and ultimately the issue of decommissioning of the dam as it ages or fills up with sediment become important. Economic and urban developments downstream of a dam can make the changes in water management effectively irreversible or only reversible on a very long (intergenerational) time scale, hence fundamentally changing the nature of the system in a single step. Figure 6.3 represents an idealized representation of cost-benefit analysis in this situation, with the benefits occurring in the short term and costs occurring to future generations. When applying cost-benefit analysis to environmental management, the rate by which we discount future costs and benefits is not only one of the most influential decisions, it is also often one of the major items of value conflict and controversy (Hanley and Splash 1993). In Figure 6.3, the area between the horizontal axis and the curve denotes the net benefits, that is, the difference between the benefits and the costs. Flood control, hydroelectricity, and irrigation are benefits

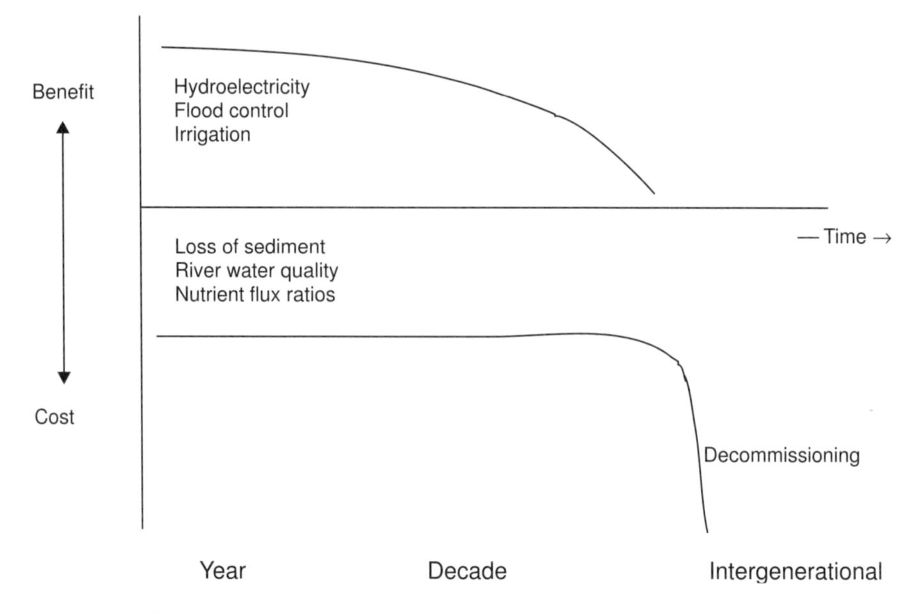

Figure 6.3 An illustration of the relative cost-benefit of damming over time.

during the life span of the dam while loss of sediments, river water quality, changes in nutrient flux ratios, construction, and decommissioning of the dam constitute the costs. Through the choice of discount rate, we are within the analysis able to make this dam seem either beneficial or not for the society.

Sudden release of contaminants (chemical time bombs; Liss et al. 1991) as a result of environmental changes represents another step change function. Other examples of step functions include loss of benthic macrophytes and changes in phytoplankton species in the Black Sea (Mee, this volume). These irreversible or nonlinear changes on long time scales represent particularly difficult challenges to the coastal management process.

Another complex long-range scale of interaction is that resulting from remedial measures in ICM themselves. The experience of phosphorus removal in the Baltic, which resulted in local environmental improvement but also promoted long-range transport of nitrogen (due to inefficient local utilization by P-limited phytoplankton) and caused environmental problems in other parts of the Baltic (i.e., proliferation of cyanobacteria), again emphasizes the imperative of considering coastal management within appropriately large physical and temporal boundaries (Elmgren and Larsson, this volume).

The interaction between scientists and managers needs to recognize the possibility that there are uncertainties, ignorance, and step function changes as well as considering the time scales of damage and recovery of systems. In the Baltic, for example, the political time scale for environmental improvement (politicians have set a target of 2020) is likely to be incompatible with the longer residence times of nutrients and contaminants in this sea (e.g., five years for N and eleven years for P) (Wulff et al. 2001) and in the soils and groundwater of the drainage area. In general, recovery times for an ecosystem will be longer than the time scales of initial perturbation, even without step function or irreversible effects (Costanza and

Mageau 1999). Rehabilitation costs are also usually larger than emission controls (Ten Brink et al. 1991). Recovery times will also vary for different physical parts of the system depending on water residence time, for example, and for different chemical components depending on chemical residence times. Biological recovery times are particularly complex and difficult to predict and depend on the scale of impact that affects recruitment of stocks. Recovery can be expedited by direct remediation measures (e.g., removal of contaminated sediment); however, such measures are only realistic for small systems. Beyond this, recovery will reflect the characteristic residence time of the contaminant of interest in the ecosystem in question. To avoid generating unreasonable expectation of success, these time scales must be conveyed effectively to managers and the public.

Economic Valuation

Conventional cost-benefit environmental economics handles long time scales and the risk of step function changes in a particular way that is conditioned by a utilitarian ethical stance. Such a perspective has been challenged by a range of commentators. In addition, no single economic instrument is suitable for all problems. The basic condition for efficiency (that the utility to all affected individuals should be accounted for when determining efficient quantity/quality) is one of the fundaments in modern environmental economics. The need to include all individuals is complicated if the rights or interests of future individuals must also be considered. The common approach in economic analyses for dealing with the effects on costs and benefits of a long time scale is to use some kind of interest rate to discount future benefits and costs, i.e., longer-term effects are given less weight in the appraisal. While this approach may be politically realistic in the short term, it is problematic in the context of environmental change over relatively long time horizons. There is no consensus on the appropriate interest rate to use, which is not surprising given the clear ethical judgements that are involved. To use environmental economics effectively in coastal management in this context, it is necessary to be explicit about the assumptions and decision criteria that have been adopted to avoid inconsistency and inefficiency (Frykblom 1998).

One of the driving forces behind anthropogenic change is clearly the existence, and sometimes the lack, of economic incentives. Economic theory has a whole battery of tools to accomplish an economically effective solution to emissions where a simple zero emissions situation is not the required outcome. When an emitter has a negative impact on the well-being or production of others, economic theory often prescribes internalization of the caused costs. That is, the emitter should carry the costs of its actions and hence a more cost-effective solution may eventually be achieved. As the zone of influence of pollutants extends beyond regional and national borders, the economical, political, and practical difficulties of implementing cost-effective measures are increasingly difficult. Pollutants crossing boundaries impose external costs that neither emitters nor nations have the proper incentives to control (Tietenberg 1992). Even if there are economic measures to deal with emissions within each country, there are often incentives to "cheat," i.e., the emitter does not fully internalize transboundary costs.

One way to solve this problem has been through international agreements. Despite not being the economically most efficient solution in theory, these can sometimes be the best that can practically be achieved, and far better than doing nothing. Agreeing on a certain reduction of, for example, nutrients has a pedagogic merit in its simplicity. However, looking strictly from an economic point of view, further significant welfare gains can be achieved if economists are allowed to participate in the design of such an international agreement. Allowing for trade of emission reduction is one such measure that can lead to a further cost-effective solution (Appendix 6.1). This holds true despite possible spatial restrictions of where emissions can be released. International agreements are sometimes difficult to establish, and to make matters even worse, they are often not strictly obeyed. The problem can be described by the classical game theory of the prisoner's dilemma (Hirshleifer 1983; Pearce and Turner 1990; Tietenberg 1992).

Integration

The complexity of the various challenges within individual scientific disciplines are substantial in themselves, as are the challenges to national and international organizations of transboundary problems. These challenges pale by comparison with the task of integrating all these issues in order to allow decision makers to balance the competing pressures in the coastal zone.

Models, ranging from conceptual to statistical to mathematical, provide a powerful means to integrate scientific understanding of the interactions among issues and drivers. The greater computational capacity now available is allowing the development of spatially explicit models that represent the biophysical dynamics of drainage basins and coastal ecosystems (Boesch, this volume) and even integrate economic drivers. When used with proper caution, concerning the propagation of uncertainties in complex models and with interactive revision based on new knowledge, models can greatly assist transboundary decision making.

STRENGTHENING THE EFFECTIVE USE AND COMMUNICATION OF SCIENTIFIC KNOWLEDGE

Setting Boundaries

It is immediately clear that coastal zone management needs to encompass all the relevant processes and to draw the management boundaries appropriately for the specific issue under consideration. Thus coastal zone management must consider processes throughout the relevant river catchments, the offshore exchange processes, and atmospheric deposition often from regions far from the coastal zone, sometimes called the "airshed." The boundaries must also include the relevant socioeconomic drivers that may be within the coastal zone catchment but can also be very remote, as in the case of tourism or economic opportunities/pressure for export to overseas markets. The boundaries for the management process will be different for different environmental concerns and can be assessed by suitable preliminary scoping exercises, which should be subject to review in an iterative process of adaptive management as discussed below. A process that fails to draw appropriate boundaries is likely to be compromised from the start. Similarly, heterogeneity within coastal zones must be considered both in terms of environmental management and in terms of the perception of the state of the environment. Setting appropriate boundaries includes the issue of setting appropriate and realistic goals. Geological as well as socioeconomic records can sometimes be used to assay the past state of a coastal environment and scales of change (e.g., Bianchi et al. 2000), although these records are always imperfect. However, it must be recognized that coastal systems are in a continuous state of flux, even without human intervention, and hence it is unrealistic to aim for a full return to a past specific state even if this were desirable.

Using Science in Adaptive Management

An important framework for the effective application of science in decision making in the face of uncertainty is adaptive management. Simply stated, adaptive management involves treating economic uses of nature as experiments, so that we learn more effectively from experience (Lee 1993). Adaptive management first requires making expectations of the outcome of policies explicit, so that methods to measure their effectiveness can be designed. Second, it involves collection and analysis of information so that actual outcomes can be compared with expectations. Finally, adaptive management must be prepared to correct errors, improve understanding, and appropriately change action plans. The notion of a continuous feedback process such as the Drivers-Pressures-State-Impact-Response (D-P-S-I-R) framework and in the GESAMP cycle for ICM discussed by von Bodungen and Turner (this volume) is akin to adaptive management (see also Olsen, this volume).

Adaptive management provides great opportunities and challenges to science. It requires science to develop models that are useful in projecting expectations based on best understanding. It places a great premium on monitoring of actual outcomes, including the development of meaningful indicators and separation of natural variation from responses to management activities. Adaptive management also challenges scientists to link monitoring, modeling, and research more effectively to improve projections of expectations and interpret observations made of these policy experiments.

In the Netherlands, targets and goals were set for the marine environment that were essentially expectations of the achievable outcomes of management (Ten Brink et al. 1991). These proved to be strong incentives for monitoring of key organisms and environmental variables thought to be indicative of the effects of pollution reduction measures. However, as Boesch (this volume) points out, the integration of modeling and monitoring and their use in iterative policy reformulation typically falls short of the ideal of true adaptive management, even in the Chesapeake Bay where these efforts are well supported and sustained.

Targets and Goals

The process of setting goals should involve all stakeholders from throughout the region. Natural and social scientists must be partners in the process to ensure that the goals are realistic in terms of achievability and in terms of being measurable. Thus, for example, restoration to some past state is often effectively impossible, while improving environmental quality and aiming for sustainability against some agreed target is possible. Similarly, a goal that requires the abandonment of shipping operations in an estuary would generally be unrealistic, though management of shipping to minimize environmental damage may well be possible. However, it must be recognized that the process of ICM will only succeed with all important stakeholders involved and committed. The primary role of the scientist in this process is to provide options, not make decisions, and to ensure that targets and timetables set are realistically attainable.

Institutions

Transboundary problems that cross geopolitical boundaries will require transboundary institutions, which we assume can and will be established. Examples of some of these are contained within the chapters in this book together with some other examples in Appendix 6.1. The success of these institutions in ICM depends in part on the individuals involved. It also requires that the stakeholders remain engaged in the process, and for this the institutions need to be seen to be successful in addressing their goal. Transboundary management that crosses geographic boundaries will require stakeholder participation from all countries affected, particularly scientific participation. Transboundary institutions can expedite this process by funding cooperative scientific interactions. The creation of the widest possible international scientific consensus on an issue is critical to achieving political agreement on action.

Scientific Input to the ICM Process

Scientists must engage as equal parties in the process to create a trustworthy network of communication involving themselves, managers, stakeholders, public, media, and politicians. Mutual and interdisciplinary scientific capacity needs to be developed across geographical boundaries. It may be necessary for scientific organizations involved in this process to acquire brokers to interface with these stakeholder groups. However, this must not be at the expense of scientists playing an active role interacting with other stakeholders but rather as a complement to this process. In general, the difficulties of creating effective ICM increase as time and distance scales increase. Local contamination problems are often more amenable to relatively simple technological solutions. An example would be bacteriological contamination of bathing waters, which can be remedied readily by sewage treatment, although even in this context complex issues of risk perception and public trust, or the lack of it, can be present. By contrast, many transboundary issues in ICM involve more complex issues that require a societal response involving changes in lifestyle or economic practice. Thus, restoring freshwater flows to coastal zones in some areas may necessitate reductions in water abstraction and hence reduced societal use of water. Some reductions in nitrate inputs to coastal waters can be achieved by improved wastewater treatment, but further reductions will require major changes in agricultural practice or reductions in atmospheric nitrogen deposition via changes in fossil fuel combustion (Figure 6.4) with associated major costs.

The initial scoping exercise and subsequent adaptive management should help focus scientific research on eliminating ignorance of key processes, reducing uncertainties, and increasing capacity building for the provision of relevant advice. Science is needed to help set sensible milestones for the adaptive management process and to identify the full range of effects resulting from particular actions, for example, the additive benefit of wetland creation for flood prevention, habitat creation, and nutrient retention (e.g., Jickells et al. 2000). It is also often helpful to associate a monetary calculus with this process using cost-benefit analysis that, where meaningful, more fully recognizes less tangible benefits, such as those related to ecosystem services (Costanza et al. 1997), together with appropriate adoption of precautionary approaches in the face of ignorance and uncertainty. The D-P-S-I-R framework (von Bodungen and Turner, this volume) may be useful in this context; however, it is necessary to recognize the wealth of information required for this process. In addition, the information

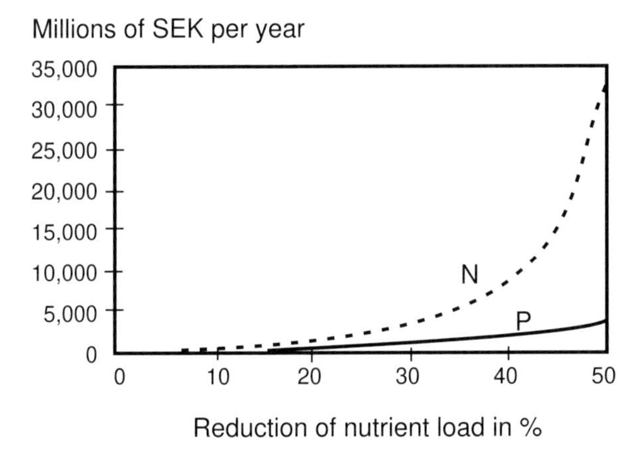

Figure 6.4 Relative cost in Swedish crowns (SEK) of reducing anthropogenic fixed nitrogen and phosphorus inputs to the Baltic. Reductions of several tens of percent can be achieved by relatively simple and inexpensive approaches such as improved sewage treatment works and minor changes in agricultural practices. Subsequent reductions become increasingly more expensive and difficult since they involve major changes in farming practices as well as societal lifestyle such that a 50% reduction can cost twice as much as a 45% reduction (Gren et al. 1997; reprinted by permission of Academic Press).

needed will often be different for natural and socioeconomic parts of the process and, therefore, requires careful translation across disciplines and on to policy makers. The goal of the scientific research should be the presentation of choices for future environmental quality and the associated costs.

In addition to this work focused on improving environmental management at a local or regional level, we wish to note two global initiatives that consider the state and functioning of coastal zones: LOICZ (Land–Ocean Interactions in the Coastal Zone; Pernetta and Milliman 1995) and GIWA (Global International Waters Assessment; Mee and O'Toole et al., both this volume). The scale of these activities makes them particularly appropriate for considering transboundary issues and, in the process, reveal new issues and help set the future research agenda by complementing and synthesizing work at local levels.

FUTURE RESEARCH NEEDS

In drawing together future research needs, we focused on the following large generic issues:

- Develop indicators that can be used to describe environmental impacts and hence facilitate comparison between regions, quantify changes within an ecosystem, and set targets for ICM. This may need to be done on a regional scale, since for most indicators, no single global approach will be appropriate.
- Improve methods to value meaningfully the cost-benefit of environmental change in the coastal zone.
- Develop/improve models that describe complex environments (e.g., atmosphere/landscape/hydrology/coastal zone), and hence allow modeling of scenarios of different management options and of climate change.

- Couple these models to economic valuation and other multi-criteria assessment models, hence allowing managers and politicians to compare management options in a realistic manner.

We recognize, however, that in any coastal ecosystem, substantial research is also necessary to quantify biogeochemical and socioeconomic processes and parameterize models of these processes.

APPENDIX 6.1

European Monitoring and Evaluation Programme (EMEP) (J. Pacyna)

More than two decades ago, an international program was established within the UN Economic Commission for Europe (UNECE) Convention on Long-Range Transboundary Transmission of Air Pollutants (LRTAP), with the aim of providing the governments of the European countries with information on source receptor relationships for air pollutants emitted in one region and deposited in another region after long-range transport within air masses. The European Monitoring and Evaluation Programme (EMEP 1999) was thus established to provide information that could be used to validate the UNECE international agreements on emission reductions of air pollutants in the UNECE region. EMEP proved to be very successful in the development of tools to trace the implementation of the international agreements through both long-range transport models and a monitoring network. The experience of EMEP is used directly by the European Marine Convention for the protection of the North Sea and North East Atlantic (OSPARCom) and the Baltic Sea (HELCOM). EMEP experience is also used within the Arctic Monitoring and Assessment Programme (AMAP). The methods used in EMEP may be appropriate for other parts of the world.

Debt/Environment Swap (J. Pacyna, M. Voss)

A number of creditor countries (the so-called Paris Club member countries) have agreed to convert a part of Polish international debt to finance projects in Poland aimed at a reduction of pollution generated in the country. One of the five areas of these activities, coordinated by the Polish EcoFund Foundation, is the reduction of Baltic Sea contamination arising from within Polish coastal waters. The above debt to environment swap arrangement has proved very successful over the last seven years and may be applicable in other parts of the globe.

REFERENCES

Addiscott, T.M., A.P. Whitmore, and D.S. Polson. 1991. Farming, Fertilisers and the Nitrate Problem. Wallingford: CAB International.

Baden, S.P., L.O. Loo, L. Pihl, and R. Rosenberg. 1990. Effects of eutrophication on benthic communities including fish: Swedish west coast. *Ambio* **19**:113–122.

Bianchi, T.S., et al. 2000. Cyanobacterial blooms in the Baltic Sea: Natural or human induced? *Limnol. Oceanogr.* **45**:716–726.

Boesch, D.F. 1996. Science and management for four U.S. coastal ecosystems dominated by land–ocean interactions. *J. Coast. Conserv.* **2**:103–114.

Boesch, D.F. 1999. The role of science in ocean governance. *Ecol. Econ.* **31**:189–198.
Botsford, L.W., J.C. Castilla, and C.H. Peterson. 1997. The management of fisheries and marine ecosystems. *Science* **277**:509–515.
Conversi, A., and S. Hameed. 1998. Common signals between physical and atmospheric variables and zooplankton biomass in the Subarctic Pacific. *ICES J. Mar. Sci.* **55**:739–747.
Cornwell, J.C., W.M. Kemp, and T.M. Kana. 1999. Denitrification in coastal ecosystems: Methods, environmental controls and ecosystem level controls, a review. *Aquat. Ecol.* **33**:41–54.
Costanza, R., et al. 1997. The value of the world's ecosystem services and natural capital. *Nature* **387**:253–260.
Costanza, R., and M. Mageau. 1999. What is a healthy ecosystem? *Aquat. Ecol.* **33**:105–115.
De Jong, F., et al. 1999. Wadden Sea Quality Status Report. Wadden Sea Ecosystem No. 9, Trilateral Monitoring and Assessment Group. Wilhelmshaven: Common Wadden Sea Secretariat.
Dobson, A.P., A.D. Bradshaw, and A.J.M. Baker. 1997. Hopes for the future: Restoration ecology and conservation biology. *Science* **277**:515–522.
EMEP. 1999. The European Monitoring and Evaluation Programme (EMEP). Report of the Sixteenth Session of the Executive Body of the Convention on Long-range Transboundary Air Pollution. Report No. ECE/EB.AIR/59. Geneva: UNECE (United Nations Economic Commission for Europe).
Frykblom, P. 1998. Questions in the contingent valuation method: Five essays. Acta Universitatis Agriculturae Sueciae 100. Uppsala: Swedish Univ. of Agricultural Sciences.
Gollasch, S., J. Lenz, M. Dammer, and H.-G. Andres. 2000. Survival of tropical ballast water organisms during a cruise from the Indian Ocean to the North Sea. *J. Plankton Res.* **22**:923–938.
Goolsby, D.A. 2000. Mississippi Basin nitrogen flux and Gulf hypoxia. *EOS* **81**:321–327.
Gren, I.-M., T. Soderqvist, and F. Wulff. 1997. Nutrient reductions to the Baltic Sea: Ecology, costs and benefits. *J. Env. Manag.* **51**:123–143.
Hanley, N., and C. Spash. 1993. Cost-benefit Analyses and the Environment. Aldershot: Edward Elgar.
Healey, M.C., and T.M. Hennessey. 1994. The utilisation of scientific information in the management of estuarine ecosystems. *Ocean Coast Manag.* **23**:167–191.
Hirshleifer, J. 1983. From weakest link to best shot: The voluntary provision of public goods. *Public Choice* **41**:371–386.
Howarth, R.W., et al. 1996. Regional nitrogen budgets and riverine N and P fluxes for the drainage to the North Atlantic Ocean: Natural and human influences. *Biogeochemistry* **35**:75–139.
Hsü, K.J. 1991. Fractal theory and time dependency in ocean margin processes. In: Ocean Margin Processes in Global Change, ed. R.F.C. Mantoura, J.-M. Martin, and R. Wollast, pp. 235–250. Dahlem Workshop Report. Chichester: Wiley.
Humborg, C., et al. 2000. Silicon retention in river basins: Far-reaching effects on biogeochemistry and aquatic food webs in coastal marine environments. *Ambio* **29**:44–49.
Jickells, T.D. 1998. Nutrient biogeochemistry of the coastal zone. *Science* **281**:217–222.
Jickells, T.D., et al. 2000. Nutrient fluxes through the Humber estuary: Past, present and future. *Ambio* **29**:130–135.
Justic, D., T. Legovic, and L. Rottini-Sandrini. 1987. Trends in oxygen content 1911–1984 and occurrence of benthic mortalities in the northern Adriatic Sea. *Estuar. Coast. Shelf Sci.* **25**:435–445.
Justic, D., N.N. Rabalais, and R.E. Turner. 1995. Stoichiometric nutrient balance and origin of coastal eutrophication. *Mar. Poll. Bull.* **30**:41–46.
Kay, R., and J. Alder. 1999. Coastal Planning and Management, p. 27. London: E. and F.N. Spon.
Lee, K.N. 1993. Compass and Gyroscope: Integrating Science and Politics for the Environment. Washington, D.C.: Island Press.
Liss, P.S., et al. 1991. What regulates boundary fluxes at ocean margins? In: Ocean Margin Processes in Global Change, ed. R.F.C. Mantoura, J.-M. Martin, and R. Wollast, pp. 111–126. Dahlem Workshop Report. Chichester: Wiley.
Mann, K.H., and J.R.N. Lazier. 1991. Dynamics of Marine Ecosystems. Oxford: Blackwell Scientific.

Mantoura, R.F.C., J.-M. Martin, and R. Wollast, eds. 1991. Ocean Margin Processes in Global Change. Dahlem Workshop Report. Chichester: Wiley.
Milliman, J.D. 1991. Flux and fate of fluvial sediment and water in coastal seas. In: Ocean Margin Processes in Global Change, ed. R.F.C. Mantoura, J.-M. Martin, and R. Wollast, pp. 69–89. Dahlem Workshop Report. Chichester: Wiley.
Nienhuis, P.H., and A.C. Smaal, eds. 1994. The Oosterschelde Estuary (The Netherlands): A Case Study of a Changing Ecosystem. Dordrecht: Kluwer.
Nixon, S.W. 1995. Coastal marine eutrophication: Definition, social causes and future concerns. *Ophelia* **41**:199–201.
Nixon, S.W., et al. 1996. The fate of nitrogen and phosphorus at the land–sea margin of the north Atlantic Ocean. *Biogeochemistry* **35**:141–180.
North Sea Task Force. 1993. North Sea Quality Status Report 1993. Oslo and Paris Commissions, London. Fredensborg: Olsen and Olsen.
Pearce, D.W., and R.K. Turner. 1990. Economics of Natural Resources and the Environment. Hertfordshire: Harvester Wheatsheaf.
Pernetta, J.C., and J.D. Milliman, eds. 1995. Land Ocean Interactions in the Coastal Zone: Implementation Plan. Report No. 33. Stockholm: IGBP (International Geosphere–Biosphere Programme) Global Change.
Philippart, C.J.M., G.C. Cadee, W. van Raaphorst, and R. Riegman. 2000. Long-term phytoplankton–nutrient interactions in a shallow coastal sea: Algal community structure, nutrient budgets and denitrification potential. *Limnol. Oceanogr.* **45**:131–144.
Planque, B., and A.H. Taylor. 1998. Long-term changes in zooplankton and the climate of the North Atlantic. *ICES J. Mar. Sci.* **55**:644–654.
Rabalais, N.N., et al. 1996. Nutrient changes in the Mississippi River and system responses on the adjacent continental shelf. *Estuaries* **19**:386–407.
Schellnhuber, H.-J., and H. Sterr. 1993. Klimaänderung und Küste. Heidelberg: Springer.
Showstack, R. 2000. Nutrient over-enrichment implicated in multiple problems in U.S. waterways. *EOS* **81**:497–499.
Ten Brink, B.J.E., S.H. Hosper, and F. Colijn. 1991. A quantitative method for the description and assessment of ecosystems. *Mar. Poll. Bull.* **23**:53–60.
Tietenberg, T. 1992. Environmental and Natural Resource Economics. New York: Harper Collins.
von Westernhagen, H., et al. 1986. Sources and effects of oxygen deficiencies in the south-eastern North Sea. *Ophelia* **26**:457–473.
Wulff, F., L. Rahm, K. Hallin, and J. Sandberg. 2001. A nutrient model of the Baltic Sea. In: A Systems Analysis of the Baltic Sea, ed. F. Wulff, L. Rahm, and P. Larsson. Heidelberg: Springer.

7

Shoreline and Land-use Development Perspectives in the Context of Long-term Decreasing Functional Diversity in the Delta Region, SW Netherlands

I. DE VRIES

National Institute for Coastal and Marine Management, RIKZ, P.O. Box 20907, 2500 EX The Hague, The Netherlands

ABSTRACT

The decreasing functional diversity of the Delta region, SW Netherlands, has been described as a result of interaction between natural dynamics (e.g., sea-level rise, marsh accretion) and increasing anthropogenic influences ("draining, digging, dredging") over 9000 years. The Delta area is in its present state one of the few remaining rural areas within the European core economic region. The increasing natural threat to the area (sea-level rise) combined with occupation pressure require a policy response. A layer model has been presented which arranges elements of the spatial structure into three physical layers with different rates of change: a ground layer (soil, water) with long intrinsic time scales (up to 1000 years), a networks layer (infrastructure) with intermediate time scales (up to 100 years), and the occupation layer (land use) with short time scales (10 years). The layer approach has been suggested as a framework for the integration of science ("understanding") and policy ("intervention"). A common spatial policy or planning approach could be based on an overall guiding principle that layers representing slow dynamics should enable and constrain the development in layers representing faster dynamics. Implementing this planning approach for the Delta region will result in spatial design and intervention strategies based on "building with nature" instead of "working against nature." An example is the re-establishment of gradients between upstream rivers, estuaries, and adjacent sea to restore the natural buffer and filter capacity of the Delta and of feedback between landscape units in general.

Science and Integrated Coastal Management
Edited by B. von Bodungen and R.K. Turner

INTRODUCTION

The Delta area (Figure 7.1) in the southwestern part of the Netherlands developed over the last 9000 years through an ongoing interaction between natural dynamics (e.g., sea-level rise, peat formation, marsh accretion) and anthropogenic influences (draining, digging, dredging). Its present state (resilience, robustness), including land use, is the result of this historic development. In the region's center, traditional agricultural land use dominates. The sand coast and dune areas of the western area and the (former) tidal waters represent an important natural value and a high potential for recreational use. The largely rural delta is surrounded by densely populated urban and highly industrialized zones in northern, eastern, and southern direction. The delta area is one of the few remaining rural areas within the European core economic region (delineated by London, Paris, Milan, Munich, and Hamburg), where regional trends indicate further future agglomeration. Pressure on the delta area by port and infrastructure development, urbanization, intensive agriculture, and recreation increases. In addition, sea-level rise and other impacts of climate change impose an increasing threat upon the area. These combined anthropogenic and natural pressures require a policy response.

In this chapter, I describe the decreasing functional diversity that results from this interaction between natural dynamics and human responses using the layer model and Drivers-Pressures-State-Impact-Response (D-P-S-I-R) framework for this analysis.

LAYER MODEL

The European Spatial Development Perspective (ESDP 1999) and the Dutch Fifth National Policy Document on Spatial Planning contain an attempt to integrate the various layers and dimensions of the spatial structure. "Thinking in layers" builds on an older debate within geography, which examines elements of the spatial structure that display faster or slower rates of change including implications for the choice of intervention strategies.

Three physical layers can be identified (Figure 7.2): The *ground layer* (soil, (ground)water, ecosystems), with ecological sustainability as a guiding principle, represents the natural resources that ultimately enable as well as constrain land use and resource exploitation. The intrinsic time scale of many processes in this layer is very long, from tens of years for stock renewal of commercial fish species to 100–1000 years for geomorphologic processes that shape the coastline. This implies that recovery from (potentially rapid) deterioration caused by mismanagement, misuse, or overexploitation is a slow process. The *networks layer* (infrastructures such as roads, railroads, waterways, urban structures, coastal defense structures), with economic functionality as a guiding principle, represents the results of public investments. The planning horizon and consequently the time scale of this layer is on the order of 10–50 years. The *occupation layer* (land use and land cover), with social equity as the guiding principle, represents the result of private investments. The rationale of short-term benefit maximization determines the time scale of this layer and can be as short as 1–5 years.

These three layers are backed up by a "fourth dimension": cultural identity and experience of "virtual space." This layer is not merely an afterthought but rather represents the immaterial aspects of the other three. The guiding principle here is that the development of global networks must not lead to a leveling off of cultural differences. Cultural identity lies in its diversity.

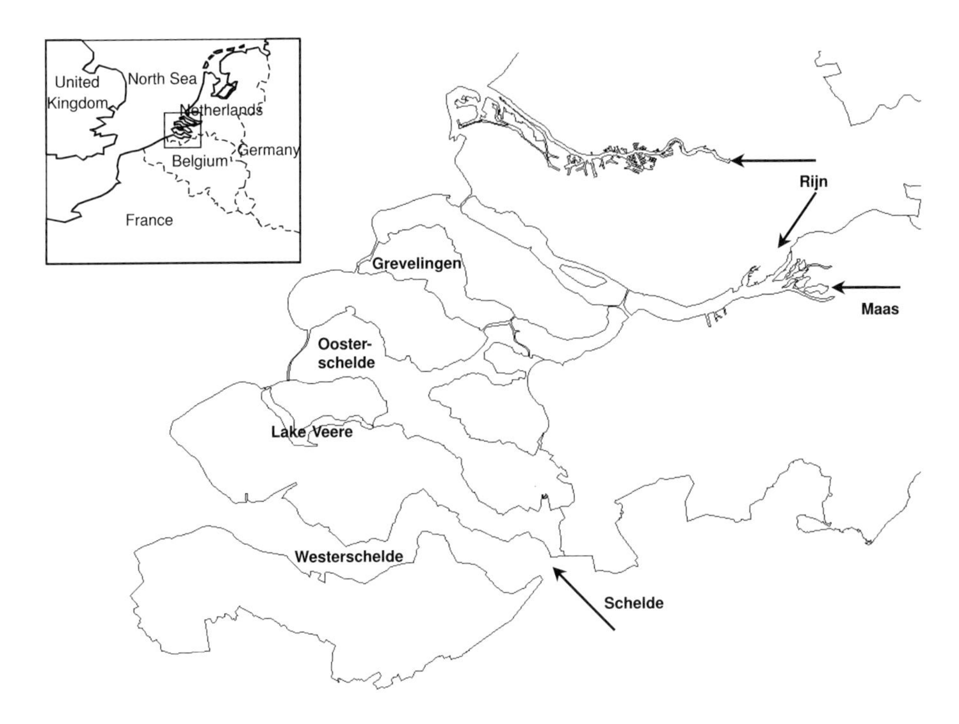

Figure 7.1 The Delta of Rhine (Rijn), Maas, and Schelde, SW Netherlands.

The three physical layers combined with the "fourth dimension" can be used as a framework for integrating the various discourses on spatial development strategies. This approach can play a crucial role in the formulation of a common planning "doctrine." Such a doctrine could be based on an overall guiding principle that layers representing slow dynamics should enable and constrain the development in layers representing faster dynamics: soil and

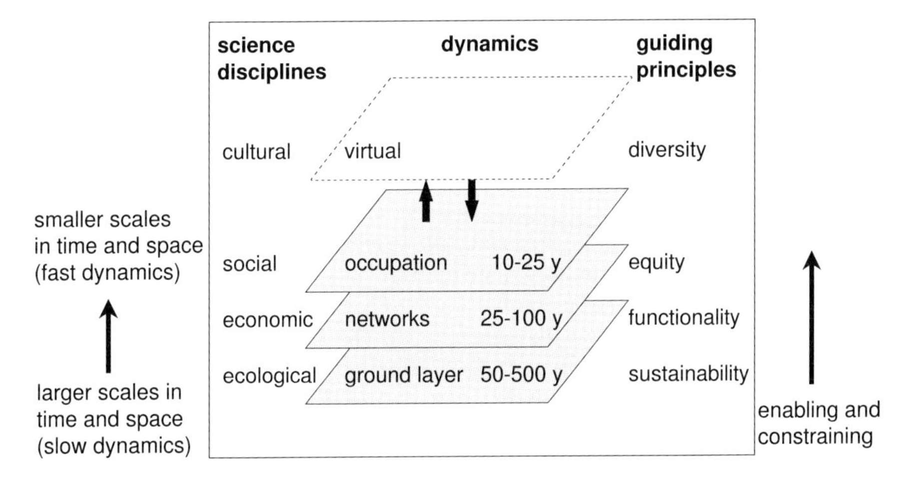

Figure 7.2 Thinking in layers.

(ground)water characteristics should condition infrastructure development which, in its turn, should be conditional for land-use development.

The layer approach could also become the framework for the integration of the professional debate, dedicated to the world of "understanding," and the policy discussion on the design of "intervention" strategies. An example is the overall target of the Dutch national spatial policy in the coming thirty years: the combination of a continued economic development and population increase (networks and occupation layers) with an absolute decrease of environmental pressure and an increase of the spatial quality of the living environment (ground layer). This target is based on the assumption and hope that it is feasible to "decouple" economic development from environmental and spatial pressure: the de-materialization of the economy.

DRIVERS-PRESSURES-STATE-IMPACT-RESPONSE

To distinguish between understanding and intervention, the D-P-S-I-R framework can be used as a complementary second concept to structure the process of understanding, i.e., the scientific analysis of land-use dynamics (see von Bodungen and Turner, this volume).

This framework describes (a) the causal sequence of driving force (e.g., economic sectors, administrative context), environmental pressure (resource exploitation), state and quality of the environment, and impact on society; and (b) feedback or "coevolution" between this causal chain or network and the (policy) response options of society.

Quality indicators should be developed for each step in the D-P-S-I-R chain. There is no need to assign monetary economic values to these indicators. Instead, indices representing ratios between indicators could be calculated:

- "Uncoupling index": the ratio of economic activity (driving force) and environmental pressure indicates the efficiency of converting environmental resources into economic goods and services. High values of this index indicate the environmental efficiency or de-materialization of the economy.
- "Environmental assimilation index": the ratio of pressure and state indicates the extent of decreasing environmental quality by the environmental pressure. This could be defined as the relative assimilation of environmental resources by the economy or the level of utilization of the environmental user space.

THE RHINE, MAAS, AND SCHELDE DELTA

The geomorphologic history of the area since the beginning of the Holocene (7000 B.C.) is characterized by subsequent periods of inundation and land accretion. The first period of inundation, ca. 6000–3500 B.C., had a purely natural cause: sea-level rise after the last ice age, when the sea level was 45 m below the present level. The Atlantic transgression (sea extension) reached its maximum in 3500 B.C. Thereafter, as a result of the decreasing rate of sea-level rise, inundation was replaced by land accretion through the sedimentation of tidal gullies and marshes and subsequent peat formation in the period 3500–500 B.C. During this period there was only scattered habitation, almost exclusively restricted to the western coastal barriers.

Large-scale habitation of the peat marshes began in mid-Roman times, i.e., at the beginning of the era. This earliest habitation had a disastrous impact on the area as natural drainage was augmented through the digging of ditches in an attempt to cultivate the waterlogged peat areas. This artificial drainage caused soil subsidence by compaction and peat oxidation. In addition, peat was cut for salt making and lime burning. As a consequence of these human interventions and subsequent erosion, the former peat lands were flooded. This self-accelerating process started approximately A.D. 300 and resulted in a total reshaping of the landscape within 50–100 years. It changed from an extensive peat area, crossed by relatively small streams, to an abandoned delta consisting of salt marshes and vast tidal areas.

It took another 1000 years of geomorphologic processes (e.g., accretion of marshes, sedimentation of older gullies resulting in "inversion ridges") before the area was suitable for the next colonization, in the Middle Ages. The safeguarding of animals and other economic assets against storm floods required the construction of artificial dwelling mounds. However, as a consequence of the continuing sea-level rise, these mounds needed to be regularly raised. Some of them eventually reached a level of 5 m or more above sea level.

Inhabitants of the area started building dikes around the marshes for protection against storm surges in the 11th to 12th century. In approximately 100 years, almost all marshes were embanked. The excavation of peat (for fuel use and salt production) was the second large-scale intervention in the landscape during this period and resulted in the complete disappearance of peat marshes. Two irreversible processes, which still continue today, were the direct result of these interventions: (1) soil subsidence, as a result of the disconnection of embankments from the natural sedimentation and accretion, and artificial drainage in the polders; and (2) increasing level of storm floods because tidal waves could no longer spread out over the marshes. As a consequence, dike bursts by storm surges had catastrophic effects and could not be repaired. Considerable parts of the embankments were subsequently flooded by storm surges (Borger and Ligtendag 1998; Vos and van Heeringen 1997).

The last catastrophe took place in 1953, when a northwest storm breached 180 km of coastal defenses, flooded 160,000 ha of polderland, and caused 1835 casualties. As a result, the Delta Project was created as an answer to the continuous risk of flooding. The primary task of the Delta Project required the closure of the main tidal estuaries and inlets, with the exception of the Westerschelde, which was needed to allow continual shipping access to Antwerp. The original plan — complete closure of the estuaries and transformation into stagnant freshwater lakes — was partly altered during its early execution to protect and conserve marine natural resources in the region. This resulted in the decision to maintain stagnant saline conditions in some of the former estuaries (Grevelingen and Lake Veere). In the mouth of the Oosterschelde, a storm surge barrier was built. The barrier guarantees protection against flooding while allowing the tide to enter the estuary freely, thus safeguarding the tidal marine ecosystem.

Notwithstanding the alterations to the original plan, the Delta Project is a reactive form of management which has fundamentally changed the hydraulic, morphological, and ecological characteristics of the region. While the embankments since the Middle Ages separated "land" and "water," the Delta Project has disconnected "estuaries" from each other and from incoming "rivers" and the adjacent "sea." The overall change can be summarized as the loss of dynamic gradients. Gradients were replaced by discrete boundaries. Because dynamic gradients are a prerequisite for natural constructive processes, the historic sequence of interventions can be regarded as "working against nature" rather than "building with nature." They have caused largely irreversible, decreasing functional diversity (see von Bodungen and Turner, this volume): "the variety of responses to environmental change, in particular the variety of spatial and temporal scales with which organisms react to each other and to the environment."

DISCUSSION

In retrospect, three consecutive intervention strategies can be discerned, and these have left their scars in the landscape and diminished the utility of the Delta region (de Vries et al. 1996).

Reactive Single Issue Management (1000–1975)

From medieval embankments until the first stage of the Delta Project, interventions focused solely on safety: disastrous flooding "should never happen again." This strategy is reflected in the embankments: isolated low-lying polders suffering from soil subsidence, saline seepage, etc.; in the stagnant freshwater lakes and in the dams and sluices as discrete boundaries between rivers and sea. These interventions did not promote sustainable development, because buffering feedbacks were destroyed between land, coastal, and inland waters and natural adaptation processes. Loss of robustness and resilience as well as increased vulnerability are particularly evident in the landscape units created by these interventions.

Protective Bio-ecological Management (1975–1990)

In the second stage of the Delta Project, protection and conservation of existing remaining values of landscape and nature were added to the main goal of safety. Nature protection arguments were behind decisions to maintain saline conditions in some of the stagnant lakes or former estuaries, as well as to build a storm surge barrier in the Oosterschelde instead of closing this estuary. However, the (revised) interventions have also some serious drawbacks, which are apparent in the Delta region. According to Kavaliauskas (1995), the bio-ecological approach focuses on threatened nature values "where they are now," without satisfying the actual needs of integrated landscape management. This approach results in isolated subsystems that are not functionally integrated in the landscape. Continuous and intensive management is needed to maintain these isolated natural values. Despite intensive care, these (sub)systems remain susceptible to external perturbation.

Constructive and Adaptive Geo-ecological Management (1990–present)

The third strategy, constructive and adaptive geo-ecological management, has not yet been implemented, but plans are being formulated and experiments conducted. Ideally, the strategy should be based on the functional relations in the landscape, focusing on restoration and development of functional natural values "where they must be" (Kavaliauskas 1995). Estuaries are seen as stabilizers of geosystems, they form a natural buffer or filter between upstream rivers and downstream marine environments. The same holds for marshes on a smaller spatial scale between land and sea. To restore these functions, gradients should be reestablished within the water systems as well as between land and sea. Examples are reclamation of salt marshes by de-poldering and reconnection of enclosed water systems to rivers and sea by sluice manipulation experiments. In general, this strategy aims at reestablishment of feedback within and between landscape units, contributing to the self-organizing capacity and the integrity of the geosystem (von Bodungen and Turner, this volume).

EPILOGUE

Water and coastal management in the Delta region, in the context of policy on spatial planning, is confronted with serious dilemmas and can be summarized as follows:

1. Overall dilemma: Choice between finding sustainable solutions or shifting the problems on to future generations.
   - *Continuation of reactive policy*: this has the advantage of fast and cheap solutions but the disadvantage of creating future problems.
   - *Shift towards anticipative policy with water and soil as resources*: here, the starting point is the robustness of the ground layer, which enables and constrains infrastructure development and increased human settlement; shifting problems on to future generations is prevented.
2. Safety policy: Vertical or horizontal solutions.
   - *Vertical*: hard technical solutions, e.g., building and heightening dikes and sea defenses, which increase drainage and pumping capacity.
   - *Horizontal*: retardation of soil subsidence by storage and retention of fresh water, "storing instead of pumping," space for coastal resilience.
3. Water distribution in space and time: Dominance of agriculture or multifunctional use.
   - *Maintaining dominance of agriculture*: this requires new interventions in water management against natural processes.
   - *Multifunctional freshwater supply*: this counters soil subsidence and extensive pastureland/dairy-farming as the only feasible agricultural practice.
4. Land–sea, freshwater–marine: Added value by gradients and multifunctional use.
   - *Rigid conditions for monofunctional use*: shifts spatial claims from land on to (new reclaimed areas at) sea.
   - *Integral spatial development perspective for coastal zone and sea*, including anticipative and tailor-made regional solutions.
5. Ground layer: Functionality or imagination
   - *Ground layer as ordering principle*: the above four dilemmas refer to the ground layer from the perspective of use (occupation) and control.
   - *Ground layer as exponent of spatial design*: keywords for imagination are pride, identity, social cohesion, admiration, freedom, and differentiation.

REFERENCES

Borger, G.J., and W.A. Ligtendag. 1998. The role of water in the development of The Netherlands — A historical perspective. *J. Coast. Conserv.* **4**:109–114.
De Vries, I., A.C. Smaal, P.H. Nienhuis, and J.C.A. Joordens. 1996. Estuarine management strategies and the predictability of ecosystem changes. *J. Coast. Conserv.* **2**:139–148.
ESDP (European Spatial Development Perspective). 1999. Towards Balanced and Sustainable Development of the Territory of the EU, Potsdam, May 1999. Luxembourg: Office for Official Publications of the European Communities.
Kavaliauskas, P. 1995. The nature frame/Lithuanian experience. *Landschap* **12(3)**: 17–26.
Vos, P.C., and R.M. van Heeringen. 1997. The Holocene geology and occupation history of the province of Zeeland (SW Netherlands). Report of the Netherlands Organisation for Applied Scientific Research (NITG-TNO), p. 59. Mededelingen: NITG-TNO.

8

The Anglian Coast

J. PETHICK

Dept. of Marine Sciences and Coastal Management, University of Newcastle upon Tyne,
Ridley Building, Newcastle upon Tyne NE1 7RU, U.K.

ABSTRACT

The mismatch of scales between institutional regimes seeking to administer coastal areas and the biogeochemical processes present generates a difficult set of socioeconomic, political, and scientific complexities. Prolonged human intervention in the coastal zone has served to increase variability, such that the coast is almost permanently in a state of disequilibrium. Despite the protection intention behind the intervention process, a state of affairs has been created in which risk to humans living and working in the coastal zone has actually increased. The coastal morphology is not being allowed to function efficiently as human intervention has reduced the available "free" coastal space. The Anglian coastline of England is a good illustration of the problems that have been created by a coastal protection and defense strategy not properly attuned to coastal functioning.

INTRODUCTION

One of the key questions facing our management of the coast is whether we can provide the scale of administrative structures which equate with the scale of coastal processes. Too often the scale of coastal interventions is constrained by political and institutional considerations and devolves down to the lowest common denominator: local pressures placed upon short lengths of coast over short time periods. The proprietorial interests shown by local people in "their" section of the coast is an extremely powerful force and one which democratic systems find difficult to pass over. Yet natural coastal systems, both physical and ecological, are driven by processes that transcend the local scale and the short time period. Wave, tidal energy, and sediment transport patterns may develop over the scale of regional seas, while the resilience of a coastal system may operate over time scales measured in hundreds of years. Human intervention in these complex and large-scale systems can have results that we do not understand fully at present. We urgently need research into these interactions, but even more important is to develop an awareness that they exist at all. Educating coastal users to consider

Science and Integrated Coastal Management
Edited by B. von Bodungen and R.K. Turner

impacts of their actions in coastal areas remote in time and space from themselves is perhaps the most important principle in any coastal management program.

Before either educational or administrative systems can be developed to deal with large-scale coastal interactions, scientists must attempt to define their temporal and spatial boundaries. It is hardly remarkable that very little scientific research has attempted to do so: large-scale spatial systems are difficult enough to observe and to measure, but long-term temporal periods extending over centuries present an impossible subject for direct observation. In this chapter I attempt to provide a preliminary examination of these difficult areas for research leading to a consideration of the administrative and institutional mechanisms that could be developed to deal with them. Although the subject of this chapter is the Anglian coast of the U.K., the concepts may apply to any developed country's coast. For the purposes of this paper, the Anglian coast is assumed to extend from the Humber estuary to the Thames estuary (Figure 8.1), an area larger than traditional East Anglia but which recognizes the wider interactions of physical and ecological systems that govern this complex coast.

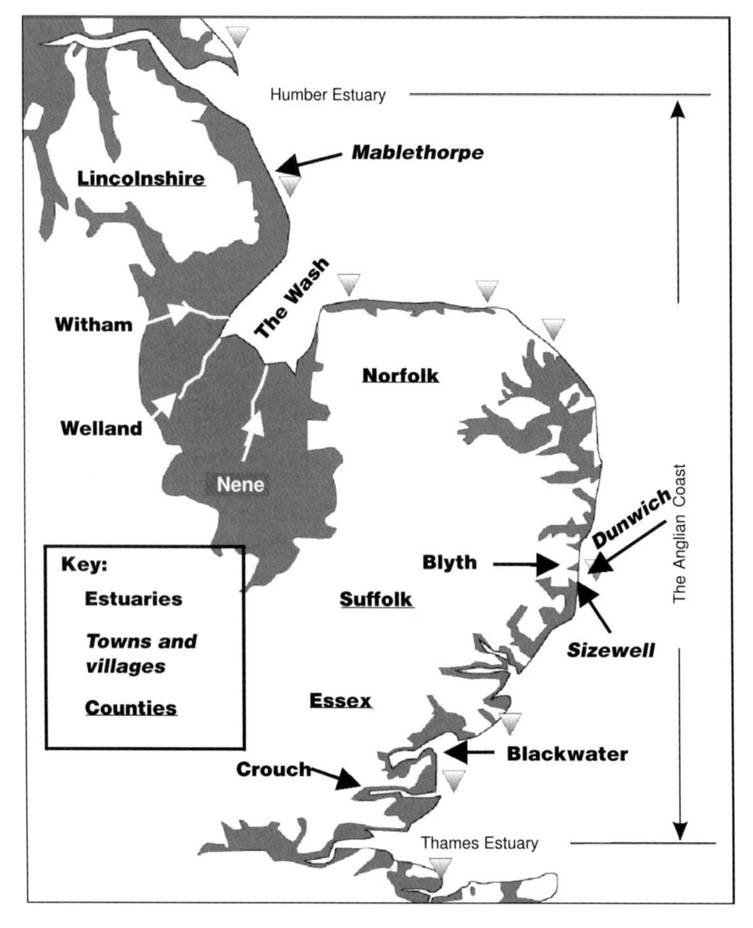

Figure 8.1 Location of the Anglian coastline as defined in the text. Shaded areas indicate reclaimed intertidal areas now at risk from tidal flooding. Triangles indicate principal areas of coastal erosion.

The Anglian coast epitomizes many of the problems facing sustainable coastal management over the next century. These problems are a compound of past management, future physical changes (particularly sea-level rise), and conflicting pressures from development agencies and conservation groups. It is argued, however, that it is the inappropriate management of this coast in the past, over periods extending back several hundreds of years, that is the major cause of our present and future problems. Furthermore, instead of learning from our past mistakes in order to secure the future, we are proceeding, largely in ignorance, to repeat them (cf. de Vries, this volume).

PAST MANAGEMENT OF THE ANGLIAN COAST

The Anglian coast is characterized, in places, by a high population density, soft-rock geological structure, low-lying coastal hinterland, and rising sea levels (due to land sinking). These characteristics are sufficient to convince coastal managers that risks, from flooding and erosion, are an inevitable adjunct to coastal use here, but they are wrong. Natural coastal systems are resilient and respond to accommodate environmental change, for example, cliff erosion results in widening abrasion platforms upon which wave energy is increasingly dissipated so that erosion rates decrease; flooding results in deposition of sediment and elevation of low-lying surfaces so that flood frequency decreases. Both these natural negative feedback mechanisms operate in our interest so that, instead of accepting risk as an integral part of a coastal location, we should ask ourselves why the coast has not adjusted more efficiently to the forcing factors of wave, tide, and sediment supply. The answer may be, quite simply, that since we live in the first half of an interglacial period during which sea level has risen by over 100 m in 10,000 years, there has been insufficient time elapsed to allow these negative feedback mechanisms to develop a steady-state coast. The alternative hypothesis, proposed here, is that our massive and continuing intervention in the coastal system has forced it to readjust so frequently that it remains almost permanently in a state of disequilibrium, a state in which risk to ourselves is maximized.

COASTAL LANDFORMS

A central proposition to the argument is that coastal forms, comprised, for example, of estuaries, marshes, dunes and beaches, are adjusted to their function. The function of the coastal morphology is to dissipate energy, principally from waves and tides. This dissipation is accomplished mainly by spreading the energy loading out over a wide area so that the stresses per unit area are reduced to below the critical threshold for erosion. This results in medium- to long-term stability in the shore morphology (although short-term episodic storm events produce minor, oscillatory, adjustments), in which wave and tidal energy is dissipated leading to low tidal range and wave height — in short to lower risks for coastal users. The main requirement for such adjustment is space — wide areas of intertidal beach, mudflat, and salt marsh are a necessity for a stable and efficient shore.

Each individual component of this coastal morphology, for example, a sand beach, occupies an energy niche on an energy gradient that reduces from high values in the nearshore to zero in the coastal hinterland. The reduction in energy produced by each coastal component means that its landward neighbor occupies a lower energy level and, consequently, exhibits a different morphology, for example, sand beaches at the mouth of an estuary give way to mudflats further landward.

This mosaic of coastal landforms functions holistically, i.e., the whole is greater than the sum of its parts and each component depends upon the others. Changes in the energy regime (e.g., as sea level rises) means that the entire mosaic shifts its absolute location but the relative positions of each component remain constant. Thus, as sea level rose after glaciation, the entire coastal landform mosaic moved landwards, the so-called Holocene transgression, maintaining the coastal function and form but adjusting continuously to energy changes.

RESOURCE EXPLOITATION: RECLAMATION

Our own entry into this complex pattern of landform mosaics and energy niches proved disastrous. We failed to see that the energy dissipation role of the coast was one that acted for our own good in reducing risks from flood and erosion, and that wide intertidal areas were a necessity. Instead, we saw coastal space as a resource to be exploited. The 18th to 20th centuries saw massive reclamation of intertidal land on the Anglian coast for agriculture, urban development, and industry. These intertidal areas of mudflat and marshes were an essential component of the landform so that removing them from the coast reduced the energy dissipation capacity of the coast.

Over 2000 km^2 of reclamation has occurred on the Anglian coast over the past 300 years, roughly 50% of the original intertidal area. The major area of reclamation was in the Fenland, where three estuaries (Nene, Witham, Welland) flowed, via The Wash, into the North Sea; all of the estuaries of this coast, including the major systems of the Thames and Humber, suffered loss of intertidal areas. The results of this loss within the estuaries were catastrophic, involving increased tidal range, wave heights and, therefore, erosion and flooding. Results from model simulations of the Humber estuary show that the loss of 500 km^2 of intertidal area due to reclamation produced a 1 m rise in tidal amplitude. If similar impacts were felt in other estuaries of the Anglian coast, including the Thames, it may explain the increase in flood events and in flood defense construction in these East Coast estuaries during the latter part of the 18th century when reclamation was at its peak.

Perhaps the most important change produced by the reclamation of the estuaries was so large-scale that we failed to connect cause and effect. The discharge from the tidal estuaries into the open sea results in interruptions of long-shore sediment transport on the open coast and the development of extensive tidal deltas (see, e.g., Fitzgerald and Penland 1987). These deltas represent a type of crossroads between the movements of sediment along the shore and the movement of sediment into and out of the estuaries. They consist of sandy sediment, which forms extensive subtidal sand waves and sand ramparts that offer considerable protection to the open coast shoreline. In fact, the deltas act as a form of natural wave break, held in place by the force of the tidal waters in the estuaries. Thus, the form of the open coast is a dynamic one, representing a balance between the force of the estuaries and of the incoming wave energy.

The size of these tidal deltas on the Anglian coast is considerable. The discharge from an estuary such as the Humber, for example, with a length of 120 km and a tidal volume of 1.42 ×

10^9 m^3, results in a delta covering 120 km^2 and extending 10 km seaward and 30 km along the shore. Similar sized delta systems characterize the Wash estuaries and the Thames. Smaller estuaries, such as the Blyth in Suffolk or the Crouch in Essex, possess deltas which, although smaller, coalesce with those of their larger neighbors. There are 15 major estuaries punctuating the 300 km long Anglian open coast. The tidal deltas of these estuaries overlap approximately 100 km of this open coastline so that, today, one third of the open Anglian coast is protected by deltaic deposits. The picture 300 years ago exhibited very likely more extended deltaic areas protecting the shorelines.

The evidence for the links between reclamation and erosion cannot be fully rehearsed here but some examples may be given. The Lincolnshire coast, south of Mablethorpe has been eroding over the past 250 years. This coast is a low-lying area of former salt marsh which was reclaimed in the 13th to 15th centuries and the question must be asked why the energy conditions, which formerly were conducive for deposition, changed to erosion in the 18th century. A predictive model of the Humber estuary delta based upon work by Walton and Goodall (1972) showed that the reduction in tidal prism in the Humber after reclamation would have reduced the extent of deltaic overlap on the Lincolnshire coast from 50 km to 23 km. This would have exposed the entire southern section of this coast to wave attack (Pethick 2001) and changed conditions from deposition to erosion. This erosion is still proceeding today, necessitating a recent £100-million beach recharge program by the Environment Agency. Similarly, the Dunwich cliff erosion in Suffolk may be related to the reclamation of the Blyth estuary that formerly possessed a delta front extending south to Thorpeness and whose protection encouraged the development of the port of Dunwich, the second largest port in Britain until the 17th century. The probability of such an extensive development taking place on an eroding coast is remote and suggests that conditions here too changed dramatically in the 18th century when the Blyth estuary was reclaimed. Again, predictive modeling suggests that an area of delta front extending over 15 km would have been lost as a result and would explain the onset of erosion on this coast.

Coastal Defense

These erosional changes were, of course, merely the natural response of the coastal landform system to imposed changes, an attempt to recover the level of energy dissipation which was lost as the intertidal area decreased. Left to itself, the coast would have recovered and, in some places on the Anglian coast, has done so. However, in most cases the coast was not left to adjust due to our further intervention.

Our reaction to the catastrophic changes that we had instigated merely compounded the problem. First, we failed to realize that the flooding and erosion were the results of our own ineptitude and we continue, to the present day, to believe that these risks are the outcome of natural phenomena. Second, we attempted to stifle the natural response of the coast to the imposed changes. The floods and erosion which marked the adjustment of the coastal morphology to estuarine reclamation were reduced by a series of hard engineering structures, the flood embankments and wood, rubble, or concrete revetments which today front over 50% of the Anglian coastline.

The effect of these engineering structures has been, if anything, even more catastrophic than the initial reclamation. The Anglian coast has been immobilized and starved of sediment supply from formerly eroding cliffs. It has no space left for energy dissipation or landward transgression. The coast has been, literally, petrified.

Inappropriate Development

The immobilization of the Anglian coast by the elaborate coastal defense system during the last 150 years, resulting from the initial reclamation era, has itself resulted in yet another human response, which further reduces the sustainability of the coast. Although the initial reclamation was almost entirely for agricultural purposes, during the 19th and 20th centuries the reclaimed land has been used increasingly for industrial and urban development. The opportunity to develop on flat land close to the navigational pathways provided by the estuaries and which also provided water and waste disposal, outweighed any consideration of increased risks in these low-lying areas. Sea-level rise and land compaction has steadily increased those initial risks so that much of today's ongoing development is at least 1 m below high tide levels. This inappropriate development includes extensive areas of London fronting the Thames, industrial areas along the Humber, and port infrastructures along most of the Anglian estuaries.

Whereas flood embankments protecting agricultural land may be regarded as temporary features of the coastal landscape, these urban and industrial sites are seen as permanent and the attitude has gradually developed that they should, indeed must, be protected against any risk from flood or erosion. This attitude was reinforced after the disastrous 1953 surge tide that flooded almost all of the reclaimed areas along the Anglian coast (Figure 8.1) and resulted in the loss of 186 lives. The response to the flood event, however, was not one that recognized the risk and moved people and property out of such a high-risk area, but instead the response was to upgrade the existing line of fixed defenses. The result has been to reinforce the locational drift towards these areas, a prime example being the massive holiday and retirement home complex along the Lincolnshire coast where some 15,000 people, mainly elderly, now live in often single-storied buildings at elevations below high water level. The beach recharge on this coast, mentioned above, in conjunction with a concrete flood embankment, offers protection against a 1:250 year flood, but the residents of this and other areas appear to believe that their protection is absolute. More important for long-term coastal planning, the fact that they, and the industrial complexes along this coast, cannot easily be relocated means that the shoreline is immobilized permanently.

CONSERVATION POLICY

In addition to these physical responses to coastal changes, we have also initiated a series of administrative measures that compound the problems of landform immobility. Recognizing that the coastal landforms, habitats, and ecology had been severely reduced by coastal defense structures and by erosion and flooding, we have decided to preserve what little natural coastal environment was left to us. The British system of nature reserves was initiated on the Anglian coast in the early years of the 20th century. These were later translated into a national network of Sites of Special Scientific Interest (SSSIs) and National Nature Reserves (NNRs) and supplemented with international designations such as the Ramsar sites. More recently, the European Union's (EU) Birds and Habitats Directives have been instrumental in the designation of a series of Special Protection Areas (SPAs) and Special Areas of Conservation (SACs), respectively. In each case, the area and type of coastal ecosystem has been defined in the designation, and stern warnings issued that this conservation status should not be allowed to deteriorate. In effect, this has meant that all such designated areas are under preservation orders that prevent the coast from adjusting to the many changes we have imposed upon it. These conservation policies for the coast will, therefore, have precisely the same effect as the hard engineering structures to which they were a response. The only difference is that the location of the conservation areas is mainly complementary to the location of the hard defenses so that the entire coast is effectively immobilized.

The loss of coastal habitat due to reclamation was paralleled by the loss of freshwater habitat along river corridors as the agricultural land area extended over the past 250 years. In the late 20th century rush to preserve dwindling habitat, it was recognized that many areas of reclaimed salt marsh, now used as grazing pasture, provided a reasonable substitute for these lost river floodplains. Consequently, these former saline marshes were designated as conservation areas for freshwater habitat, despite the fact that they lay below sea level and were difficult if not impossible to sustain as freshwater areas over the long term.

The absurdity of this approach to coastal habitat conservation is demonstrated on the North Norfolk coast where attention of many coastal management organizations has been centered recently. Here, seaward of an 18th century flood defense, an area of open coast marshes and intertidal sands is designated as a Special Area of Conservation under the EU Habitats Directive, legislation which effectively prevents "deterioration" of such a site. On the landward side of the flood defense an area of freshwater reed bed has been designated as a Special Protection Area under the EU Birds Directive, conferring similar protection against "deterioration." Rising sea levels in this area are forcing the intertidal area to migrate landward and, in order to prevent its deterioration, the flood defense should be moved landwards to facilitate this. Moving the flood defense landward would, however, mean loss of designated freshwater habitat, so that an impasse has been brought about in which, if no action is taken, both habitats will suffer irreparable loss. It is clear from this one example that a conservation policy based upon preservation of a dynamic coastal system is not sensible. Moreover, the conservation dilemma outlined here should also make it clear to coastal managers that any coastal policy based upon static defenses is unsustainable.

FUTURE DEVELOPMENT

As a result of this history of coastal resource exploitation, we have now, on the Anglian coast a largely immobile coast that is unable to respond to changes and yet we now face perhaps the biggest change in coastal inputs since the early Holocene. The predicted rise in sea level due to global warming will be in addition to the already rapid rise in the rate of sea-level rise on the Anglian coast. The coast must be able to respond to these changes if risks to our own infrastructure and to the natural environment are to be minimized. This response is one of relocation of the coastal landform mosaic, a movement that would allow each landform component to take up its correct energy niche. Without such adjustment, landforms and habitats will disappear and flooding and erosion potential will increase. We will, of course, not allow this potential risk to be realized and will construct ever higher flood defenses and develop more effective erosion control. Such a response has not worked in the past and will not work in the future: it is unsustainable. In the words of the Ministry of Agriculture, responsible for coastal defense in England it "locks future generations into a unsustainable spiral of increasing investment in coastal defense."

Two examples of the future implications of this immobilization of the Anglian coast can be discussed here. The landward transgression of estuaries as a response to sea-level rise was first suggested by Allen (1990) for the Severn estuary and has subsequently been the subject of research in the Blackwater estuary, Essex, (Pethick 1997, 1998) and the Humber estuary (Environment Agency 1999). Estuarine transgression appears to enable an estuary to maintain its relative position within the tidal and wave energy frame. It entails a vertical upward movement, keeping pace with sea-level rise, as well as a landward movement that maintains its position in the longitudinal energy frame. The research has demonstrated that the process of transgression consists of sediment erosion from the outer estuary, principally from upper intertidal areas, and its transfer to the inner estuary where it is redeposited on the upper intertidal marshes. The increased elevation in these inner estuarine areas keeps pace with sea-level rise and allows the estuary margins to migrate landward over previously supratidal areas. Research in the Humber has shown that the rate of such migration is in the order of 10 m of horizontal transgression for every 1 mm of sea-level rise.

The problem in the Humber estuary, however, as with all of the Anglian estuaries, is that the inner estuary intertidal area has been reclaimed and flood embankments now prevent the redeposition of sediment on these areas. As a result, the estuary is prevented from landward migration and thus energy levels increase, leading to accelerated intertidal erosion and amplification of the tidal range. This means increased risk for the considerable infrastructure — industrial, urban, and agricultural — along the estuary shores, an increase in risk which is always ascribed to sea-level rise but which is, in fact, due to mismanagement of the mobile coastline.

The second example of the difficulties facing our future management of the coast due to past mismanagement is from Norfolk. The North Norfolk shoreline is characterized by particularly complex wave refraction patterns, the result of nearshore bathymetric variations over the tidal deltaic deposits of the Wash. Here the shoreline lies at an oblique angle to the nearshore contours and the resultant wave refraction pattern shows a series of wave foci spaced at approximately 10 km intervals. Figure 8.2 shows the pattern for a northerly wave with an 8-second period. The wave foci are areas where wave rays are compressed, causing increased wave heights and energy levels, while the areas between foci experience diminished wave height and energy levels as the wave rays are spread out over wider areas of shoreline. The landform response to this sequence of energy gradients is also shown in Figure 8.2. Areas of high wave energy are characterized by coarse sediment beaches, such as those at Hunstanton, Wells, and Blakeney. Between these wave foci, areas of low energy are occupied by salt marshes, such as those at Thornham and Stiffkey.

These landform sequences merge into one another and their boundaries are blurred by the variations in location of wave foci as wave approach angle changes during successive storm events. Nevertheless, the overall sequence is extremely sensitive to the imposed wave energy gradients and, therefore, to changes in sea level. The impact of sea-level rise is demonstrated in Figure 8.3, in which sea level is assumed to have increased by 1 m (i.e., taking place over

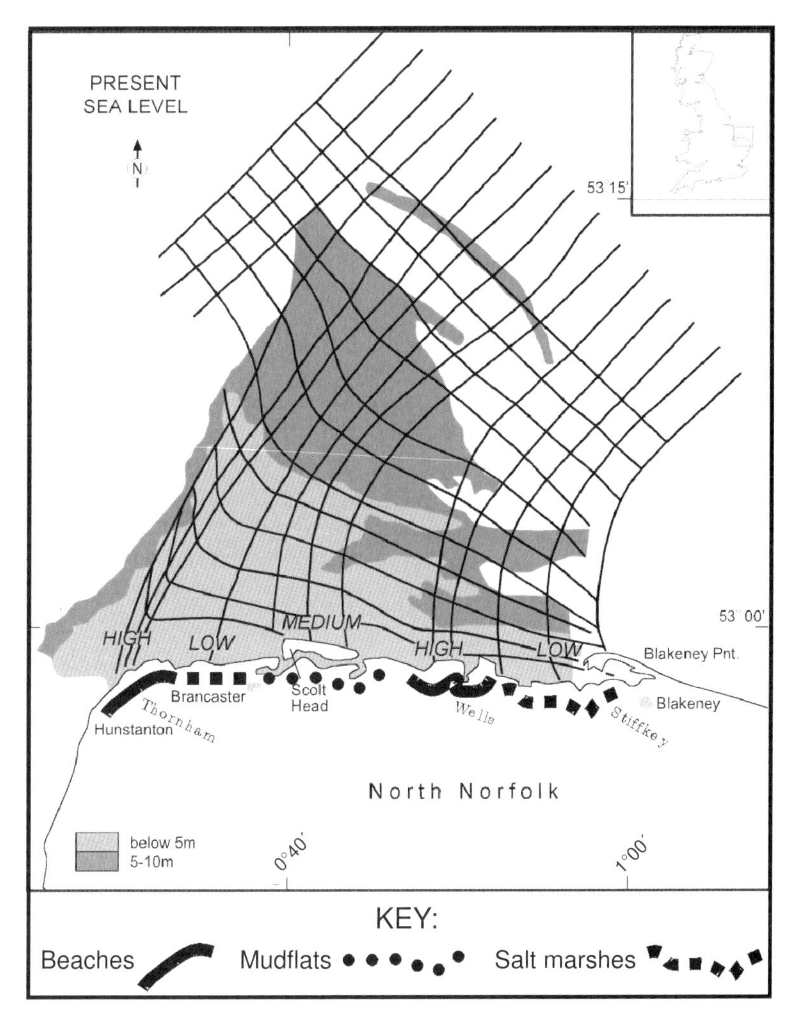

Figure 8.2 Wave refraction and coastal landform units on the North Norfolk coast. Areas characterized by sandy beaches are associated with higher waves due to the presence of wave foci. Areas characterized by salt marshes are associated with lower wave heights between the foci. Modeled wave is 8 s period from the northeast. Sea level is for present day.

the next 150 years at an average rate of sea-level rise of 6 mm per year). The wave foci are shown to have migrated along the shore by 8 km, at a rate of 53 m per year, so that locations that previously experienced minimum now experience maximum wave energy. The response of the landform sequences along this coast is predicted to be equally dramatic. Sand beaches, such as those at Wells, are predicted to be replaced by mudflats and salt marshes, while sand beaches will replace the designated salt marsh and sand dune areas of Thornham, Scolt, and Stiffkey.

Two outcomes of this predicted migration of landforms must be of great concern to coastal management. First, the inertia of existing infrastructures means that development based upon coastal landforms, for example, the recreational industry based on the proximity of sandy

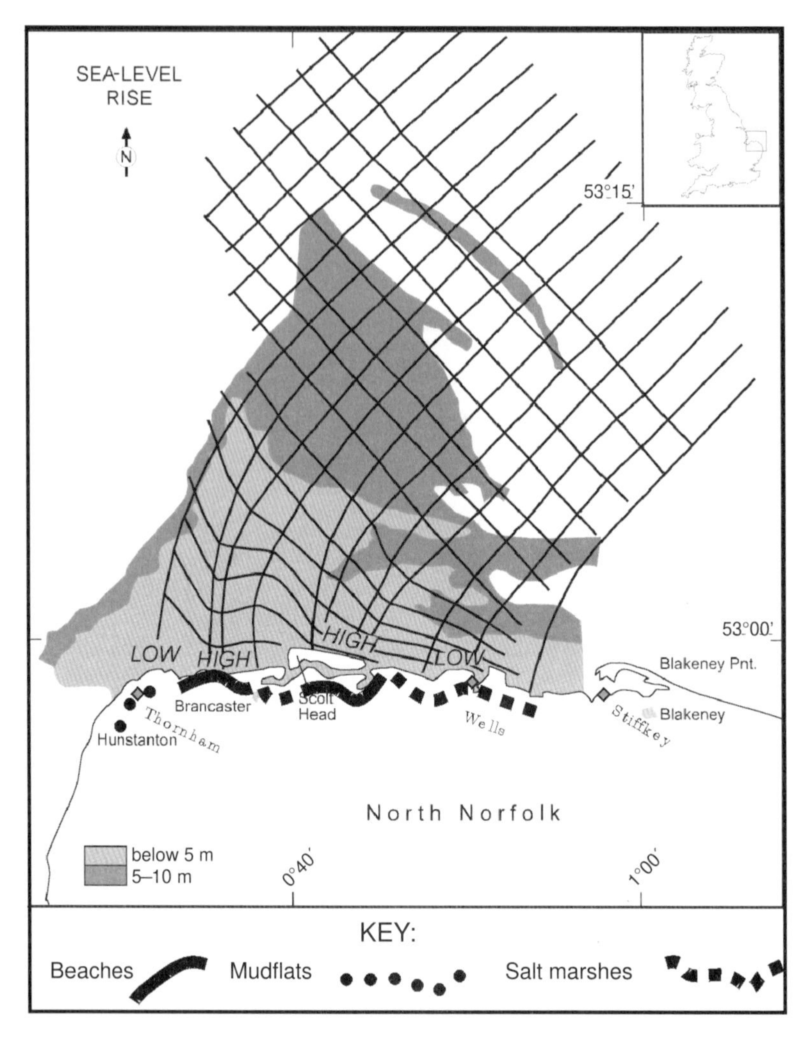

Figure 8.3 Wave refraction and coastal landform units as for Figure 8.2, but with a 1 m increase in sea level. Note the migration of the wave foci along the shore as compared with Figure 8.2, a movement predicted to be associated with a similar migration of the landform units.

beaches or on bird-watching habitats, will not be able to move along the coast in response to landform migration and will deteriorate. Second, the existence of coastal defenses, even on this relatively undeveloped coast, means that the predicted migration of landforms cannot take place in its entirety so that major losses of, for example, salt marshes, will result, again threatening jobs and industries which depend on the coast and its landscape for their livelihood.

These two examples demonstrate that the physical legacy of reclamation and the attitude it engendered of treating the coast as a static line, rather than a dynamic zone, have reduced our options for the sustainable management of the Anglian coast. Where we have constructed towns or industries in high-risk coastal sites we may have no option in the future but to increase defense heights. The protection that such defenses appear to provide encourages further development: initiating a positive feedback process that cannot be sustained.

SOLUTIONS

Solutions to the problems posed by the series of errors committed on the Anglian coast over the past three centuries become increasingly difficult to find, since we continue to compound the errors by inappropriate developments of all kinds. At first sight, the obvious approach appears to be restoration of the Anglian coastline in order that its natural self-sustaining processes might be resumed. Attractive as such a proposal might be to conservation bodies, it is quite impractical. The problem of existing development on reclaimed land, however inappropriate, cannot be ignored, while the wider implications of a complete restoration program are of even greater importance. Restoration of all of the Humber estuary reclaimed land, for example, would result in a 1 m reduction in tidal range which would disrupt navigation to the ports along the estuary as well as causing massive loss of existing intertidal habitat, all of which is protected under the EU Habitats Directive. Similarly, the increase in tidal discharge caused by total intertidal restoration of any of the Anglian estuaries would lead to severe channel erosion, again leading to loss of intertidal habitat and threatening urban areas and other infrastructures, including pipelines.

Clearly total restoration is untenable, at least in the short term, but a program of phased restoration could be advantageous. Thus, the Humber estuary research has shown that restoration of a 300 ha area of intertidal land in the inner estuary would result in a 0.3 m fall in high tide level, thus offsetting the predicted sea-level rise for the next 50 years. Small-scale, phased, restoration would increase intertidal sedimentation, diverting sediment from navigation channels and reducing the necessity for maintenance dredging. Increased sedimentation would benefit water quality by locking up both nutrients and heavy metals in the increased intertidal areas. The major advantages, however, would be, first, in decreased flood risk as tides and waves are dissipated over the extended intertidal area and, second, in conservation enhancement due to increased habitat. More radically, proposals have been made to create new tidal deltas by forming small inlets adjacent to important infrastructures. One such proposal suggests that the Sizewell nuclear power station on the Suffolk coast, currently threatened by shore erosion, might benefit from the opening of a new tidal inlet into an adjacent freshwater marsh so that the sandbanks associated with its tidal delta might protect the power station.

All of these arguments are being forwarded in support of the managed retreat program that is currently being assessed in the U.K. Nevertheless, despite a general willingness on the part of government to consider the program in theory, little action is being taken, although a number of managed retreat trials under way in the Essex estuaries are proving successful. The urgent need for full-scale implementation of a managed retreat program was presented in the recent House of Commons Select Committee report on coastal management (House of Commons 1998). The report argued that our present approach was unsustainable, stating that *"We are of the opinion that flood and coastal defence policy cannot be sustained in the long term if it continues to be founded on the practice of substantial human intervention in the natural processes of flooding and erosion."*

The report continued by emphasizing the need to allow the coast more space in order to carry out its primary function of energy dissipation, thus conferring benefits to all users. It was, however, careful to point out that any retreat program should be carefully phased so as to avoid some of the problems outlined above: "*Greater priority should be given within national policy to managed realignment and sediment control. In each case, the total area of land which would be affected represents only a tiny fraction of the national land surface, and the associated costs could be diminished by implementing managed realignment of the coast over long time scales.*"

The building of coastal defenses encourages development, which, in turn, creates the need for further defense. The need to break this "development spiral" was recognized by the report, in which the House of Commons Committee states that: "*We believe that a clear presumption should be made against future development in flood plain land.*" Second, and most controversially, the report proposed a long-term policy to move those industrial or urban structures that would impede the natural coastal function. The report urged "*...the formulation of long-term adaptive policies, for example, encouraging the gradual managed abandonment of certain coastal areas, possibly over the course of many decades, and conferring residual life on defence works currently protecting assets which are untenable in the long-term.*"

Perhaps most important, the report advised that the system for coastal management in the U.K. should be reviewed in order to provide larger-scale administrative structures. This would overcome the problems of local-scale pressures that ignore wider interactions at the coast and have been responsible for all of the problems of the Anglian coast that have been outlined in this paper.

CONCLUSIONS

The problems of coastal management on the Anglian coast have been attributed, in this paper, to a misunderstanding of the true nature of coastal risk. Risk from flood and erosion is not, contrary to widespread belief in the U.K., due to natural agencies against which we feel we must wage a constant war using the militaristic nomenclature of "defenses" and "strategies." Instead, the risks we face in choosing a coastal location appear to be of our own making. For the Anglian coast at least, this was due to our initial mistake, made in the 18th century, of assuming that intertidal areas were a useless area of coastal land that could be exploited with impunity. Our second error was to completely miss the connection between the loss of these intertidal lands and the flooding and erosion that occurred as a result in areas and in time remote from the initial interference. Part of this problem was due to the local-scale approach to coastal management that was totally unable to contemplate coastal erosion, for example, as being the result of intertidal reclamation in areas over 100 km away and several centuries ago. We continue with this attitude today with our system of defenses coupled with widespread urban and industrial development and, in so doing, provide an unsustainable coastal system for future generations.

The solution to these self-inflicted wounds to our coastal system is partly practical and partly institutional. We should continue to develop the rather tentative managed retreat program in order to create a more functional coast capable of self-sustenance over the long-term. At the same time, we should create a more suitable administrative structure that looks at the wider implications of coastal development, both in space and time. Coastal policies should address a regional seas scale, for example, the southern North Sea and the Anglian coast constitute a single coastal system and must not be subdivided down into local units which ignore impacts extending over tens, or even hundreds, of kilometers. Similarly, in developing coastal policies we must look forward not just to the coming century but also the next, being aware that it is on this time scale that large-scale coastal landforms, such as estuaries, respond to changes we impose upon them. This is demonstrated by the Anglian coast that is still reacting to 18th century reclamation. Policies to phase out inappropriate development over the next century, in order to allow restoration of the coast, would avoid the upheavals inherent in short-term measures and would break the spiral of development and defense which has bedeviled the Anglian coast over the past three hundred years.

REFERENCES

Allen, J.R.L. 1990. The Severn estuary in southwest Britain: Its retreat under marine transgression and fine sediment regime. *Sediment. Geol.* **66**:13–28.

Environment Agency of England and Wales. 1999. Humber estuary: Geomorphological studies. Internal report. London: Environment Agency of England and Wales.

Fitzgerald, D.M., and S. Penland. 1987. Backbarrier dynamics of the East Friesian Islands. *J. Sediment. Petrol.* **57**:746–754.

House of Commons. 1998. Flood and Coastal Defence. House of Commons Agriculture Select Committee. London: HMSO.

Pethick, J. 1997. The Blackwater estuary: Geomorphological trends 1978 to 1994. Report to the Environment Agency. London: Environment Agency of England and Wales.

Pethick, J. 1998. Coastal management and sea level rise: A morphological approach. In: Land Form Monitoring, Modelling and Analysis, ed. S. Lane, K. Richards, and J. Chandler. Chichester: Wiley.

Pethick, J. 2001. Coastal management and sea-level rise. *Catena* **42(2–4)**:307–322.

Walton, F.D., and H.G. Goodall. 1972. Sedimentary dynamics under tidal influences, Big Grass Island, Taylor County, Florida. *Mar. Geol.* **13**:1–29.

9

The Great Barrier Reef, Australia

Partnerships for Wise Use

C.J. CROSSLAND[1] and R.A. KENCHINGTON[2]

[1] LOICZ International Project Office, Netherlands Institute for Sea Research (NIOZ), P.O. Box 59, 1790 AB Den Burg, Texel, The Netherlands
[2] Great Barrier Reef Marine Park Authority, G.P.O. Box 791, Canberra ACT 2601, Australia

ABSTRACT

The Great Barrier Reef is an ecosystem under active management as a large-zoned, multiple-use area. More than 98% is within a marine park, and overall management is through a single authority in keeping with the World Conservation Union (IUCN) tenets. Conservation of natural heritage is a fundamental objective and a range of human uses is permitted. Management issues are identified and approaches to resolving problems are described, including institutional arrangements, community consultation and participation, and information needs. Scientific research information is a vital ingredient for effective management, as is the development of partnerships, trust, and communication between reef users, management, and researchers. Emergence of these partnerships and the delivery of accessible and useful science has not been easy or clear-cut; some history and experience is described to the point where "reasonably good" working alliances are in place. The urgent need for greatly enhanced capacity and information about socioeconomic aspects is highlighted. In our experience, the involvement of policy users in the establishment of research objectives and ongoing liaison over work in progress provides a clear focus for the research without compromising its excellence; it also provides an ownership and policy acceptance for the outcomes.

BACKGROUND

The Great Barrier Reef (GBR) covers about 350,000 km^2 and stretches along more than 2000 km of the northeastern coast of Australia (Figure 9.1). It comprises a complex and extensive

Science and Integrated Coastal Management
Edited by B. von Bodungen and R.K. Turner

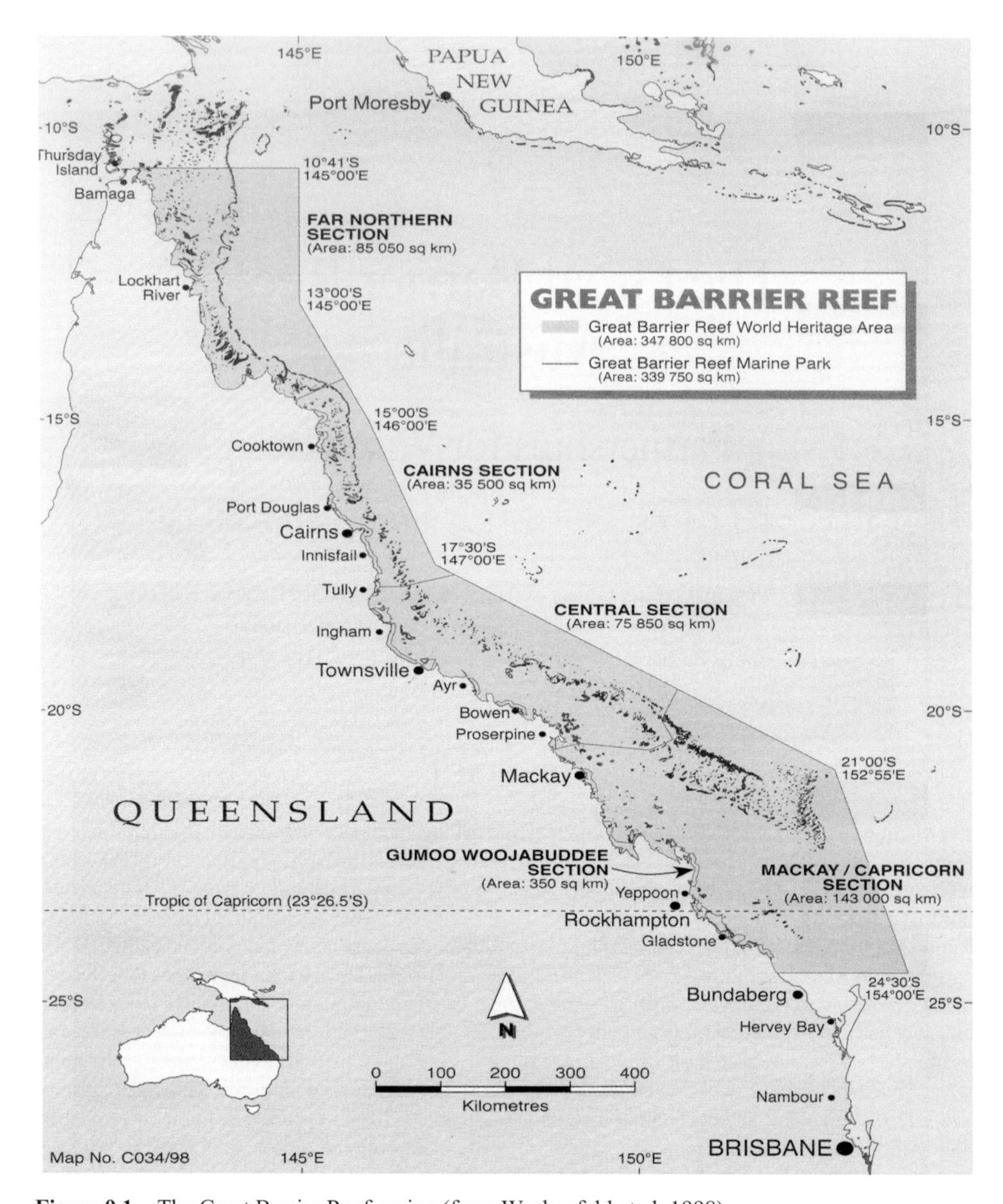

Figure 9.1 The Great Barrier Reef region (from Wachenfeld et al. 1998).

array of interlinked tropical marine ecosystems, including mangroves, seagrasses, coral reefs, soft-sediment communities, fish, bird, and marine mammal populations. The GBR is a large globally iconic ecosystem. The GBR Marine Park comprises more than 98% of the Great Barrier Reef Region (GBRR). All of the GBRR and the islands and intertidal areas within it are contained within the Great Barrier Reef World Heritage Area (listed in 1981). Overall management is provided through the GBR Marine Park Authority, which was established by Australian federal legislation in 1975 (for a recent history and overview, see Harvey

1999). The GBR Marine Park Authority performs its functions in association with the Queensland State Government primarily through the Parks and Wildlife Service, which has responsibility for adjacent land areas, internal waters, and islands above low tide mark. Day-to-day management of the GBR Marine Park is conducted by Queensland State Government agencies subject to the GBR Marine Park Authority.

The initial phase of the GBR Marine Park Authority's operations concerned establishing and implementing the GBR Marine Park. There is a clearly stated and accepted goal for management:

> *to provide for the protection, wise use, understanding and enjoyment of the Great Barrier Reef in perpetuity through the development and care of the Great Barrier Reef Marine Park.*

This goal is supported by a strong political (and community) will.

The GBR is managed as a large-zoned multiple-use area (Table 9.1). Human activity is generally subject to permitted use in different zones of the area (Kenchington 1990; Kelleher and Kenchington 1992). Conservation of the natural heritage is a key objective, and ecologically sustainable human activities are managed. In the generally accepted terminology of IUCN, the World Conservation Union, the entire GBR Marine Park is a Category VI Protected Area — Managed Resource Marine Protected Area. Within it, the zones address the objectives of other IUCN categories.

The GBR Marine Park Authority is scheduled to review the park zoning, by section, every seven years. While this work has fallen behind, smaller and local areas planning within the zones has progressed, especially in the areas of strong human use (e.g., the Whitsunday and Cairns plans of management; see www.gbrmpa.gov.au/corp_site/info_services/library/resources/electronic_publications.html). Consultative and participatory processes underpin this work.

The main human activities occurring within the GBR Marine Park include tourism and recreation, fisheries (recreational, commercial trawl, and line), shipping (transit and ports), aboriginal traditional hunting, and research. Oil drilling and mining are prohibited. Activities occurring outside the GBR Marine Park, but actually or potentially affecting it through coastal runoff and increased demand for access and use of the GBR Marine Park, include the

Table 9.1 Major zones for the Great Barrier Reef Marine Park.

GBR Zone	Attribute	% of Marine Park Area	IUCN Category
Preservation and Scientific Research	No access; strict nature reserve	< 0.5	Ia
National Park	"No take" zone; ecosystem protection and recreation areas	4.5	II
Habitat Protection	No trawling, but other fishing allowed; habitat/species management marine protected areas	15	IV
Recreation	Limited fishing (1line/hook)	< 0.5	V
General Use	All reasonable activities (including trawling) allowed; protected seascape marine protected areas	80	VI

full range of coastal and catchment land use and development of the adjacent Queensland coast.

PROBLEM IDENTIFICATION

Conservation, Biodiversity, and Heritage

The fundamental problem that must be addressed in managing the GBR Marine Park is to provide for conservation and reasonable use. This is the core issue of ecosystem-scale sustainable management. To be effective, management must be capable of identifying and addressing change — actual and perceived — as it affects the biophysical systems of the GBR, and the social and economic systems underlying the human activities that use or impact upon those systems. A key issue is to ensure that the most protected zones of the GBR Marine Park contain an adequate and comprehensive representation of the full range of habitats or communities of the GBR.

The biophysical elements of the problem relate to the cost-effective collection and interpretation of data on the state and dynamics of the ecosystem. A major issue is achieving time series data from which the self-repair capacity of the system can be assessed or predicted, in the face of the combined impacts of natural and human-induced phenomena. The understanding of natural versus anthropogenically driven change in environmental status is a continuing conundrum, especially for science.

The social elements of the problem relate to identifying and reconciling the range of values which people place on the natural system — from the aesthetic and cultural to the pragmatic and economic. Issues relating to indigenous people remain a substantial gap in both effort and knowledge; these are likely to have increasing prominence.

The economic elements of the problem relate to the benefits that can be derived from using the natural system and the extent to which people are prepared to accept the immediate or tactical costs of measures to protect the natural system in the strategic or longer term. Community values and perceptions are vital ingredients, and scientific data and information are building blocks towards resolution. The building of community awareness and understanding of the related issues and knowledge is difficult.

The GBR has high global value ranging from the cultural and aesthetic (Lucas et al. 1997) to direct economic return. The annual value acquired from human use of the GBR is about A$ 3 billion.

Tourism

Tourism is the major industry with most activities being day-trip (often to reef site pontoons) and mainland based; there are about 20 resort islands. In the last decade, visitor numbers have expanded more than tenfold to about 2 million per year. Modern, fast high-tech vessels underpin the industry, and technological improvements will likely lead to, or provide pressure for, greater areas of access in the GBR. Almost all of the development of the GBR tourist industry has occurred since the GBR Marine Park was established. Activities within the GBR Marine Park are subject to environmental impact assessment and permitting. Research and compulsory monitoring have demonstrated that environmental impacts are slight and are limited to areas a few meters from the point of mooring or anchorage. Equity of access and the desired setting in terms of style and number of visits to popular sites can present a challenging socioeconomic dimension.

Fisheries

Commercial fisheries include bottom-trawl over wide areas of the soft-bottom GBR lagoon. The effort and the extent of area trawled have increased dramatically over the last 15 years, reflecting fisheries management practices and technological improvements. Other fisheries include reef line fishing (multispecies fisheries; undergoing changes from an increased emphasis on the live-fish trade), coastal netting, and recreational charter vessel activities. The problem for the GBR Marine Park, as for virtually all marine areas of the world, is to develop a system of management that contains the fishing effort and impacts within levels which can be demonstrated to be sustainable. This must be considered both in terms of the stocks that are targeted and of the ecological systems that sustain or coexist with the target species. Active management with constraints to remove or reduce unintended or accidental impact on endangered species, such as dugong and turtles, are another source of conflict with commercial fisheries; restrictions on coastal fisheries were recently agreed. Recreational fishing is a major pastime and often enters into conflict with commercial fishing, where the same stocks are targeted by commercial and recreational fishers. Turtle and dugong hunting is a limited and restricted traditional fishing activity for indigenous Australians in some areas.

Water Quality, Coastal Land Use, and Development

Port development and transit of predominantly commercial cargo and ore carriers have increased over the last two decades. There is an inevitable element of risk, and the management problem is to devise and implement measures to reduce that risk to the lowest possible level. Pilotage is compulsory in the northern region through the inner reef route. There is an oil spill response plan incorporating wildlife rescue measures. This is rehearsed regularly. Further potential for risk reduction lies in global measures to achieve better training of ships' crews and better maintenance of vessels.

Water quality and the effects of changes in adjacent land use (e.g., urbanization, agricultural development and activities, dams, wetland alienation, sediments) and consequent land-based sources of marine pollution are key issues that concern reef managers. In general, the GBR World Heritage Area remains relatively pristine, with elevation of nutrient and associated pollutants and sediments localized near towns and cities, and mainly associated with ephemeral seasonally flowing rivers, their estuaries, coastal deltas, and discharge plume paths (Wachenfeld et al. 1998). Here, potential pollution and eutrophication issues are coupled with need for appraisal of socioeconomic dimensions. Point-source terrestrial discharges and sewage outputs (terrestrial and associated with transit vessels) are under active management.

Effective environmental monitoring tools and programs as well as assessment of biodiversity are priority elements. Knowledge of reef processes and ecology, including teleconnections (e.g., for selection of reserves as ecological replenishment areas), are fundamental to areas of policy and management.

Table 9.2 Key issues for management.

Land runoff	Mainland catchment and land-use changes provide increased discharges of nutrients and sediments to rivers, overland flow, and groundwater.
Urbanization	Urban infrastructure and development contributes discharges from stormwater and drainage, sewage, landscaping, landfills, transportation, and road surfaces; atmospheric transmission from industrial and suburban areas; drainage and groundwater modification; coastal habitat modifications often through loss of wetland and mangrove "natural filtration" ecosystems.
Ports and shipping	Discharges to atmosphere, water, and sediments from industries and support services, port bulk terminals and loading facilities; dredging and sediment disposal, including landfill and coastal reclamation; modification of coastal hydrodynamic regimes; operational and other discharges from ballasting and cargo loss in accidents.
Fishing	Benthic habitat modification/disruption; resuspension of sediments and changes to substratum oxygen/redox regimes; resource and biodiversity depletion from fish harvesting; and destruction/disposal of bycatch.
Tourism	Transportation, food, and support services; demographic shifts and changes of loci for population pressure; marine operations, including small vessel/yacht waste discharges.

The use and activities in adjacent mainland catchments are the principal pressures under assessment and management and require enhanced scientific information (Table 9.2). The relatively small population and localized population centers along with active regulation and management regimes (from integrated catchment management on land to controlled/permitted activities in GBR waters) combined with directed public awareness programs, have served to minimize the actual and potential impacts of people's activities.

SOLUTIONS DEVELOPMENT AND POLICY MAKING

In 1998 the GBR Marine Park Authority underwent reorganization to provide a structure better placed to anticipate and manage issues relating to the major factors of reef management and to enhance the consultative and participatory processes that have developed. The new organization is built around key policy sections which are represented as four critical issues groups that focus respectively on:

- conservation, biodiversity, and heritage,
- fisheries,
- tourism and recreation,
- water quality, coastal land use, and development.

The four policy sections are supported by a number of service delivery units:

- information services (including research management, information management, library),
- program delivery (including permits, planning and major project coordination, indigenous cultural liaison),

- education and communications,
- corporate services,
- legal service.

The development of solutions to problems and the establishment of policy have a number of approaches, dependent on the issue. The outcomes range from legislation and enforcement to development of industry codes of practice and accreditation, to modification of people's behavior by education. A strong political will and legislative framework underpins the management of the GBR. Operationally, the development of remedial and strategic policy is driven from the GBR Marine Park Authority and pursued within a consultative framework to ensure that policy is based on best information, is "owned" by users and the community, and will thus be effective. This is not a trivial process in time, effort, and resources.

Scientific research and acquired information is a vital partner and plank in management of the GBR. While the "precautionary principle" underpins the management approach, scientific research innovation and information, tools and models, as well as assessments and advice on marine systems processes and uncertainties remain the major management demands. The response of the scientific community has been variable through time, as have the scientist–manager interactions on priorities and access to unambiguous data and information (Kelleher 1996).

The new structure is based on the concept of the critical issues groups working closely with the information services group to identify and prioritize the types of information they need to address policy issues. If the information exists, it is accessed through library and information systems. If not, it is the role of the information services group, working with the critical issues group, to define and broker research or information collection tasks, which are generally carried out by expert agencies on contract.

Marine science capacity has tended to parallel the increased significance to the community of the GBR. Major developments at James Cook University and the establishment of the Australian Institute of Marine Science occurred in the early 1970s. These and an array of Australian national and state research agencies contribute to the science enterprise; indeed, Australian reef science and GBR management approaches attract a high level of recognition and application elsewhere. Links between GBR research and management have not always been close; in the early days of management, the road was decidedly rocky. However, in the last decade there has been a deliberate and reasonably successful alliance established.

The last 25 years have seen a trend towards close interaction develop between researchers and managers. Scientists have realized the user and community demand for interaction, accountability, and application of knowledge beyond academe; managers have recognized the time scales and uncertainties associated with much research. Here, communication, trust, partnerships and recognition of roles, and improved institutional dimensions have been and remain vital facilitating elements. In recent years, the science–management interaction has moved to incorporate GBR user interests, contributing further to reduced conflict, profitability, and "wise use" outcomes based on "data not dogma" (Crossland 2000).

In 1993, the Cooperative Research Centre for the GBR World Heritage Area (CRC Reef; see http://www.reef.crc.org.au) was funded under a national program, and the Centre has added a unique conduit to the alliance. The Centre has recently received funding for a further seven years with a program that builds on the initial success. The new program further

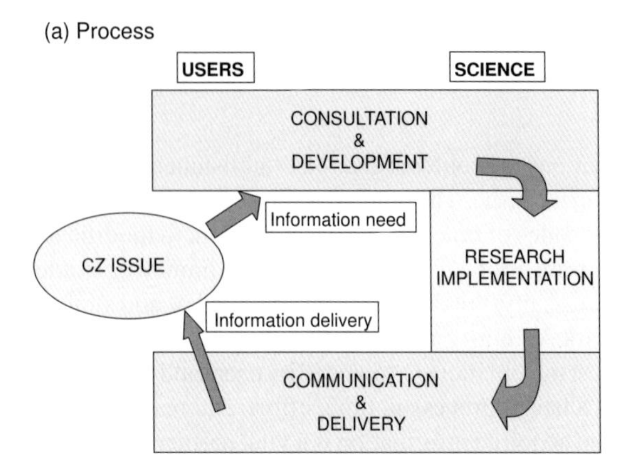

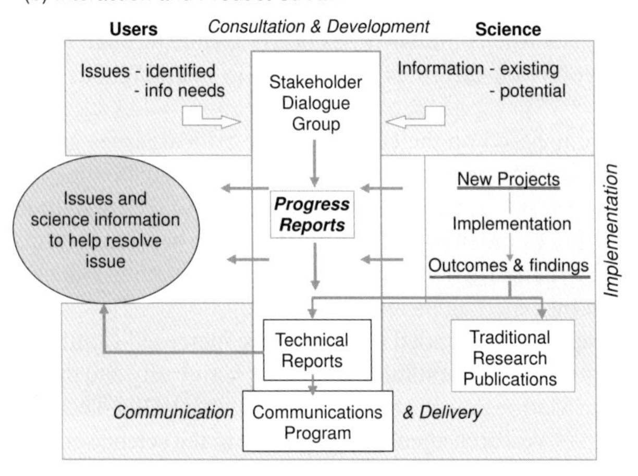

Figure 9.2 Idealized CRC Reef model for consultation and science interaction with users. This framework has been evolving since 1993 and, generally, has had good success (from Crossland 2000).

develops the important role of management users in design, continuing coordination and eventual interpretation and application of the outcomes of research (Figure 9.2). A key component of the program is research to develop effective performance criteria for all aspects of sustainable management. This includes the performance of the user groups as well as that of the managers and the condition of the managed ecosystem.

Key questions for improved science knowledge include fundamental science elements and outcomes that separate the signal of what can be managed from the background noise of natural variability (Table 9.3).

Recognizing that environmental management is about managing people and their actions, a high priority must be placed on greatly improving socioeconomic research, data, and information. The social, cultural, and economic values are influential and vital in decisions affecting the management of the GBR. To date there is a dearth of capacity and information about

Table 9.3 Key issues for science.

Environmental Aspects

- Systems/process understanding (especially soft-bottom communities; scaling to a systems view)
- Teleconnection (e.g., reef larvae, dugong) and transboundary (e.g., catchment–coastal) processes
- Ecology and biodiversity of mobile populations (turtles, seabirds, mammals)
- Management and effects of fishing (especially trawl)
- Time-series data (trends and long-lived species)
- Time scales and context for evaluating changes in systems (natural vs. anthropogenic changes)
- Development of rigorous and sensitive indicators of environmental change (for monitoring)
- Modeling and visualization techniques (planning, forecasting, education)

Socioeconomic Aspects

- Trends in use patterns: industry sector demands and forecasting, economic patterns and risk
- Motivation and perception of reef visitors, recreational users, and industry (and changes)
- Behavior patterns of visitors and other users
- Valuation and benefit/cost tools and models, and risk assessment approaches
- Societal change and impact assessment of coastal development
- Integrated decision support systems
- Development of rigorous and sensitive indicators of the success and efficiency of management
- Development of rigorous and sensitive indicators of the sustainability and efficiency of resource use
- Modeling and visualization of intersectoral and cumulative social and economic impacts of use mix and change scenarios
- Administrative/planning/legal options for managing private use of common and linked resources
- Relative efficiency and effectiveness, and appropriate mixtures of self-regulation, government regulation, and education in achieving and maintaining sustainable use

the socioeconomic elements of the GBR and its use, as well as about the human uses within the contiguous land areas. Limited as we may perceive it to be, our understanding of the biophysical systems of the GBR far exceeds our understanding of the socioeconomic context within which it must be managed. From a heretical viewpoint, it could be argued that we can make some good guesses about the natural systems and their functions; for socioeconomic issues, there is but a tiny coherent platform for assessments (see Drimyl et al. 1997; Pearce et al. 1997). However, it is clear that the uncertainty about the function of biogeochemical systems limits the data available for risk assessment and, in association with needed socioeconomic information, constrains development of acceptable and rigorous decision support systems.

Increasingly, the resolution of problems involves multidisciplinary and interdisciplinary team approaches. The building of these partnerships across disciplines of science remains a difficult, practical and often sociological challenge but continues to have success as communication, trust, respect, and "common language" develop within the teams. The successes of this approach (e.g., crown of thorns starfish outbreaks, dugong conservation) are most notable where managers and relevant users are full and contributing members to the team in science evaluations and analysis, ensuring mentoring of information across the continuum of problem-to-policy. The information brokerage concept within the new GBR Marine Park Authority structure appears to be working well and to provide a means of increasing the effectiveness of communication between scientists, social scientists, and policy makers.

POLICY IMPLEMENTATION

An approach of education rather than regulation and community consultation with participatory management underpins the stewardship actions of the GBR Marine Park Authority. This is backed by legislative and enforcement capabilities.

In 1994, more than 60 interest and user groups (policy, management, industry, community, research) agreed on a common vision and 25-year strategic plan for management direction and outcomes (GBRMPA 1994). This was a vital step, determining all users' expectations and a strategic template for management operations, including:

- conservation,
- resource management,
- education, communication, consultation and commitment,
- research and monitoring,
- integrated planning,
- recognition of aboriginal and Torres Strait islander interests,
- management processes,
- legislation.

The plan has 25-year objectives and 5-year goals (initially for the first 5 years). It provides a set of strategic and operational targets for management, reflects combined (and agreed compromise) agendas and expectation of all stakeholders, and provides a framework for guiding science and resource priorities. As a broad source reference, this plan is relevant not only to the GBR but also can be viewed as encapsulating issues relevant to global reef areas — the local or regional concerns will differ by degree, reflecting the scale and intensity of human pressures and effectiveness of management regimes.

The plan development process was extremely valuable in identifying and developing the linkages between groups and disciplines involved in use, management, and study of the GBR. While many of the targets (particularly for industry and community groups) appear to have been overambitious, they have provided a robust context for the development and review of management objectives and performance criteria. For the GBR Marine Park Authority the plan has provided the basis for the restructure and outcome-based programming systems that were introduced in 1998.

Use-conflict resolution is an important part of management. Working groups to address elements of critical issues, public consultations, and workshops are continuing tools in policy development and implementation. Recognizing that the vital management tool for the GBR is the zoning plans, the development of special (small) area management plans for certain areas is an ongoing exercise which influence regional community use and access. Regional Marine Resources Advisory Committees were set up regionally to act as conduits for localized community inputs and review of plans for relevant policy. Science involvement is an element in these initiatives.

Another feature of the GBR Marine Park Authority reorganization has been the establishment of an advisory committee for each of the four issues groups. Each advisory committee has an independent Chair and membership comprises a range of expertise and community interests relevant to the issue. The intention of the system is to provide for early, balanced, and effective consideration to identify and address issues of policy before the GBR Marine Park Authority embarks on widespread public consultation.

Industry consultations and "experimentation" in self-regulation of industry use (e.g., tourism) are part of the policy implementation thread. Attitudinal shifts in most user industries are resulting in the awareness of the commercial and practical need for a "clean and pristine" GBR that has seen adoption of goals of world's best practice for in-house self-regulatory approaches being put in place. This is consistent with the developing practice of industry, governments, and accountancy considering the triple bottom lines — economic, environmental, and social.

POLICY EVALUATION

The GBR Marine Park Authority is a statutory authority that reports to the Australian Commonwealth (or federal) parliament through the Minister for Environment. The GBR Marine Park Authority is an independent four-person board with a Chair who is also Chief Executive Officer. The Minister may give directions to the board but, if this is done, such directions are required to be published in the Annual Report of the GBR Marine Park Authority. The intention of such arrangements is to remove the work of the GBR Marine Park Authority from most immediate pressures of day-to-day political dynamics. There is also the GBR Consultative Committee (representing a range of interest groups) which is appointed by and reports to the Minister on performance of the GBR Marine Park Authority, and can raise issues and perceived problems associated with the GBR.

Public reporting on the environmental state of the GBR has been instituted. In 1995, the State of the Environment Report for Australia (Taylor 1996) determined that the GBR is in "Excellent-Good" condition. The State of the Great Barrier Reef World Heritage Area report (SORR) for 1998 "allows for cautious optimism about the state" of the area with respect to human impacts, noting uncertainties in trend data, limited knowledge about some species and processes, and need for further management action (Wachenfeld et al. 1998). This condition is not accidental but has resulted from major national commitment of political will, funding, people skills, and the development and application of management approaches to minimize human effects while maintaining access and use. Managers and scientists are actively engaged in further development of a suite of sensitive and effective criteria for measuring "change" (both natural and anthropogenic).

This favorable picture of the environmental status of the GBR is probably right; however, it has significant levels of uncertainty. Both the SOTE and SORR were derived from expert description and pieces of monitoring data; there is a long way to go in establishing a comprehensive and effective monitoring program for the entire GBR. Time-series information is limited for most attributes, so patterns of natural variability and palaeoclimate effects are sparse. Similarly, the design of effective performance indicators for management is a field in its developmental infancy. The SORR (Wachenfeld 1998) is based on the Pressure-

State-Response model for environmental reporting (OECD 1994). Despite a high level of effort (and there are several relevant areas for which scientific information is still lacking and for which rigorous indicators of environmental state or change are absent), performance criteria for pressure and response and new methodologies are needed.

The evolving management strategies and policy continue to develop within a tapestry of large and small user/interest group conflicts, a variety of agenda, and perspectives of opportunities and societal demands, all couched within the local variable winds of political and institutional changes.

TO THE FUTURE

Clearly, key scientific research and management actions must be sustained. It can be argued that we have a good idea about the mechanisms and how to do these things; however, dealing with the dynamics of environment, perceptions, and new knowledge remain challenges.

What we must be clever about is making sure that science is an active and responsive partner in the total process. Preliminary successes (and failures) in the GBR arena demonstrate that this requires energy, time, and institutional awareness and support. The development of an effective and mutually responsive relationship between science and management is not a serendipitous process. It goes beyond the traditional realm of training of scientists and requires some changes in mind-set in the scientific environment. The way forward will require increased attitudinal shifts, discovery of additional and effective ways to be society-friendly, with researchers gaining and applying new abilities, and perhaps dispelling some of the popular media-generated suspicion of science as the producer of threatening change.

Importantly, effort is needed for continuing to build management–user–researcher partnerships. Such participation establishes a "common language," often leading to reduced conflict between users and science disciplines. Multidisciplinary research team approaches are common, but we need to learn how to be more interdisciplinary in effecting their actions; training of individual researchers with strong cross-disciplinary skills is happening in places and should be strengthened.

Greatly improved communication is an imperative. Some significant (and successful) steps have been taken linking science with managers and other users, and science with the media; action and resources should sustain and expand the successes. Improved information access and effective tools, including more coherent data bases and visualization techniques, and mentoring of data would greatly assist end-users to obtain fast, relevant information from an "overload" situation.

Institutional structures and networks need to evolve actively in order to meet the political (and community) will. Consultative mechanisms and active, public communication of research outcomes and information are vital ways of gaining an effective approach and wide ownership of sustainable environmental development and wise use in the Great Barrier Reef.

REFERENCES

Crossland, C.J. 2000. Improved strategies for linking coastal science and users. In: Socio-economic Aspects of Fluxes of Chemicals into the Marine Environment, ed. J.M. Pacyna, H. Kremer, N. Pirrone, and K.-G. Barthel, EUR 19089. Brussels: European Commission.

Drimyl, S.M., T.J. Hundloe, and R.K. Blamey. 1997. Understanding Great Barrier Reef economics. In: Great Barrier Reef Science, Use and Management: A National Conference, Proc. vol. 1, pp. 335–342. Townsville: GBRMPA (Great Barrier Reef Marine Park Authority).
GBRMPA (Great Barrier Reef Marine Park Authority). 1994. The Great Barrier Reef, Keeping It Great: A 25-Year Strategic Plan for the Great Barrier Reef World Heritage Area, 1994–2019. Townsville: GBRMPA.
Harvey, N. 1999. Australian integrated coastal management: A case study of the Great Barrier Reef. In: Perspectives on Integrated Coastal Zone Management, ed. W. Salomons, R.K. Turner, L.D. de Lacerda, and S. Ramachandran, pp. 279–296. Berlin: Springer.
Kelleher, G. 1996. Case study 2. The Great Barrier Reef, Australia. In: The Contributions of Science to Coastal Zone Management, GESAMP (Joint Group of Experts on the Scientific Aspects of Marine Environmental Protection) Reports and Studies No. 61, pp. 31–44. Rome: FAO (Food and Agriculture Organisation of the United Nations).
Kelleher, G., and R.A. Kenchington. 1992. Guidelines for Establishing Marine Protected Areas. Gland: IUCN (The World Conservation Union).
Kenchington, R.A. 1990. Managing Marine Environments. New York: Taylor and Francis.
Lucas, P.H.C., T. Webb, P.S. Valentine, and H. Marsh. 1997. The Outstanding Universal Value of the Great Barrier Reef World Heritage Area. Townsville: GBRMPA (http://www.gbrmpa.gov.au/corp_site/info_services/publications/wha).
OECD (Organisation for Economic Co-operation and Development). 1994. Project and Policy Appraisal, Integrated Economics, and Environment. Paris: OECD.
Pearce, P.L, G. Moscardo, and B. Woods. 1997. Understanding the tourist market. In: Great Barrier Reef Science, Use and Management: A National Conference, vol. 1, pp. 343–352. Townsville: GBRMPA.
Taylor, R. ed. 1996. Australia: State of the Environment 1996. Melbourne: CSIRO (Commonwealth Scientific and Industrial Research Organisation) Publishing.
Wachenfeld, D.R., J.K. Oliver, and J.I. Morrissey, eds. 1998. State of the Great Barrier Reef World Heritage Area. Townsville: GBRMPA.

10

Shoreline Development on the Spanish Coast

Problem Identification and Solutions

R. SARDÁ

Centre d'Estudis Avançats de Blanes (CSIC), Camí de Santa Barbara, s/n,
17300 Blanes, Girona, Spain

ABSTRACT

This chapter identifies the need for strategic regional coastal plans as a managerial solution to the problems of the Spanish Coast. The inclusion of these plans in a national strategy for the conservation of the Spanish Coast is advocated. An overview is provided of the development of the Spanish coastline and the related policy that governs it. Largely uncontrolled growth, late management/conservation measures, and heterogeneity of competent authorities leave a difficult legacy for further effective management of the coast. Tourism is identified as the most dynamic sector in Spain, which contributes considerably to the most serious problem along the coast: erosion and urbanization pressure.

INTRODUCTION

Spain, with its important marine tradition, is situated in a strategic location receiving waters from the Lusitanian, Mauritanian, and Mediterranean biogeographical regions. Environmental conditions in Spain's coastal zone are highly variable, resulting in a large variety of habitats which host the richest and most varied biodiversity in Europe (Warwick et al. 1996). Spain's coastline stretches approximately 3200 km along the Mediterranean, 1200 km along the Cantabrian coast, and 3500 km along the Atlantic (including the Canary Islands). The sandy beaches along the Mediterranean and Atlantic coasts are ideal summer resort areas and they attract thousands of visitors annually. As a result, natural resources in these coastal areas have rapidly declined. Although less degraded, the natural resources along the Cantabrian coast have, in the past, been exploited and are in danger from serious pollution. Adverse

Science and Integrated Coastal Management
Edited by B. von Bodungen and R.K. Turner

impacts on natural coastal systems usually result from industrialization, urbanization, and agriculture carried out on an industrial scale (Ros 1994). Although there are different, heavily impacted environmental black-spot locations along the Spanish coast due to industrial practices (e.g., Huelva, Bilbao, Portman Bay, Barcelona, Valencia, Tarragona), the most important pressure on the natural systems of the Spanish littoral comes from the tourist industry and urbanization and all their associated impacts (Sardá and Fluvià 1999).

Located along the Spanish coastline are 478 municipalities, representing a constant population of more than 13 million people (around 30% of the Spanish population). The population, however, easily triples during the summer months. The Mediterranean coast and the two archipelagos, Balearic and Canary Islands, are the main destinations of foreign tourists visiting Spain. The environmental impacts of the resultant overcrowding on the fragile coastal ecosystems are the primary cause of past and present damage and deterioration. Human economic activities in the coastal zone of Spain are, however, extremely important for the economy of the region. The coast is the basic site of the most dynamic and productive sector: tourism. With annual revenues exceeding 26 billion Euros, tourism represents 10.6% of the country's gross national product and serves to finance a large part of its trade deficit. According to the latest tourism input–output estimates, the tourist industry (both from residents and nonresidents) currently generates around 1.15 million jobs, which represents 8.1% of the active population (MEH/MMA 1999). Most tourist statistics are related to "sun and beach tourist activities," and these activities dominate the littoral edge of our coasts.

As in other Mediterranean countries, Spain faces a difficult dilemma regarding its coastal zones: it is being forced to combine increased public awareness for the coastal environment with the attempt to better the coastal activities in an effort to improve its economic indexes. For example, in 1998, the regional authorities of the Balearic Islands approved measures limiting the opening of new tourist facilities and activities; local managers subsequently repositioned established tourist destinations in the islands. However, at the same time, regional tourist managers encouraged, welcomed, and were highly enthusiastic about the increased number of tourists to the islands during the season. This paradox led to a social demand for the implementation of measures to help balance these two extreme positions. Here, I discuss ways in which the desired sustainability of the coastal zone can be secured for the future.

OVERVIEW OF THE DEVELOPMENT OF THE SPANISH COAST AND ITS POLICY

It was not until the middle of the 20th century that the Spanish coast was developed. Until then, industry and transport activities were mainly connected to densely populated areas (large cities). The rapid transformation observed in the rest of the coast was due to the transformation of agriculture practices to meet the needs of the tourism industry and in places its replacement by tourism facilities.

Tourism as a potential source of revenue was recognized early in the 20th century. However, it was not until after the Spanish Civil War that tourism per se was enthusiastically pursued. From the 1950s until the 1980s, tourism produced a period of euphoria in which economic recovery and social modernization took precedence and obscured any recognition of future consequences. This resulted in changes never dreamt of by the players involved (Goytia 1996). Although the 1960s are considered to be the initial decade when tourism

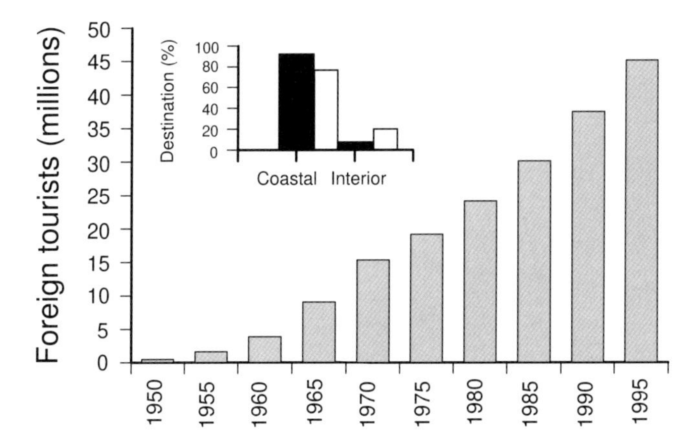

Figure 10.1 Number of foreign tourists entering Spain since 1950. Each gray bar represents the mean annual number of foreign tourists that visited Spain during the selected years. The percentage of tourists who visited the coastal and interior zones during the last decade are plotted on the inset of the figure (filled bars = foreign tourists; open bars = Spanish tourists) (adapted from SGT 1990 and Valls 1996).

began to grow, the trend actually began in the 1950s and steadily grew thereafter (SGT 1990; Valls 1996) (Figure 10.1). Today, Spain is second in the world in foreign tourist arrivals and fourth in income generated by tourism (WTO statistics). Most tourists who visit Spain (around 80%) select the coast as their final destination (Figure 10.1).

During all these years, legislation was enforced; however, it was not enough. Activities carried out on the coast were based primarily on particular interests rather than on the general public interest. Under the umbrella of the old Coastal Law of 1969 and the land-use legislation of 1976, unplanned growth and overdevelopment, abusive building of second homes (often on vulnerable natural areas), unsustainable use of natural resources, and abusive practices of enrichment based on the appropriation of public goods, were carried out everywhere. The comparison of the construction of houses along the shoreline of the Costa Brava against the average construction in Catalonia (Figure 10.2) shows a pattern repeated all over the Spanish littoral, especially in the Mediterranean areas. The management of the Spanish coast was extremely difficult and many political, economical, and social conflicts arose everywhere.

The rapid spread of tourism and the coastal policies developed from 1960 to 1988 were the cause of many problems still present in the Spanish littoral today:

- The demographic increase led to excessive and uncontrolled urban pressure.
- The coastal management was not integrated into sectoral policies and lacked homogeneity between local, autonomic, and central administrations.
- Coastal plans delivered by the autonomic administrations were almost nonexistent.
- The activities carried out by the Public Works Ministry, the main office responsible for the application of these laws, were almost entirely based on the defense of coastal public property, regeneration efforts, and the concession of authorization permits and were oblivious to environmental issues.
- Seawater quality declined as a consequence of freshwater pollution and uncontrolled sewage discharge from urban and industrial sites.

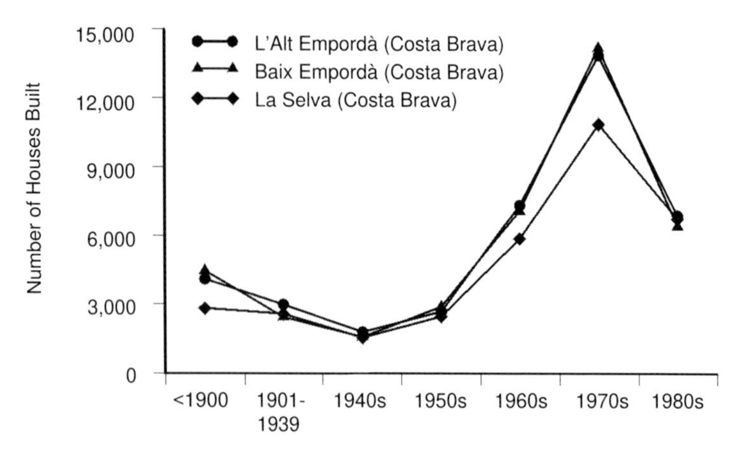

Figure 10.2 Total number of new houses built in the periods indicated on the X-axis in the three regions of Costa Brava (adapted from Sardá and Fluvià 1999).

- Humans have greatly influenced beach erosion processes by varying the sediment supply, modifying the regimes and water discharges of the rivers, and constructing coastal infrastructures (e.g., ports, jetties, breakwaters, marinas).
- Very few Marine Protected Areas were preserved along the Spanish coast during these years (Palma Bay, Tabarca, Mouro, and Chafarina Islands). In addition, Spain rapidly lost ecologically vulnerable and sensitive areas, such as salt marshes, coastal lagoons, and seagrass meadows. Furthermore, most of the Spanish coastal dunes were destroyed due to the construction of hotels and infrastructures for tourism (Sanjaume and Pardo 1992; Ramos-Esplá and McNeill 1994).

The new Spanish Coastal Law of 1988 was enacted to protect the coastal zone. Although this law arrived late in many parts of the Spanish territory, it has served to reaffirm the public domain and to define the rights and limitations on property. To summarize, the Spanish Coastal Law: (a) defines the precise form of coastal public property, (b) establishes easements and property limitations on neighboring land, (c) regulates the use of the coastal public property, (d) establishes that works in coastal public property shall be financed from the State budget, (e) defines infringements of this coastal legislation, and (f) determines the responsibilities of the various administrations involved in the coastal area. The Coastal Law of 1988 was developed to guarantee the public nature of coastal public property and to conserve its natural characteristics. It has sought to balance economic development against the necessity of preservation (Montoya 1991).

Nevertheless, the overall application of this law was insufficient as administrations, autonomous communities, and municipalities that have the competency to act on a much wider coastal band did not manage the territory properly. The law was reaffirmed by the government as the only way to have a more ordered coastal environment, alternating spaces already urbanized with other natural spaces in a planned fashion to remove the urban pressure from the coastal zone. During the implementation of the law, many coastal towns tried to favor the urbanization process of the coast through various means. In particular, they tried to delimit the coast as urban soil, land that did not fall under legislated restrictions, thus reducing the protected area from 100 m to 20 m. Despite the many problems associated with the implementation of the Coastal Law, social demand for maritime recreation as well as for environmental protection was partially fulfilled.

The entrance of Spain into the European Union (1986) and the ratification of programs developed by the Rio Conference (1992), marked the beginning of a new era in the management of the coastal environment in Spain. The European Union determines a series of community directives, which delineate the objectives and policies for the environment, including the coastal zone. The European Commission's (EC) fifth environmental action program, "Towards Sustainability," confirmed a long-term strategy for future community actions in the field of environment and provided the means to move towards sustainable development. It also marked a qualitative step forward in addressing the forms of future economic growth management. In terms of coastal habitats, the program was extremely important since it considered, for the first time, littoral areas as a priority. It called for new management strategies and had a clear objective: the sustainable development of coastal areas and their resources should depend on the environmental carrying capacity.

Following the Spanish Coastal Law of 1988, a Coastal Management Plan (1993–1997) was created. This plan, however, followed previous initiatives in which coastal management was held to be synonymous with coastal engineering and efforts to immobilize the coastline by engineering structures (García Mora et al. 1998). Waterfront protection, beach replenishment, and the creation of marine walks account for the majority of the budget allocated to the management of Spanish coasts (Barragán 1994).

In 1996, the Spanish Ministry for the Environment was created. The Ministry was divided into three main departments. The Department of Water and Coasts was selected as the main office for the management of the Spanish Coast. In 1997, this department designed a new concept in coastal management. Their main objective was to reach a sustainable development of the coastal resources through the implementation of an Integrated Management Plan (MMA 1998). The designed strategy outlined the necessity of introducing environmental principles to activities on the coastal environment made by the sectoral office involved. The recognized environmental principles applied in the plan were: sustainable development, precautionary principle, prevention principle, sustainable management of common natural resources, inland management, land-use management, the externality concept recognition, and the polluter-pays-principle. The department also recognized that erosion of the coast is the main problem for the Spanish littoral, especially in the Mediterranean and South Atlantic zones. During the last few years, the General Directorate of Coast, under the Department of Water and Coasts, has been conducting different activities based on these environmental principles: (a) an ambitious easement plan, (b) punishment policy, (c) more sustainable urban planning, (d) managing the use of coastal public property, and (e) establishing projects of public work.

It should be noted, however, that Spain is not a homogenous country. It is made up of autonomous communities, and these communities actually manage the activities that influence the structure and dynamics of the coastal ecosystem. Thus, current management of the Spanish littoral is divided among national, autonomic, and municipal administrative offices which makes the coordination of efforts very difficult and, in many cases, results in duplication. This style of management represents the main obstacle to effective coastal management today.

PROBLEM IDENTIFICATION: THE PRESENT STATUS OF THE COASTAL PROBLEMS IN THE SPANISH LITTORAL

As stated above, the Spanish coast is dominated by tourism. Current environmental threats stem primarily from the various impacts of the tourist industry (Sardá and Fluvià 1999). Since 1989, a large-scale coastal survey has been carried out each autumn by the Coastwatch Europe network. This survey provides general information on the coast, effluent discharges, litter, pollution, and destruction of habitats, which are shown in Figure 10.3 for several regions (Peris et al. 1998). In the Mediterranean provinces (Catalonia, Valencia, Baleares, Murcia, and Andalucia), erosion (primarily of anthropogenic origin) and urbanization are the two main factors, while pollution-related problems are more prevalent in the Northern Atlantic provinces (Galicia, Asturias, Cantabria, and Pais Vasco).

Physical Occupation of the Territory: Coastal Erosion and Beach Management

Anthropogenic coastal erosion is one of the most serious and widespread problems in the coastal zone. For many years, the increased concentrations of settlements, transport facilities, recreational developments on the shore, the proliferation of ports and marinas, the absence of sediment transport by rivers due to the construction of dams, loss of beach stability by redistribution onshore, offshore, and alongshore in zones where sand is extracted, as well as the destruction of natural protection systems such as dunes and coastal lagoons have altered the natural equilibrium of erosion and accretion. Long-term erosion impacts in the coastal zone and beach regressions are crucial problems in the management of coastal areas in Spain.

Undoubtedly, the beach is a resort's basic asset. Most of the fifty million tourists are particularly attracted by the extended beach areas. Since the appearance of the Coastal

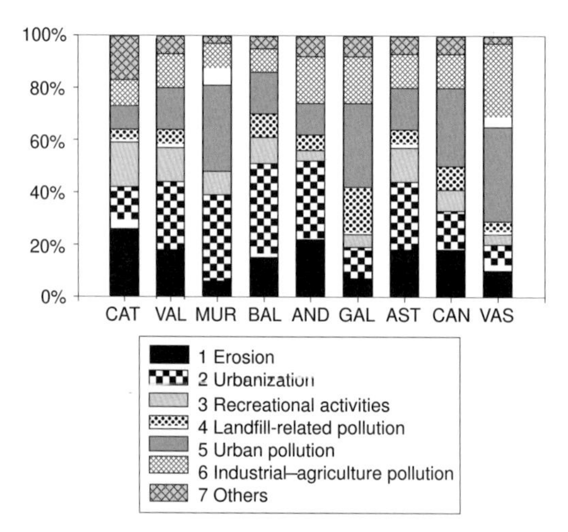

Figure 10.3 Main coastal environmental problems considered in the Coastwatch Report of 1997 for the Iberian peninsula (CAT: Catalonia; VAL: Valencia; MUR: Murcia; BAL: Islas Baleares; AND: Andalucia; GAL: Galicia; AST: Asturias; CAN: Cantabria, and VAS: Pais Vasco) (adapted from Peris et al. 1998).

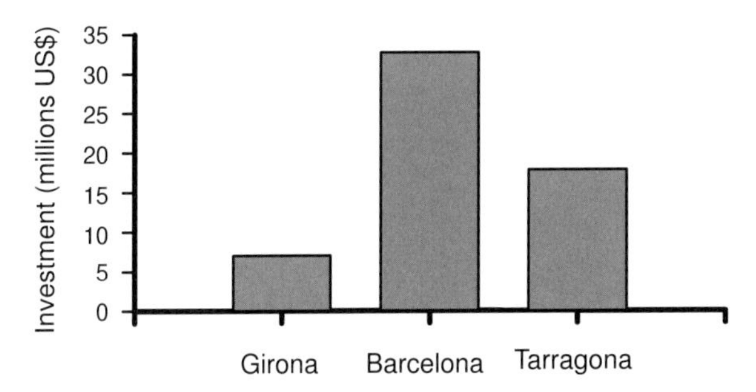

Figure 10.4 Investment efforts on beach nourishment in the three coastal provinces of the Catalan coast (northeastern Spain) until 1996 (adapted from MOPTMA 1996).

Management Plan (1993–1997), the various offices in charge of beach management have had the responsibility of recovering the physical integrity of many beaches (quantity) and ensuring the free access of the public to these sites for recreation. Engineering solutions have been adopted to protect these natural beauties: from beach nourishment techniques to protection by breakwaters and other infrastructures. However, these are costly solutions with high environmental impact (Essink 1997), and they yield a very artificial location that needs periodic maintainance. In the Catalan coast, 35 beaches of Tarragona (28% of the total), 18 beaches of Barcelona (26% of the total), and 5 beaches of Girona (4% of the total) were nourished before 1997, accounting for an investment of around 9000 million PTAs (Figure 10.4). Autonomous administrative bodies and municipal authorities need to develop yearly plans to guarantee beach cleanness in order to maximize the quality of the tourist attraction as well as to minimize the environmental impact generated by anthropogenic pressure on the coastal environment.

Water quality at the Spanish beaches has greatly improved since the enforcement of the EC's Directive on bathing water (76/160/EC). In 1997, the seawater of 379 coastal towns was monitored. From 1588 sampled points, 76.1% were judged to be of excellent quality, an increase of 0.7% over the previous year (MMA 1998). Water treatment plans of littoral municipalities are being scaled to the maximum population size at the tourist season, and this has contributed to the improvement of seawater quality near coastal towns.

Physical Occupation of the Territory: The Urbanization Pressure

In 1998, the Spanish State enacted a new land-use law. The aim of the new regulation was to increase the offer of developable land, in an effort to reduce land and housing prices. The consequences of this new law for the Spanish coastline have been outlined by Doñate (1998, pp. 64–65):

> The new law states that nondevelopable land must cease to be considered as leftover land, being newly defined to be included in this category due to it being the object of special protection incompatible with its transformation, or that which the general planning considers necessary to conserve or considers unsuited for urban development. With this ruling, the residual land area is developable, with the exception of cities and towns without planning policies, in which land not "developed" will be "nondevelopable." The new law establishes a new land appraisal system by which the methods used to determine land's true value will be those employed in the real estate

> sector or the ones that are commonly applied in determinations of cadastral value. The law attempts to pave the way for a greater offer of land, making it possible for the land which has not yet been developed (and for which there are no reasons for its conservation) to be considered potentially developable. This represents a setback for the balanced urban organization of Spanish coastline overpopulation.

Although the law aims to reduce land prices, those who will most likely profit most from it will be developers and the construction industry (Doñate 1998). However, since Spain gives most of the urban competencies to the autonomous communities, these autonomous regions will have the opportunity to moderate the predicted negative impacts on the environment.

Apart from problems related to urban development, the second critical problem concerning the physical occupation of the terrestrial systems in the coastline is related to the past development of the tourist industry as a whole and to the needed incorporation of the environmental factor in Spanish tourist planning.

In 1998, a Sustainable Tourism Program Plan was jointly developed by the Spanish Ministry of Economic Affairs and Inland Revenue and the Spanish Ministry of the Environment. The plan aims to contribute to protect natural areas, to raise competitiveness in the Spanish tourism sector, and to reduce seasonal fluctuations in Spain's present tourist industry (MEH/MMA 1999). The plan contains several specific programs: (a) tourism and planning, (b) tourism and environmental management, (c) tourism and natural areas, (d) training in sustainable tourism, and (e) cooperation in international forums.

In addition to this plan, many new initiatives are being promoted: the Balearic Islands tourist moratorium, the Calvià Local Agenda 21, the Action Plan of Sustainable Tourism in the Girona regions, the Costa del Sol Excellence Plan, the Costa Blanca strategy of repositioning, the process of change in the Canary tourist model, or the Lanzarote's Island Territorial Regulation Plan and Local Agenda 21. This proliferation of initiatives indicates that tourist managers are aware of the need to incorporate the environment into day-to-day business operations.

Coastal Environmental Pollution

The Spanish littoral still has environmental "hot spots" of industrial pollution. The most acute problem was in Portmán, a small fishing harbor in the southeastern Mediterranean. For 30 years, mining waste was dumped without any consideration into the bay. More than 90% of the total industrial waste dumped into the Spanish littoral happened at this site. It is estimated that around 50 tons of waste, with high concentrations of heavy metals, are still in the vicinity of the bay. Although the Spanish Ministry of the Environment has an ongoing project to remedy the present situation, the Portmán case can serve as an example of extreme environmental degradation. Other black spots related to industrial activities are Huelva, Tarragona, and Bilbao.

Urban waste management has greatly improved over the last decade. Autonomic governments manage sewage treatment and plan to provide all municipalities larger than 2000 inhabitants with biological treatment systems by the year 2005 in accordance with the 76/160/EC Directive. By contrast, large metropolitan areas, such as Barcelona, have problems with waste treatment. The large urban population and industries produce huge amounts of treated wastewater and sludge. The mouth of a submarine pipeline, located 4 km off the coast at a depth of 55 m, discharges on average 40 metric tons of sludge per day. This has resulted in a highly contaminated deposit covering a surface of approximately 1 km^2 and with a thickness of 2.5 m (Checa et al. 1988). A second treatment plan that utilizes another submarine pipeline has been designed to work within a few years. There is no actual plan to treat such enormous quantities of sludge on land.

Exploitation of Natural Resources: Management of Marine Protected Areas and Other Conservation Measures

Declaration of marine protected areas (MPAs) is a relatively new concept in Spain. The first MPA was declared in 1982 (Palma de Mallorca Bay). However, the economic benefits of such areas are now recognized and, in some provinces, the protected littoral is becoming increasingly important. In the province of Huelva, over 60% of the coastline has been included in the network of protected areas (García Mora et al. 1998).

MPAs are classified into different categories: national parks, hunting refuges, marine reserves (sometimes biosphere reserves), fishery preserved zones, marine–terrestrial parks, and natural sites (Table 10.1). The use of MPAs as a management tool for the protection of marine habitats and species is gaining impetus in Spain (Ramos-Esplá and McNeill 1994).

Other conservation measures and plans have been enacted to protect species and habitats, such as the Spanish Law of Conservation of the Natural Heritage, Flora, and Fauna of 1989.

Table 10.1 Marine protected areas in Spain until 1998.

Location	Type	Date	Area (km^2)	Reasons
Bahía Palma	FPZ	1982	2000	Restocking (fish), preservation of *Posidonia*
Sonabia	FPZ	1982	100	Restocking (fish, crustaceans)
Chafarinas	HR	1983	—	Protection of Monk seal
Tabarca	MR	1986	1400	Marine resources and communities
Columbretes	MRNP	1990	4000	Marine resources and communities
Medes	MR	1990	550	Marine resources and communities
Cabrera	MTNP	1991	8164	Marine communities
Ses Negres	MR	1993	80	Educational
San Antonio	MRNP	1993	85	Marine resources and communities
Cabo Gata	MRNP	1995	12,200	Marine resources and communities
Cabo Palos	MR	1995	1898	Marine resources and communities
Graciosa	MR	1995	70,700	Marine resources and communities
Restinga	MR	1996	750	Marine resources and communities
Cabo Creus	MNRP	1998	8769	Marine resources and communities

FPZ = fishery protected zone
HR = hunting refuge
MR = marine reserve
MRNP = marine reserve and natural park
MTNP = marine–terrestrial national park

To protect particular species, such as seagrasses, autonomous communities have legislated specific guidelines.

SOLUTION DEVELOPMENT AND POLICY MAKING

Reducing the effects of anthropogenic activities on the coastal environment in Europe and recovering degraded sites need to be resolved quickly. Coastal erosion in the Spanish littoral (including the alarming process of beach regression) and the increased pressure of urban development and construction obscure other dangers, such as the effects of pollution or losses in biodiversity.

Environmental impacts on the coast face another problem: the lack of cooperation between many different administrations on coastal issues, each one addressing different managerial aspects related to the coastline and often with opposing objectives and strategies. Furthermore, managerial practices are even more problematic due to the administrative division into autonomous communities, which are responsible for applying environmental policy within the limits of their territories. The statutes of autonomy of these communities define the distribution of competencies. These competencies can be different, and the designation of administration to manage the same issues may vary regionally. Such a structure makes coastal management extremely haphazard. Figure 10.5 illustrates how responsibilities on coastal issues are distributed in the Catalan region.

All of this has contributed to a disharmonious development and ongoing environmental degradation of the coastal zone. The urgent need to change this status has now been recognized. Over the last years, several Plans of Excellence (best practice guidelines for local

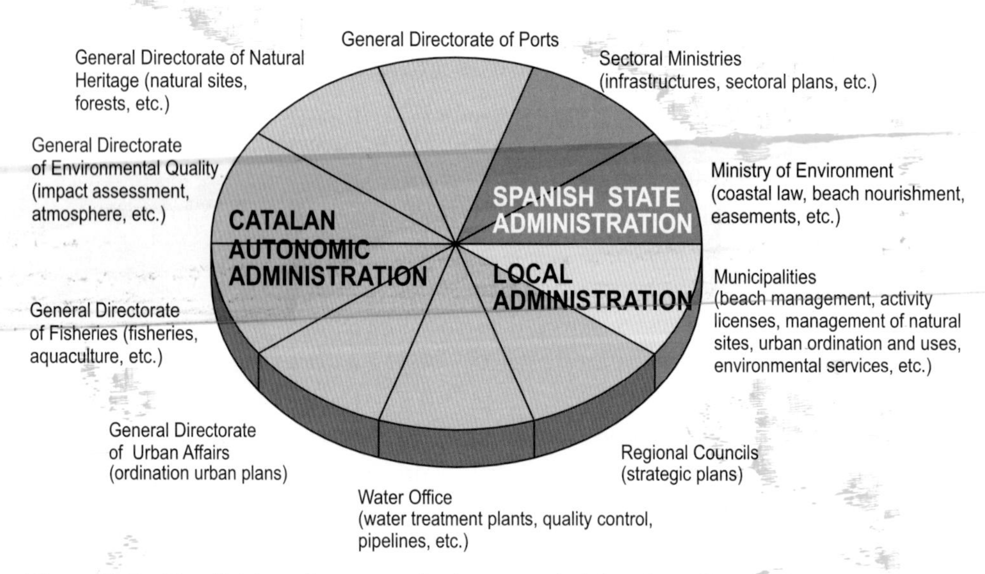

Figure 10.5 The division of competencies between administrative offices in the management of coastal issues on the Catalan coast.

sustainability) have been enforced through the implementation of Local Agendas 21 (the most well known is the Local Agenda 21 for Calvià, Mallorca). Other coastal towns are considering this new management approach.

The Spanish littoral needs a national strategy towards a sustainable development of the coast. This strategy could be incorporated in the Integrated Management Plan launched recently by the Ministry of Environment. In recognition of Spain's singularities, such a national strategy should be formulated through the implementation of strategic regional coastal development plans. These strategic regional plans should have delimited natural regions as territorial units and should be based on the analysis of future scenarios, the enforcement of Local Agendas 21, and the delineation of long-term objectives. The legal context should be contained in the proposed directive on Strategic Environmental Assessment of the European Community. The objective of such a directive is to provide high-level protection of the environment by ensuring that environmental consequences of plans and programs are identified and assessed before adoption. This objective should be completely compatible with the requirements needed to eliminate the disharmonious management scenario presently in place.

Strategic regional coastal plans are not new managerial tools in Spain. In 1986, a coastal plan for the Barcelonés region was developed in Catalonia. The main goal of this plan was to emphasize the natural values of the coastline and to coordinate initiatives to promote a new model of urbanization on the littoral edge (Bretón 1996). Recently, the Coastal Management Program of the Basque Country (Castro et al. 1998) made recommendations for the management of this coastline. Many other coastal plans have been developed for particular local sites. Although these efforts are important and have had positive effects, strategic regional coastal plans should be included in a national strategy and, if possible, be based on similar standards. We propose that a strategic regional coastal plan be divided into two phases: a preliminary and an advanced phase.

Preliminary Phase (Regional Administration Management)

a) The compilation of existing information. Territorial management requires up to date information. A description of all the information on the state of the environment as well as the socioeconomic structure is required. Past and present information related to natural resources and coastal ecosystems should be compiled (Figure 10.6, top graph). From this information the temporal evolution of observed changes should be assessed. The collection of information regarding sectoral activities and demographic trends should also be emphasized (Figure 10.6, bottom graph). The final outcome should be to characterize the present status of the regional littoral in terms of the environmental standards, economic activities, and their trends.

b) The analysis of existing information. Analysis of the acquired information should be studied under a conceptual framework of reference. The Drivers-Pressures-State-Impact-Response (D-P-S-I-R) framework (Turner et al. 1998) could be introduced for this purpose. The D-P-S-I-R framework could be used to link the information to environmental pressures and then to socioeconomic structural changes. The two obtained data sets (environmental and socioeconomic) could be analyzed together and future scenarios projected for different sites. The final product would be a general framework of "drivers-pressures-state-response" for the

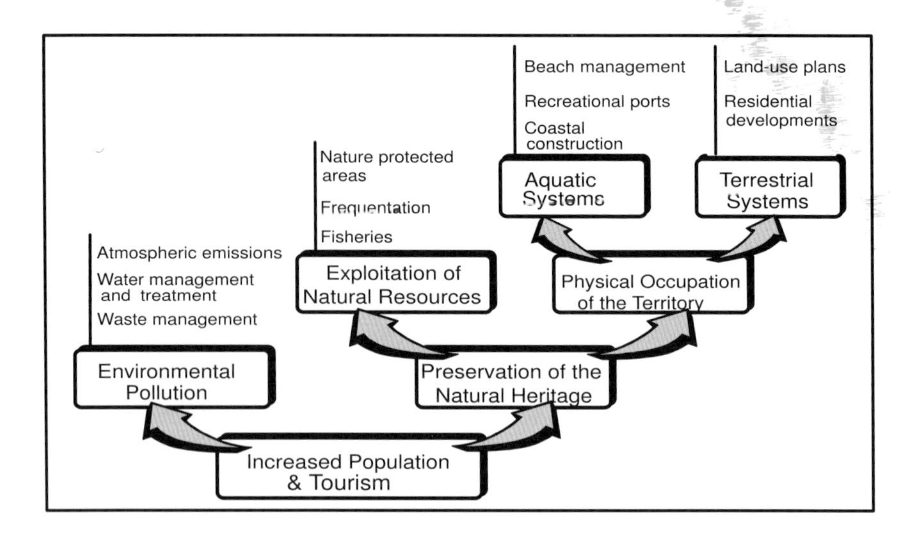

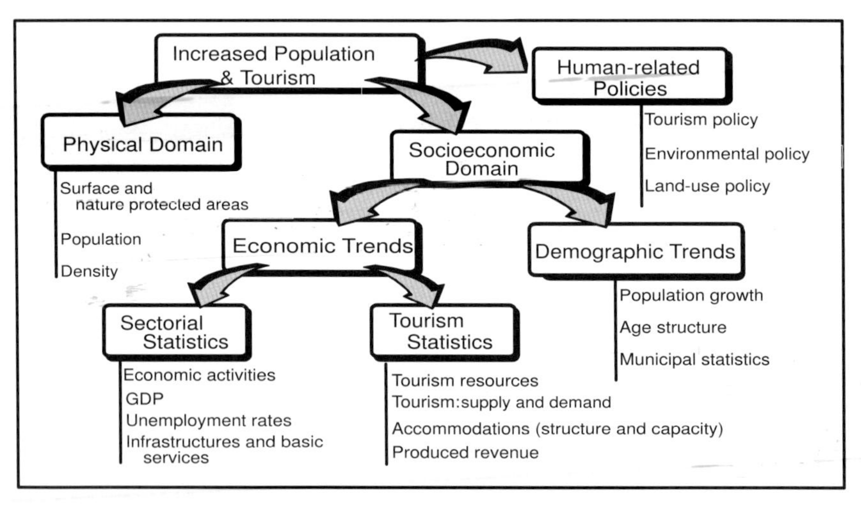

Figure 10.6 Hierarchically organized conceptual scheme to classify the existing regional information on environmental (top graph) and socioeconomic data (bottom graph). As an example, conceptual schemes are developed using tourism as the main economic activity in an area.

regional area. The document should be the essential reference for the development of the strategic plan.

Advance Phase (Advisory Council Management)

a) The construction of the managerial structure of the plan. An advisory council should be created. All regional stakeholders should play an important role and the relationships between decision makers, sectoral managers, and local residents should be analyzed. The initial goal should be to find a wide societal agreement concerning the development of the coast, based on sustainable development principles, and using the document formulated in the preliminary phase. Four main points should be emphasized by the Advisory Council:

1. The development of Local Agendas 21 should be promoted. Participation and cooperation between local authorities as well as discussion with their citizens, local organizations, and private companies are paramount to develop a local program according to the Rio Agenda 21.
2. The development of sustainable plans for sectoral organizations, which are the main agents involved in the territorial activities, should be encouraged. Introduction of environmental policies, environmental management systems, and environmental practices by industry should be recognized and awarded.
3. The protection of the natural heritage and the restoration of degraded sites should be promoted. Although vulnerable ecological areas should be the first to be protected and restored, other natural sites, even at a local scale, should be considered for protection. The economic benefits of protected areas should also be emphasized.
4. Information-based instruments must be developed to analyze the progress of the plan:

   - Geographical Information Systems (GIS) can be used to manage coastal zones. The use of GIS will be possible using the data obtained in the preliminary phase.
   - Second generation indicators should be developed to measure whether a region is moving in the right direction as proposed by the document outlined in the preliminary phase.
   - An environmental quality seal should be awarded as recognition of the plan's progress.
   - A final document of the Strategic Regional Plan should be prepared.

b) The program and monitoring aspects. Following the steps previously mentioned, each strategic plan should document its actions. These actions should contain clear initiatives that can be measured through time using environmental indicators. Although each plan should be implemented primarily at a local (municipal) level, the introduction of this local process in a structured regional plan should facilitate a harmonization of efforts and regional coordination.

A review of the current situation on the Spanish littoral suggests that such planning is needed to prevent further degradation and to foster mitigation. Integrative approaches, combining social and natural interests, should be promoted to facilitate an exchange of ideas and information. The development of coordinated strategic regional coastal plans could be a valid way of developing the coastline in a sustainable manner, thus balancing the economic development and ecosystem preservation.

ACKNOWLEDGMENTS

I am grateful to the organizers of the Dahlem Workshop for inviting me to present a paper to the working group on shoreline development. I want to thank Klaus Schwarzer, Claudio Richter, Conxita Avila, and Muntsa Solá for their useful comments. This paper is a contribution to the project 2FD97-0489.

REFERENCES

Barragán, J.M. 1994. Ordenación, Planificación y Gestión del Espacio Litoral. Barcelona: Oikos Tau.

Bretón, F. 1996. El litoral: Bases per al planejament i la gestió integrada d'un espai dinàmic i vulnerable. In: El Sistema Litoral: Un Equilibri Sostenible? Quaderns d'Ecología Aplicada 13, pp. 45–100. Barcelona: Diputació de Barcelona.

Castro, R., A. Borja, and J. Bald. 1998. Impact assessment on marine environment in the coastal management program for the Basque Country (Spain). In: Sustainable Waterfront and Coastal Developments in Europe: Socioeconomics, Technical and Environmental Impacts, ed. M. de Prat, pp. 59–62. Barcelona: Suport Serveis, S.A.

Checa, A., et al. 1988. Contaminación por metales pesados en los sedimentos. In: Efecto del Depósito Submarino de Lodos de la Zona del Rio Besós Sobre la Zona Costera de Barcelona (Proyecto Spio). Barcelona: Corporacion Metopolitana de Barcelona and Ministerio de Obras Públicas y Urbanismo.

Doñate, I. 1998. La protecció del litoral mediterrani i l'ordenació territorial. *Medi Ambient, Tecnología i Cultura* **21**:60–67.

Essink, K. 1997. Risk Assessment of Coastal Nourishment Techniques — RIACON. Final evaluation report. Haren: National Institute for Coastal and Marine Management/RIKZ.

García Mora, M.R., J.B. Gallego, and A.T. Williams. 1998. The coastal area of SW Spain: Between conservation and tourist development. In: Sustainable Waterfront and Coastal Developments in Europe: Socioeconomics, Technical and Environmental Impacts, ed. M. de Prat, pp. 177–185. Barcelona: Suport Serveis, S.A.

Goytia, A. 1996. Back to a sustainable future on the Costa Brava. In: Sustainable Tourism Management: Principles and Practice, ed. B. Bramwell et al., pp. 121–145. Tilburg: Tilburg Univ. Press.

MEH/MMA (Ministerio de Economía y Hacienda and Ministerio de Medio Ambiente). 1999. España: Un Turismo Sostenible. Madrid: Centro de Publicaciones de Ministerios.

MMA (Ministerio de Medio Ambiente). 1998. Medio Ambiente en España, 1997. Madrid: Centro de Publicaciones de Ministerios.

Montoya, F.J. 1991. An administrative regulation pattern of coastal management for Mediterranean Sea: Spanish Shores Act, July 1988. *Mar. Poll. Bull.* **23**:769–771.

MOPTMA (Ministerio de Obras Públicas, Transporte y Medio Ambiente). 1996. Guía Oficial de las Playas de España: Tomo III. Las Playas de Cataluña. Madrid: Centro de Publicaciones de Ministerios.

Peris, E., et al. 1998. Informe General y Base de Datos Sobre el Estado del Litoral Español, Coastwatch 1997. Valencia: Univ. Politécnica de Valencia.

Ramos-Esplá, A.A., and S.E. McNeill. 1994. The status of marine conservation in Spain. *Ocean Coast. Manag.* **24**:125–138.

Ros, J. 1994. La salud del Mar Mediterraneo. *Investigación y Ciencia* **215**:66–75.

Sanjaume, E., and J. Pardo. 1992. The dunes of the Valencian coast (Spain): Past and present. In: Coastal Dunes. Geomorphology, Ecology, and Management for Conservation, ed. R.W.G. Carter, T.G.F. Curtis, and M.J. Sheehy Skeffington, pp. 475–486. Proc. Third European Dune Congress, Galway, 17–21 June 1991. Rotterdam: Balkema.

Sardá, R., and M. Fluvià. 1999. Tourist development in the Costa Brava (Girona, Spain): A quantification of pressures on the coastal environment. In: Perspectives on Integrated Coastal Zone Management, ed. W. Salomons, R.K. Turner, L.D. Lacerda, and S. Ramachandran, pp. 257–276. Berlin: Springer.

SGT (Secretaría General de Turismo). 1990. Libro Blanco del Turismo Español. Madrid: Centro Publicaciones Ministerios.

Turner, R.K., W.N. Adger, and I. Lorenzoni. 1998. Towards Integrated Modelling and Analysis in Coastal Zones: Principles and Practices. LOICZ (Land–Ocean Interactions in the Coastal Zone) Reports and Studies No. 11. Texel: LOICZ International Project Office, NIOZ (Netherlands Institute for Coastal and Marine Research).

Valls, J.F. 1996. Las Claves del Mercado Turístico: Como Competir en el Nuevo Entorno. Bilbao: Ed. Deusto.

Warwick, R., R. Goñi, and C. Heip. 1996. An Inventory of Marine Biodiversity Research Projects in the EU/EEA Member States. Maarssen: MARS (European Marine Research Stations Network), Netherlands Institute of Ecology.

Standing, left to right: Job Dronkers, Rafael Sardá, Ies de Vries, Karsten Reise, Merrilyn Wasson, Chris Crossland
Seated, left to right: Klaus Schwarzer, Angela de Luca Rebello Wagener, Jane Taussik, Edmond Penning-Rowsell

11

Group Report: Shoreline Development

K. SCHWARZER, Rapporteur

C.J. CROSSLAND, A. DE LUCA REBELLO WAGENER, I. DE VRIES, J. DRONKERS, E. PENNING-ROWSELL, K. REISE, R. SARDÁ, J. TAUSSIK, and M. WASSON

INTRODUCTION

Shoreline development[1] refers to human activities that interfere with the natural evolution of the shoreline as well as with the natural evolution of the ecosystem at the land–ocean interface. Shoreline development is a global issue, modifying land–water interaction, coastal habitats and their biota, and affecting resources on which the majority of the world's population depend (Bijlsma 1993). Decisions on shoreline development are often (almost) irreversible and engage life support conditions for many generations. They imply choices that have to be based on value judgments across a great variety of factors. These factors cannot be expressed in a single valuation system nor predicted with any certainty (Turner and Adger 1996).

Many parties are involved in the decision-making process. Often these parties have different and opposing interests (Green and Penning-Rowsell 1999). The ideal situation of an objective, well-informed arbiter capable of defining clear, univocal criteria for decision making generally does not occur. In practice, it is just as difficult to formulate the relevant questions as to provide the adequate answers.

The above applies to many environmental management issues. In the case of shoreline development, the situation is particularly complex as land- and sea-related issues are both involved. Moreover, the shoreline is a region with a highly dynamic character, not only because of these multiple interacting socioeconomic issues, but also because of the intrinsic

[1] Numerous definitions related to coastlines or shorelines exist. These definitions depend on the point of view of different disciplines (economy, engineering, biology, geology) as well as on national legislation. In our discussion, we used "shoreline" in the sense of a feature that is highly dynamic in space and time and which therefore covers an area.

Science and Integrated Coastal Management
Edited by B. von Bodungen and R.K. Turner

variability of the natural system related to hydrodynamic, morphodynamic, sedimentary, and biogeochemical processes.

Conditions for implementing an effective ICM (Integrated Coastal Management) approach are sometimes expressed in terms of the three W-factors, Wisdom–Willingness–Wealth (Boudreau 1999), where:

- "Wisdom" includes scientific data and information availability,
- "Willingness" involves government will and empowerment of decision making, and
- "Wealth" represents the financial capability to invest in optimal management options.

The three Ws are not equally distributed among coastal nations worldwide. For example, in developing countries data and scientific information are often scarce by reason of access or capacity compared to relatively data-rich areas (e.g., Europe, North America). Similarly, the capacity for major investment in recovery or rehabilitation of shoreline development reflects national gross domestic product (GDP) and economic priorities (ability to pay). However, effective communication of science will affect the willingness to change and, thereby, the wealth available to ICM.

This chapter focuses primarily on the factor "Wisdom," which is more broadly interpreted as effective use and communication of science in the ICM process. Throughout, we use *science* and *scientists* to refer to both natural and social science disciplines, unless otherwise specified.

Even in regions where data and scientific knowledge are relatively abundant, they are often site and issue specific. Universal rules or models are very seldom available (Dronkers and de Vries 1999). Moreover, knowledge is scattered, not only geographically, but also over scientific disciplines and over institutions. Much of the existing information has not yet been synthesized in order to draw lessons for shoreline management. These are all reasons that hamper the effective use of science. Therefore, integration of science in shoreline development is the challenge. Much of the mismanagement of shorelines in the past and the present can be traced to the lack of effective integration of science into management (Sardá and de Vries, both this volume).

We identify four areas that are critical elements in both integrated management plans and integrated research programs: institutions, time scales, carrying capacity, and uncertainty. Science–management communication on these issues is essential. Science–management interaction modes are highlighted and impediments to communication and effective use of science are identified and discussed. Later, communication strategies are proposed that may improve science–management interaction in the ICM process. The result of the discussion is summarized in a number of simple prescriptions, which we named the "Golden Rules."

KEY ELEMENTS TO BE ADDRESSED BY ICM

In the case of ICM, the dynamics of shoreline development can be schematized broadly as an interaction between human activities and the shoreline environment. Geographically, the field of interaction is larger than the shoreline region itself, as the shoreline system can be affected by activities based on land (e.g., access routes to the shore, dams, sand mining in rivers, land-based shore protection measures, use of agrochemicals) and on sea (e.g., oil spills,

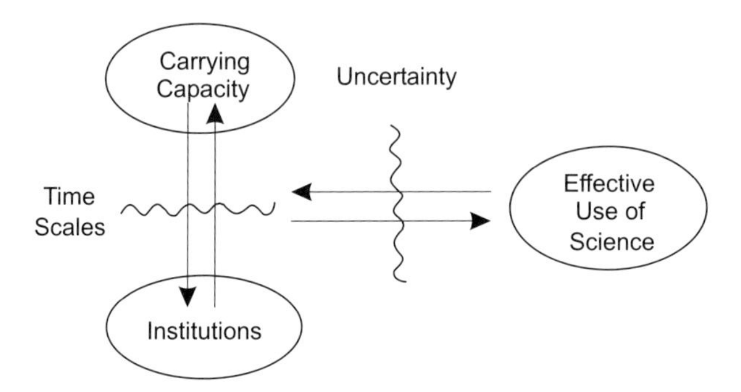

Figure 11.1 Schematic representation of the critical elements for integrated management plans and integrated research programs. Shoreline development (left) is characterized as an interactive process relating *institutions* (the human dimension) to *carrying capacity* (the physical dimension), an interaction that is complicated by the problem of dissimilar time scales. The *effective use of science* in the ICM process (right) requires communication between scientists, coastal managers, and users about their mutual understanding or perception of shoreline development, i.e., the interaction between the human and physical dimensions of the coastal system. One of the important impediments to this communication is uncertainty, i.e., lack of knowledge, statistical uncertainty, or intrinsic unpredictability.

shipping, dredging, navigation channels). According to this scenario, shoreline environment represents a resource, or a set of resources, while human activities that interact with this environment represent resource users.

Human activities related to shoreline development are determined by institutions. *Institutions* are social constructs that govern the way humans interact with each other and with their environment. Humans interact through explicit and implicit rules and through formal and informal organizations that are established to implement these rules. Institutional dynamics involve a range of *time scales* related, for instance, to political agendas. The dynamics of the shoreline environment involves a range of time scales as well, but these are generally very different from the time scales of human activities. This mismatch of time scales constitutes one of the most important potential threats to the *carrying capacity* (defined as the degree to which shoreline resources can be used without being drastically or irreversibly reduced) of the shoreline environment and is therefore a critical issue for ICM. Carrying capacity is not an intrinsic and static element but rather a dynamic property of the shoreline environment and its utilization. Another critical issue relates to the mismatch of spatial scales, which is addressed in the chapter on "Transboundary Issues" (Jickells et al., this volume). These three notions, institutions, time scales, and carrying capacity, are discussed in this section because communicating their significance is crucial to the effective use of scientific knowledge in the ICM process. A fourth notion relates to *uncertainty*, as it is also considered a critical element in communication between scientists and other parties in the ICM process. The relationship between these different notions is schematically shown in Figure 11.1.

Institutions

The key to both the effective communication and the use of science is to identify who, in reality, "manages" shoreline development. Because of the multiplicity of shoreline uses, management decisions based on the plethora of institutional arrangements will determine how the shoreline is developed. Discerning precisely which institutions exert effective change is an essential prerequisite to ensuring communication and effective use of scientific knowledge.

Shoreline development may be driven by institutions at both national (local, regional) and global scales. For instance, world trade conventions influence regional markets which, in turn, trigger local shoreline development for resource exploitation. Institutions exist at all scales of society, from local to international. Formal institutions include the constitutional and legislative arrangements of a nation, international legal regimes, and recognized property rights regimes. However, informal institutions, such as politico-commercial networks and traditional customary users of land, coasts, and fisheries, often exert the most influence on coastal resource use. Kiser and Ostrom (1982) and Ostrom (1990) identify three interacting levels of institutions:

1. Operational level, at which day-to-day decision making occurs.
2. Organizational level, where rules are made about how a resource should be used.
3. Constitutional level, where rules are set as to who is eligible to make decisions about the use of resources.

Formal and informal institutions can and do overlap in all levels of government and society. It has become an axiom that informal institutions dominate policy and resource allocation in the developing nations of the "South," while formal institutions dominate the political processes of the "North." This, however, is an oversimplification. A more accurate statement is that national institutions can be placed on a continuum in terms of the relative influence of formal and informal institutions in the process of government. In the process of resource allocation, there is a dynamic shift between the two over time in both developed and developing nations, but decisions affecting the use of shorelines will continue to be made by the interaction of private sector interests with formal planning agencies.

Much of the pressure on the shoreline results from the conflict between public and private property rights. Property rights are social institutions that have evolved as a means of enforcing claims to the right to use environmental resources (Edwards and Steins 1998). They can be classified by the levels of access exclusivity that they provide:

- Private property limits resource access to individuals or organizations, with enforceable powers to exclude other users. Rights can be traded or otherwise allocated to others.
- Public or state property arises where access rights for the public are held in trust by the state or other public body.
- Common property resources involve the sharing of rights of use within a defined group with co-equality of access by members of that group.
- Common pool resource regimes occur where access for use is shared between a number of identified individuals and/or groups but others are excluded.
- Open access resources may be exploited by any group or individual because there is no exclusion of use.

Use of resources occurs not only through formal (*de jure*) institutions or property rights, but also through assumed (*de facto*) property rights adding to the conflict potential. Individual resources may also be subject to varied and competing uses from multiple property rights. For example, a mudflat, swamp marsh, or mangrove forest may be owned by the state, contain moorings for fishing or recreational vessels (private property), and can be used by fishing communities (common property or pool resource). Therefore, scientists must clarify, and communicate to authorities, who is, in reality, managing the shoreline through its use and change.

Because shoreline development interferes with the offshore marine environment, it also needs to conform to international institutions. The UN Convention on the Law of the Sea (UNCLOS III, see Box 11.1), for example, spells out access rights to ports and communication cables which cannot be ignored in shoreline management. Of special relevance to the communication of scientific knowledge is the requirement delineated in Part 5 of UNCLOS III (see Box 11.1) that estuarine fish habitats be managed using the best scientific knowledge to conserve the fish stocks.

There is a multiplicity of institutions that operate at different spatial scales, from local to national and international, affecting the use of the shoreline. This further highlights the

Box 11.1 Exclusive Economic Zone (EEZ) and the Law of the Sea: Implications for Shoreline Development

The EEZ is intended to be a tool for the management of the coastal zone (Article 56). It gives the State control of an area up to 200 nautical miles from the 12-mile limit of the territorial sea in which the state has the right and duty to explore, exploit, conserve, and manage the zone. It also gives the state the right and duty to conduct marine research within the zone.

The rules governing the extent and the use of the EEZ are contained in Part 5 of the United Nations Convention on the Law of the Sea (UNCLOS III): The Convention aims to codify the rights and duties of states with respect to the use of the territorial seas, coasts, and high seas.

Article 61, The Conservation of All Living Resources:
This article states that the management of living resources in the EEZ must be done in a way that maintains or restores populations of harvested species, taking into account the best scientific practice.

All fish stocks that breed within rivers and estuaries along the shoreline, and shoreline habitats such as mangroves, should be taken into careful consideration in the planning of shoreline development. Complying with this requirement of the EEZ is now attracting considerable attention as fish stocks diminish globally.

The impact of Article 71 is to allow the national or regional coastal manager to suspend the "right of innocent passage of ships" along the coast in the interest of protecting vulnerable habitats and vulnerable developments such as tourist resorts. The only requirement is that shipping be given due notice of changed navigational regulations under Article 71. The danger of ship-borne damage to shorelines from oil spills is well known. Less well known but a greater problem is bilge and waste dumping from ships nearing ports. Although this practice is illegal (unless the territorial state agrees), the temptation for ships to offload ballast close to the port is enough to make it a serious problem, especially given the growth in "flag of convenience shipping."

Bacterial pollution from ballast and bilge dumping can and does affect fish stocks in estuaries and shoreline habitats. It becomes a health and economic problem for shoreline and coastal management as it enters the food chain. With ballast water and attached to ship hulls, transoceanic dispersal of alien species often has transformed estuarine ecosystems.

The powers granted to coastal states are intended for integrated coastal management and are essential for shoreline management. It is suggested that managers are made more aware of these powers to enable the protection of sensitive economic and ecological developments.

importance of identifying the key institutions of management. Additionally, the institutions that relate to the management of shoreline development are likely to vary between and within nations and to operate at different spatial scales.

Strengthening the role of science in the ICM process may require the creation of new institutions in which scientists are incorporated to cope with the multiple institutional pressures. The Great Barrier Reef is an example where informal institutions have been established as partnerships between users and scientists (Crossland and Kenchington, this volume; Box 11.2). This has enabled a more effective multiple use of the region.

Regional planning for shoreline development and conservation may be essential if use of the coastal zone/shoreline is to be optimized without damaging the robustness of ecosystems. Failure to do so may diminish their functionality and potential for use, as has been the case in East Anglia (Pethick, this volume; Box 11.3).

Temporal and Spatial Scales

The time-scale issue is well illustrated by the layer approach, a common methodology in geography which has been adopted, for example, in the recently published European Spatial Development Perspective (ESDP 1999). The layer model (Figure 11.2) is used to demonstrate the hierarchy of time scales in shoreline development. At the ground layer, geology, hydrology, and climate dynamics are factors to which ecological systems adjust. Shoreline infrastructures, developed gradually over the course of human intervention, result in shoreline architectures that are often unsustainable from a present perspective (Pethick; de Vries; Colijn and Reise, all this volume). Initial development of infrastructure determines the further development direction in the occupation layer for a long period and initiates feedback

Box 11.2 Example from the Great Barrier Reef

The Great Barrier Reef is managed as a large zoned, multiple-use area providing for conservation and reasonable use. The need for user-driven research complementing existing research was identified by the broad user community (government, management, industry) to give an improved scientific basis for management and regulatory decision making and for enhanced viability of reef-based industries (Crossland and Kenchington, this volume).

To assist in bringing science capabilities more closely to bear on management and industry issues in the Great Barrier Reef, a networked research center (CRC Reef) was established in 1993. The CRC Reef is a joint venture between reef users (more than 1100), reef managers (national and state governments), and research agencies (notably Australian Institute of Marine Science and James Cook University). The networked partnership collaborates to carry out research, training, and extension to enhance opportunities for sustainable use and to provide information to reef management. It is run by an independently chaired board with representatives of key users and research. Public and private sector funds bring together research and training skills (knowledge-providers) and ensure participation and ownership in identification of problems, setting of research goals, and the use of research outcomes by management and industry (knowledge-users). A strong communication program transfers research findings in a readily understood form to the users and the community, actively involving the public media.

The center has gained wide "ownership" by the research and user communities, has an extensive inventory of derived scientific findings being applied to management, policy, and commercial decisions, and recently gained an additional 7 years of public–private sector funding.

Box 11.3 Example from the East Anglian Coast

The shoreline of Eastern England (Pethick, this volume) exemplifies the problems of antecedent conditions as a context to current shoreline development. Centuries of "mismanagement," based on a misunderstanding of the dynamics of the shoreline and based on ignorance of the need to conserve the shoreline so as to protect the land (rather than allowing it to be developed, creating hazards), mean that policies that are currently being pursued are destined to fail. This is because the sea needs "space" in which to move, and the shoreline is part of this space that buffers this sea–land interaction. Without this buffering, because the space has been used by humans, hazards increase. Only the very recent attempts to pursue managed retreat are the signs that this misunderstanding is being addressed and more sustainable solutions being developed. Inadequate understanding of the complexity and dynamism of the coast led to, what now may be seen as, mismanagement (although, no doubt, the decisions were honestly made against the information available at the time). The results of this have seriously foreclosed options for the shoreline management strategies that can be pursued today and into the future.

loops strengthening the initial infrastructural patterns. This imposes increasing constraints on the ground layer. The resulting ground layer changes, in turn, affect and compromise opportunities for spatial occupation and resource exploitation in the future, as discussed by Pethick (this volume). Hence, maintaining and elaborating shoreline infrastructures may entail ever-increasing economic burdens.

The solution to the problems caused by the mismatch between natural processes and the infrastructure that has developed over centuries and up to millennia requires the integration of science and management devoted to increased sustainability. The precautionary principle may provide the best guide by which science and management respond to changes over large temporal and spatial scales. Similarly, organizations concerned with management at different temporal and spatial scales along the shoreline need to integrate their efforts over regions and over longer time horizons. The task for management and science is to communicate and

Shoreline System

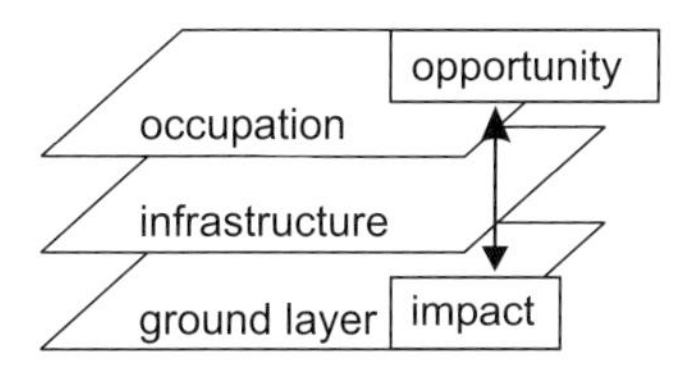

Figure 11.2 The layer approach, identifying three physical layers: The *ground layer* (soil, (ground)water, ecosystems) represents the natural resources that ultimately enable, as well as constrain, land use and resource exploitation. The intrinsic time scale of many processes in this layer is very long, from tens of years for stock renewal of commercial fish species to 100–1000 years for geomorphologic processes that shape the coastline. This implies that recovery from (potentially rapid) deterioration caused by mismanagement, misuse, or overexploitation is a slow process. The *infrastructure layer* (e.g., roads, railroads, waterways, urban structures, coastal defence structures) represents primarily the results of public investments. The planning horizon, and consequently the time scale of this layer, is of the order of 10–50 years. The *occupation layer* (land use and land cover) represents the result of private investments. The rationale of short-term benefit determines the time scale of this layer on a scale of 1–10 years.

discuss development goals for small-scale as well as large-scale shoreline areas, which are acceptable to the population in the short term as well as ecologically sustainable for the coastal environment in the longer term.

Natural and social sciences should work together to define appropriate scales that reflect the management regimes related to the expected extent of management and over which communication is necessary. Wide-ranging networks of science already in existence may facilitate the awareness of local stakeholders and managers for the long-term effects, in both space and time, of their decisions. For sustainable development, it is crucial that institutions are capable of responding adequately to the time scale of changes that occur in the natural environmental system. This is discussed below in more detail.

Response–delay Time

To protect interests engaged in shoreline development and to safeguard the carrying capacity of the coastal system, organizations and individuals will try to respond when confronted with adverse trends. Often such trends originate from human interventions. Scientific knowledge can be invoked to find the most adequate response. It is important to realize that the effective use of science depends on the timing of the response. The right response too late will not work. Like companies that compete in a free market economy, the success of ICM strongly depends on the response–delay time, i.e., the time it takes to recognize trends in the evolution of the coastal system (both socioeconomic and natural components), to develop a response strategy, and to implement it. If this response delay exceeds a critical value, the result may be a chaotic coastal development, a situation that often occurs in practice. Suppose, for instance, that exploitation benefits are still pushing ongoing shoreline development while the investment return rate has already started to decrease; as the cost curve (for maintenance, operation, restoration of environmental impacts) lags behind the benefit curve, a situation of negative net benefits will occur. This situation quickly deteriorates as costs are still increasing while revenues are decreasing (Figure 11.3).

If the response–delay time lag is too long, then the result will be a catastrophic collapse of coastal economy or natural systems. Examples include the collapse of fisheries due to increases in the fishery fleet while fish stocks are being depleted and the collapse of tourism development due to increased resort development while the coastal landscape is becoming less attractive to tourists (see Box 11.4). Avoiding crises in shoreline development requires, therefore, early awareness and a timely response to changes in the coastal system. Accumulated negative impacts on the ground layer (e.g., environmental degradation) require particular attention, because of the long recovery time (see Boxes 11.6 and 11.7). Hence, there is a need for adaptive shoreline development to reduce the time over which policy responses to emerging problems are made. A learning-by-doing policy works only if learning is fast enough. Otherwise, shoreline development will continue to be ruled by crisis management.

We note that the phenomenon described in Figure 11.3 is generally too simplistic. In reality, a multitude of feedbacks is often mobilized when crisis situations arise. However, the underlying principle seems to be valid. (Examples of delayed response situations have been discussed in literature for a long time; see, e.g., Hutchinson [1948] and Cunningham [1954].) It stresses the importance of adjusting organizational time scales to the dynamics of the natural resources being exploited.

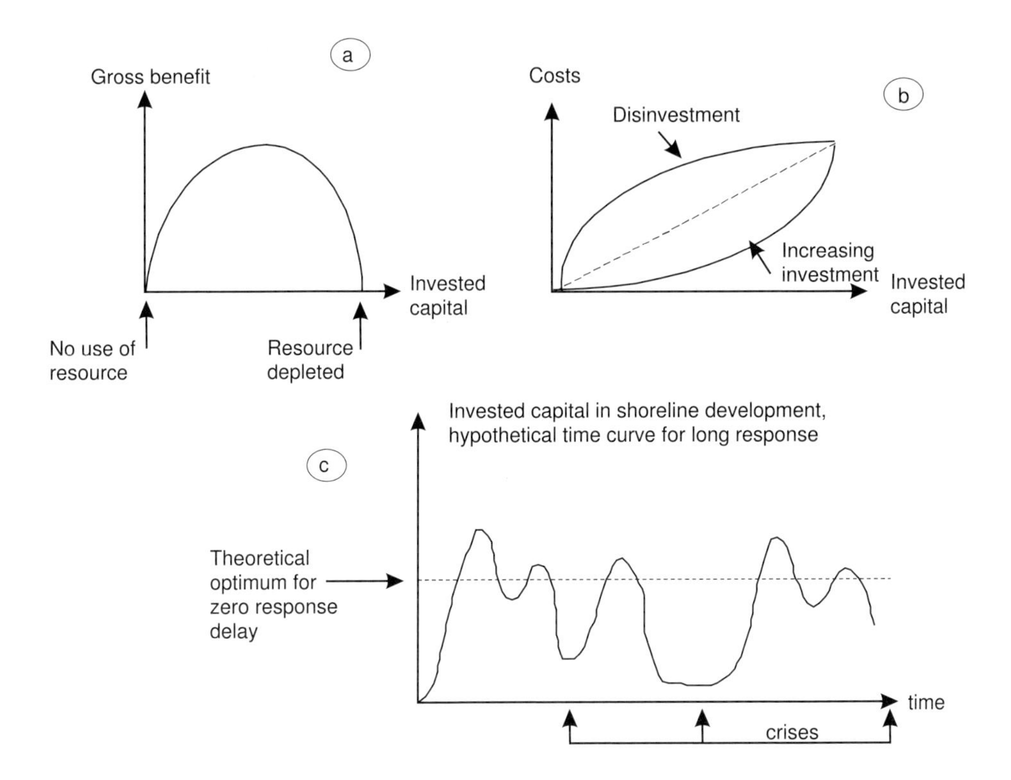

Figure 11.3 The impact of the response–delay time on shoreline development.

(a) A typical curve of gross benefits as a function of capital invested in shoreline development: Zero investment means that coastal resources are not used and do not yield any economic benefit. With increasing investment and increasing exploitation pressure, the coastal resource will become depleted; the gross benefit therefore reaches a maximum, decreasing thereafter until it is completely exhausted.

(b) The exploitation cost curve would increase more or less linearly with the amount of capital invested if there is no response delay. In practice, the cost curve often lags behind (aging of infrastructure, repayment of loans, sanitation of polluted soil), leading to the hysteresis response curve.

(c) The foregoing means that there is no stable optimum for shoreline development; when the theoretical net benefit for zero response time reaches its optimum, the net benefit for nonzero response time is still increasing. This pushes the development beyond its optimum, eventually into a regime where gross benefits decrease while costs increase. Shoreline development may then collapse and start fluctuating in a chaotic manner.

Thus research on time-scale issues should address the following questions:

1. What is the critical response–delay time for a sustainable shoreline development?
2. What adequate early warning indicators of change can be identified?
3. What are the most important time-lag factors in the cause–response chain (for instance, harmonization of organizational and natural time scales)?

Box 11.4 The Impact of Tourism on the Spanish Coastline

As in other Mediterranean countries, Spain has based a large part of its economy on the development of a robust tourism industry associated mainly with coastal areas. This shoreline development allowed economic growth and social modernization without recognition of the future consequences. Unplanned growth and overdevelopment, inappropriate development of secondary residences (often on vulnerable natural areas), unsustainable use of natural resources, and abusive practices of enrichment based on particular interest rather than on public interest and the appropriation of public goods, were widespread. The rapid spread of tourism and the coastal policies developed during the first decades of this process were the origin of various problems in the Spanish littoral. Some of the problems were solved after Spain was incorporated into the European Community. What was learned in the process is that if tourism is to contribute to the sustainable development of coastal areas, then it must be economically viable, ecologically sensitive, and socially and culturally appropriate. The dissemination of this Spanish experience should help government decision makers and local people to work in partnership with the tourism industry to promote sustainable tourism practices worldwide.

Carrying Capacity

To measure the carrying capacity of a particular territory, the desired quality of the used system must be defined. In that way, carrying capacity will depend on the acceptance of the "state" of the environment that fulfils the expectations of the different stakeholders included in a negotiation process. In this negotiation process, scenario projections should play a role. The carrying capacity will then depend on the development accepted for a particular territory and will be supported by strategic decisions in the decision-making process.

From an ecosystem perspective, the concept of carrying capacity is defined as the number of individuals of a population that a territory can maintain without reducing its ability to develop such populations in their functional, desirable levels (Ehrlich et al. 1989). Carrying capacity can also be defined from the perspective of human utilization of natural resources. Then it designates the potential for human use offered by natural resources (all kind of uses may be considered, including noneconomic uses such as amenity). Preservation of this potential is often referred to as sustainable use. A related concept is "resilience." This is the capability of the system (in our case the shoreline system, including both natural resources and human infrastructure) to absorb disturbances and gradual changes without undergoing profound restructuring into a different state. Both human interventions and natural processes may alter the carrying capacity of the shoreline system that will be thereafter characterized by a new level of carrying capacity. Sea-level changes, for instance, have been responsible over the geological times for profound alterations in shore areas, resulting in the evolution of new habitats associated with new biota. Natural trends are usually characterized by smooth slopes, while anthropogenic processes have been propelled by the exponential human population growth observed particularly during the 20th century. This is especially true for the occupation of shore areas mostly due to migration to coastal cities. Here, occupation predates adequate planning and management systems, which could conserve adequate levels of shoreline carrying capacity as has happened, for example, to the Spanish coastline (Sardá, this volume; Box 11.4).

From the ICM perspective, carrying capacity is generally not understood as an intrinsic property of the system, but rather as a characteristic that is affected by human interventions. Indeed, the notion of carrying capacity raises the question: carrying capacity of what? The coastal zone may carry different ecosystem types and it may carry different goods and services. However, in general, a choice has to be made: different types of ecosystems are often mutually exclusive and the same holds for many goods and services. For example, in the original tidal basins along the Dutch coast, very distinct ecosystems have developed after (partial) closure, depending on the remaining degree of marine influence (de Vries, this volume). The carrying capacity is therefore related to the development option that has been chosen. Ideally, this development option results from a negotiation process in which all coastal stakeholders have been involved. The ICM process must be sensitive to the dynamics of society and respond to changes in environmental perception and valuation. This requires adaptability of shoreline development to different options of carrying capacity, including options corresponding to the restoration of past development stages. Options of shoreline development in general affect the characteristics of the coastal zone ground layer and infrastructure layer (Figure 11.2). This puts restrictions on adaptability, because of the slowness of ground-layer and infrastructure-layer dynamics. Scientists who are familiar with ground-layer and infrastructure-layer dynamics of the coastal system can therefore play an important role in ICM, by providing information on consequences of different shoreline development options for carrying capacity and advice on issues related to time scales (adaptability) and risks. Boesch (this volume) addresses the effectiveness of scientific input to the ICM process and discusses how this effectiveness can be improved.

Those dealing with coastal zone issues need to respond to new choices, even if this means returning to a former level of carrying capacity. There are several examples of such decisions, for instance, the practice of restricting fisheries during reproduction periods to ensure reestablishment of stocks. The incorporation of scientific knowledge at an early stage of defining the desired carrying capacity underpins use of appropriate practices and sustainability of resources, services, and quality of life.

In developing nations, the trend to empower communities that depend closely on the shoreline for their existence has resulted in management regimes that often show a better fit between ecosystem health and human activity than other models of management (Talaue-McManus, this volume). To a considerable extent, traditional knowledge is preferred to scientific assessments of carrying capacity for fisheries. The exception may be in the restoration of habitats, where both local communities and national governmental institutions may seek scientific advice. However, the fact that the local community may lack the legitimacy of formal institutions means that shoreline management by local communities may be easily overridden by formal national and international regimes.

Environmental perturbations of coastal areas related to population density and energy dissipation can be measured in terms of waste production. Part of the waste flow is released into the environment, which has an assimilative capacity forming one of the major economic functions of the system (Pearce and Turner 1990). Assimilative capacity is an integral part of the carrying capacity and is a finite resource that, when exhausted, impairs the economic functioning of the environment. Overexploitation of environmental resources can result in serious losses for users. In the Aral Sea, for example, water withdrawal for irrigation surpassed the natural replenishing capacity and resulted in a sharp decrease in coastal water area

Box 11.5 Changes of the Carrying Capacity of the Ecosystem; Example from the North Sea

Anthropogenic activities in the North Sea have transformed the shoreline considerably. At several locations, an extended transition zone between the land and sea has been replaced by a fixed dikeline, resulting in a loss of valuable habitats and changed areal extent of habitat types. Consequently, ecosystem function, such as the production, accumulation, and remineralization of organic matter, has changed. New types of habitats were created, such as artificial rocky shores of harbors and coastal defence structures within an otherwise sedimentary environment. By introducing new species from distant coasts, the ecosystem carrying capacity was affected. In the North Sea, the introduction of the cordgrass *Spartina anglica* has created a new habitat at the seaward border of salt marshes. At the upper end of the food web, hunting and fishing may extinguish mammal, bird, and fish species with cascading effects along the food chains. Thus, in the course of human occupation of the shoreline, the carrying capacity of the ecosystem has undergone a history of changes.

with serious socioeconomic consequences. In the North Sea (Boxes 11.3 and 11.5), shore protection measures led to loss of valuable ecological habitats.

Much work on carrying capacity relates to new development in shoreline locations where anthropogenic impact may be substantial and detrimental. Parts of the coast have already been significantly changed by past development decisions (see Pethick, this volume) and they may have little environmental quality to retain, though may provide opportunities for rehabilitation. Coastal policies should ensure that best use is made of previously developed or "spoilt" land before undeveloped land is used, particularly in regions exhibiting great richness of natural quality (Taussik 1999; Box 11.6).

Maintenance of appropriate economic shoreline functioning under the current scenario of growing population density requires increasing investments in effective pollution abatement and occupation planning. In developing countries suffering from fragile economies, only a few actions can be taken before there is a drastic reduction of environmental capital (Box 11.7).

The natural and social dynamics underlying carrying capacity require further investigation. Human interference rarely produces linear and straightforward reactions or interrelations. Synergistic and antagonistic reactions and different stages of ecosystem development

Box 11.6 The Reuse of Spoilt Land at the Coast

Excellent examples of returning previously developed land to effective use are found in the wealth of historic dock regeneration areas. Rendered redundant by technological change of the 1960s and 1970s, many dock areas (including those for defense purposes) changed from being economic foci to being derelict. Much of the land was heavily contaminated. The opportunities presented by these sites have resulted in a veritable explosion of regeneration programs throughout Europe and the rest of the world. European examples include Barcelona, Gothenburg, Genoa, Cardiff, and London. The scale of such programs offers enormous capacity for the location of shoreline development and can relieve pressure on the undeveloped coast. It can, also, improve and enhance the environmental quality of the area, not just in terms of development and "hard end uses" but, also, in terms of the natural environment.

Box 11.7 Destruction of the Guanabara Bay Environment

Guanabara Bay, located at the foot of Sugar Loaf in Rio de Janeiro (Brazil), is a remarkable example of an extremely altered system. About 7×10^6 inhabitants live in the drainage basin with a ratio of population density to run-off superior to 33,000 inhabitants m^{-3} s. The bay receives 20 $m^3 s^{-1}$ of untreated sewage, industrial waste from several industries, oil, litter, and toxicants related to shipping activities (JICA 1994). Once a special site for fisheries, reproduction of valuable species, tourism, and recreation the bay is now heavily contaminated with fecal bacteria, heavy metals, oil, TBT, and other chemicals, and it exhibits strong siltation derived from deforestation and soil denudation.

and maturity play an important role in determining the degree and critical level of assimilative capacity in systems simultaneously receiving physical (e.g., land development), chemical, and biological inputs (e.g., industrial wastes, domestic sewage). These synergies make it difficult to forecast how systems will adapt to changes in their carrying capacity. Guanabara Bay provides an interesting example of synergism between contaminants. There, high levels of copper contamination should impair primary production in the bay; however, due to sewage-derived organic material, most copper is rendered inactive to biota (van den Berg and Rebello 1986) and net primary production reaches 2 g C m^{-2} d^{-1} (Rebello et al. 1988).

Systems that are especially sensitive to changes and which play important roles in providing services, for example, coral reefs and mangroves (Boxes 11.2 and 11.3), must be treated with the highest concern. Critical habitats require special attention; however, one should not forget other parts of the shoreline where the best environmental practices must always be applied.

Communicating Uncertainty

Environmental processes are subject to natural forces whose timing and intensity are difficult or impossible to forecast precisely. Scientific development has substantially improved the capacity of forecasting and, by applying models, it has become possible in several cases to distinguish between anthropogenic and natural variability. The large spectrum of possible combinations of causes and effects imposes variable, and often substantial, degrees of uncertainty on the outcomes.

A planner/manager needs to make decisions based as much as possible on hard proven evidence. In many cases, the pressures are such that decision making must go forward in the absence of full information. In such circumstances certain rules, known as precautionary principles, become important (Cicin-Sain and Knecht 1998):

- not to make decisions that have irreversible consequences,
- not to make decisions that could seriously threaten the resource base over the long term,
- not to make decisions that could foreclose options for future generations to utilize coastal and ocean resources.

If the science involved in this decision making is uncertain, it is likely to be ignored. However, uncertainty is an essential element of objective knowledge transfer. In media communication, journalists are trained to present two "sides" to every issue. They interview scientists as they would interview politicians, in an attempt to understand and underscore the differences. This highlights the extent of uncertainty.

Much of the management of shoreline development is directed towards minimizing and mitigating risk of natural processes on human activity and of human activity on natural systems. Risk is now considered to be one of the organizational paradigms of society (Beck 1992). Traditionally, certain areas of risk mitigation associated with the shoreline have been undertaken by public agencies (Pethick, this volume). Global moves towards more participatory systems of governance have repercussions on the responsibilities for risk. The basis of decision making changes from representative to participatory, thereby including the wider public in decisions involving common resources. In situations where this involves the transfer of some of the responsibility for decisions related to risk to the wider community, communication of uncertainty must be actively extended to the general public. This further emphasizes the importance of the communication of sound science to shoreline development management decision making.

Uncertainty creates particular problems for communication because of the variety of audiences. A communication strategy must, therefore, establish both the extent of agreed information and the way that the issue of uncertainty can best be conveyed.

COMMUNICATION BETWEEN SCIENCE AND MANAGEMENT: THE NEED FOR, AND NATURE OF, COMMUNICATION

The Role of Communication in the ICM Process

In the previous section, we raised a number of issues that coastal scientists feel are particularly important for ICM: *the role of institutions, time scales, carrying capacity,* and *uncertainty.* Conveying an understanding of these key concepts to managers and ensuring that they are integrated in decision making on coastal policy, as well as ensuring that best use is made of the scientific information that is available, means that science must be an equal player in the process of managing shoreline development. Effective communication of scientific information is fundamental to this relationship. Ineffective, or nonexistent, communication of scientific information will reduce the attention paid to scientific understanding and will undermine the quality of the management of shoreline development.

Communication has several purposes. It can inform or influence. It can be used to identify differences in understanding or alternative interpretations or to elicit conflicting views. It can also empower (Talaue-McManus, this volume), raise general awareness, and educate. Depending on its purpose, it may be one-way (e.g., informing), in which case the parties involved are either "givers" or "receivers," or it may be two-way, where the process of information exchange generates mutual benefit (see Table 11.1). It is important that the purpose is clearly defined, whatever it is. Purposeless dissemination of information has little benefit.

The need for integration in the management of shoreline development between sectors, between disciplines, and across the land–sea interface means that the range of material to be communicated is extensive. It originates in both the scientific and decision-making fields. Background papers for this workshop (this volume) drew on case studies on a wide range of natural sciences to management issues, such as flood risk (Meybeck, this volume),

Table 11.1 Examples of one- and two-way communication in management.

Nature of Communication	Giving Information	Receiving Information
One-way	Distributing reports	Reading reports
	Giving a lecture	Sitting in a lecture
	Publishing newspaper articles or advertisements	Reading newspaper articles or advertisements
	Distributing brochures	Reading brochures
	Broadcasting (radio or television)	Listening to radio or watching television
	Sending letter	Reading letter
	Publishing web pages	Looking at web pages
	Publishing an academic paper	Reading an academic paper
Two-way	Exchanging views at meetings; round table discussion	
	Discussion at an exhibition	
	Electronic news and discussion groups	
	Consultation	
	Conferences, seminars, and workshops	
	Certain types of questionnaire surveys	

maintenance of fish stocks (O'Toole, this volume), or pollution (Elmgren and Larsson as well as Mee, both this volume).

Social and natural sciences are of equal importance in shoreline management. Social science contributes to understanding human behavior as related to, for instance, policy generation and implementation, land-use development, and recreation activity. Understanding the interaction between human activity and the natural world necessitates communication between different sectors of the scientific community and between different disciplines. The effective transfer of knowledge from transdisciplinary research and the successful application of interdisciplinary techniques will improve decision making and, thereby, improve management of shoreline development (see also Boesch, Olsen, McGlade, all this volume). This is also illustrated by the management of the Great Barrier Reef (Box 11.2) as well as by the example of Chichester Harbour Conservancy, which shows how transdisciplinary and science–management integration may be achieved (Box 11.8). On the other hand, Talaue-McManus (this volume) highlights management failures that have resulted from poor integration of scientific information in the management process.

Whether science is enabled to become a party in decision making will depend on the relevant constitutional level of institutions. Importantly, effective use of scientific information should be made from the earliest stage of the management process and continued throughout. This requires an ongoing involvement of scientists (though not necessarily the same scientists) throughout the process and depends on effective and appropriate communication of scientific information to managers and of management needs to scientists. Examples of problems arising when this does not happen (Pethick; Sardá; Talaue-McManus, all this

Box 11.8 Integration of Science into the Management of Chichester Harbour

Chichester Harbour is a large estuary on the coast of central Southern England. It is managed by Chichester Harbour Conservancy, an unusual coastal organization in British coastal management that has extensive management powers including for the environment. Its philosophy is that management should be based on sound science and it has established close links with a number of higher education establishments in the region. It organizes its own research agenda. It has established a program of seminars bringing together scientists working on Harbour issues and coastal managers, including members of the Conservancy Board and local politicians. This has been very successful in broadening knowledge in the scientific community and in improving the scientific awareness of decision makers, thereby improving the basis of decision making.

volume) highlight the importance of ensuring that access to science is facilitated and that barriers are removed.

Impediments to Effective Communication

Over recent years, the volume of scientific research related to the shoreline has increased substantially so that a considerable body of knowledge currently exists. There is, however, concern about the general availability of this information and about its particular availability to those managing the shoreline. It is widely considered that management decisions make insufficient use of scientific information and that they are insufficiently based on scientific understanding. Although there is increasing recognition that there must be close interaction between scientists and managers, as indicated above, it is felt that there are a number of problems, or impediments, which hamper communication and exchange. Some of these relate to styles of scientific operation; others to institutions related to scientific research. These are outlined before proposals for improved communication are made in the next section.

- Scientists may not recognize that involvement in ICM is a task for science.
- Contributing to ICM may not confer the scientific status sought by scientists, making them less willing to become involved. This is exacerbated by the need for transdisciplinary science and techniques in ICM. The scientific community rates such work less highly than research in a single discipline (Box 11.9).
- The scientific community is governed by its own set of standards. For example, statements require empirical evidence and results will not be disseminated within, or beyond, the scientific community until a high degree of certainty is achieved.
- Managers need to base their decision making on clear goals, soundly grounded in relevant and appropriate knowledge. The difficulty of science to produce clear messages is an obstacle to its use in ICM. This may reflect both the complexity of, and uncertainty about, the real world in which ICM operates. This reduces the situations when it can be said that something is true or false and increases the importance of conveying uncertainty without sending conflicting messages about the environment (see section on **Communicating Uncertainty**)
- The scientific world is characterized by antagonistic styles of operation where statements are constantly challenged and new arguments proposed. Facts supported by empirical evidence must be capable of being replicated by other scientists. Such codes

Box 11.9 The Research Assessment Exercise in England

One way that scientific status is recognized in England is through the Research Assessment Exercise (RAE). This centrally administered monitoring system aims to classify universities by the quality of research undertaken. Criteria include research funding, academic papers published, and numbers of doctoral students. Departments and research units are required to submit returns of their research in relation to one of a number of Units of Assessment — each unit includes clearly recognized and established disciplines. Academic papers are graded by the quality of the publication; international, peer-reviewed journals central to a discipline obtain the highest rankings. Transdisciplinary research is immediately at a disadvantage because it belongs under no clear discipline, or Unit of Assessment. Those who undertake transdisciplinary research, even at the highest levels, may be excluded from the RAE because colleagues are concerned that the nature of their research will dilute the research ethos of the department or group. This is an example of how institutional arrangements, related to the evaluation of scientific research, can undermine the quality and status of transdisciplinary research required for ICM.

are vital to the quality and existence of the scientific community but can be an obstacle when communicating with nonscientists.

- Information to be transferred in the ICM process ranges from original material that is very specialized and costly to acquire, to processed or generally available material. It includes ideas and opinions as well as facts and figures. The modes of operation of the different "players" present constraints on how information may be conveyed or transferred. Millard and Sayers (1999) suggest that many coastal operators have difficulties obtaining necessary information, not because information is not available, but because it is inaccessible. Barriers to effective use of coastal data include when it is "privatized" and subject to restricted access.
- Subject-specific language may inhibit understanding and, therefore, use of science.
- Many people fear that they will not be able to understand science. While this may be more psychological than real, it reduces their receptiveness to new understanding.
- Scientists may have inadequate understanding of management needs. Little attention has been given to the need for managers/policy makers to communicate with scientists about the goals and objectives of shoreline communities and resource users or about their requirements and requests for assistance and support.
- Policy makers may lack sufficient awareness of the complexity of coastal issues to identify their need for inter- and transdisciplinary research.
- Many scientists are not prepared to listen.
- Knowledge is power. This may undermine the general availability of knowledge.

Improving Communication

Information transfer should not originate solely in the scientific field. Matching scientific developments more closely to the needs of management will ensure its greater use and is more likely to provide a vehicle by which science is drawn into the management process. It requires that policy makers develop sufficient awareness of coastal issues to recognize their complexity and the subsequent need for inter- and transdisciplinary research. This awareness will allow them to create appropriate groups and mixes of scientists from both the social and natural sciences to contribute to the resolution of coastal management problems. Over the longer term, the importance of such work could be reflected in revised institutional arrangements for science which give its higher status.

Management of shoreline development generates complex communication requirements for inter-sectoral, inter- and transdisciplinary and trans-land-sea integration. When so many players from such diverse backgrounds are involved, it is vital that all parties communicate their modes of operation and the rule frameworks (institutions) by which they must conduct their business to other players. Failure to do this may cause false expectation of how other players are likely to respond to problems and issues.

The integration of different types of players is a prerequisite for successful ICM. However, the very complexity and plurality of the parties involved may hamper progress because the ways of thinking and sets of values, or mind-sets, of individuals and groups are so deeply embedded. Developing awareness of and communicating in ways that "fit" the mind-sets and systems of operation of others may result in mind-sets being changed. This is part of the process of cultural development. For example, the different ways of thinking of scientists and decision makers exemplify the two dimensions of knowledge creation distinguished by Nonaka and Takeuchi (1995): ontological and epistemological (see Figure 11.4). The ontological dimension requires that knowledge created by individuals is shared within and between organizations before it is utilized whereas the epistemological traverses the tacit to explicit forms of knowledge and vice versa. The distinction and exchange between these two forms of knowledge are central to understanding and organizing the knowledge creation process. The knowledge of management is shown as tacit, personal, and context-specific. It is, therefore, difficult to communicate, whereas the explicit or "coded" knowledge of science, like procedures, handbooks, and models, can be transmitted much more easily. Effective communication of science could assist in making the decision-making realm more objective and "less individual" as explicit knowledge becomes internalized by decision makers. At the same time, increased emphasis on the needs of managers for scientific research could make science more relevant and responsive to social goals.

Partnership is an example of working together that can change the way people think. In the U.K., it has been described as "the orthodoxy of the 1990s." Bailey et al. (1995) define a

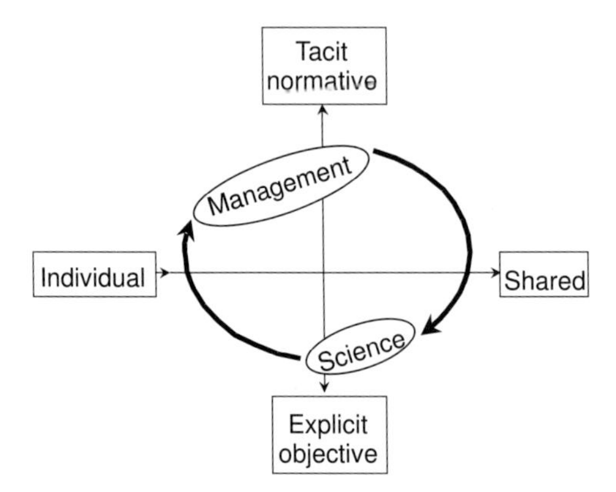

Figure 11.4 Creating and assembling knowledge, according to Nonaka and Takeuchi (1995).

partnership as "a coalition of interests drawn from more than one sector in order to prepare and oversee an agreed strategy." Partnership arrangements as communication mechanisms offer a number of benefits. While they may be created to synergize or access resources or to generate policy, of greatest interest in this context is the perception of partners that partnerships provide opportunities to change the ways of thinking of other partners. They may hope to change the views of others while not changing their own views (one-way transformation) or they may accept that the changes of view will be mutual (mutual transformation). Integrating science and/or science and management in shoreline development through partnerships, therefore, offers opportunities for changing mind-sets or ways of thinking.

Millard and Sayers (1999) propose a number of policies and mechanisms that can be adopted to improve the use of coastal data (Box 11.10). Within the scientific community, scientists from various disciplines — each with its own set of terminology and mode of

Box 11.10 Maximizing the Use and Exchange of Coastal Data

The publication by Millard and Sayers (1999) for the Construction Industry Research and Information Association (CIRIA) was produced because it was recognized that a considerable amount of existing coastal data is either unknown or inaccessible. Such data can thus make no contribution to effective management of coastal resources. The report provides a guide to best practice aimed at improving the management of data by those who operate in the coastal zone. It sets out ways in which data suppliers can provide more accessible data and how data users can best utilize existing data sets. It provides information on mechanisms that will help to achieve better exchange and consistency of data handling.

The five principles of good coastal data management (CIRIA) require that all individuals, organizations, and government institutions should:

1. Recognize, understand, and describe all data used, needed, and available.
   - Appraise data requirements to satisfy information needs.
   - Recognize all organizations as potential data suppliers or data customers.
   - Communicate data requirements and data availability with other organizations.
   - Ensure all data are accompanied by appropriate metadata.
2. Understand legal issues and execute responsibilities related to data management.
   - Ensure conditions for the use and exchange of data are clearly defined between customer and supplier.
   - Ensure all staff in an organization operate a "duty of care" towards the management of data.
3. Identify and specify organization processes and procedures.
   - Document and describe all organization procedures associated with the data-processing chain relevant to a specific organization.
   - Identify, specify, and implement organization procedures that incorporate management of data life cycles.
4. Identify and implement appropriate technologies for data.
   - Identify technologies that are compatible with data suppliers and data customers.
   - Monitor changes in data-transfer and data-storage technologies and apply those appropriate to business processes.
   - Appraise the level of security and protection attributed to data storage.
5. Audit and monitor the processes for data use and exchange.
   - Identify indicators to monitor the success of data exchange.
   - Review and monitor.

operation — must be able to communicate with each other, for example, to work effectively in multidisciplinary teams. In addition, scientists must also be able to communicate with various policy-making bodies, each of which also utilizes its own terminology and which operates on a range of scales and modes. Both groups, in turn, must be able to interact with the stakeholder community of NGOs, interest groups and individuals, business and "individual" coastal managers, i.e., those making individual decisions about development in the shoreline area. Some of these will also be providers of scientific information. This relationship can function at a number of levels: a town, a bay, an estuary, a delta, or a regional sea. At any time, communication between any pair of these players will be influenced by each party's ability to receive and digest information. The means and form (e.g., language) of communication must be selected with those abilities in mind. Consider, for example, communicating ideas on coastal defence in East Anglia (Box 11.3; also Pethick, this volume) to politicians, the general public, and government officers. Very different means and language must be used for each group. Mechanisms that will assist in improving the dissemination of ideas and concepts include:

- use of transdisciplinary teams,
- wide and cross representation at meetings,
- use of devices that enable individuals to operate in environments of others,
- development of networking arrangements of different types, and
- attempts to bridge "language" problems by using particular communication skills.

The range of scientific representation on transdisciplinary teams depends on the context of any project; however, both natural and social scientists should be involved. This will improve scientific information in the shorter term and enable an understanding of the *modus operandi* of other disciplines, which will be of great benefit for the longer term. The value of such inter- and transdisciplinary teams is demonstrated by experience on the Great Barrier Reef (Crossland and Kenchington, this volume; Box 11.2).

Communication Mechanisms and Strategies

Above, we emphasized the value of establishing a communication strategy at the onset of any scientific or coastal management project. Such a strategy should establish the goal/purpose of communication and develop targets and means of communication in the context of each project. It should incorporate quality control (e.g., how the pedigree of the scientific information is to be recognized and protected), resource implications in terms of time and cost, and evaluation mechanisms to ensure that communication is effective.

Communication about shoreline development is required at a range of scales, from local to international. The players, and the means of communication, will vary according to the circumstances. Opportunities to involve different mechanisms for enhancing the communication of science in a range of contexts are shown in Table 11.2. The range of mechanisms available, and the range of contexts in which information transfer must take effect, reinforces the importance of incorporating communication strategies at the inception of any shoreline development project.

Table 11.2 Application of mechanisms for enhancing the communication of science in different contexts.

Mechanisms for Enhancing Communication	Examples of Contexts Requiring Improved Communication				
	Within the scientific community	Between science and management	National policy level	Local project level	Wider community
Integrated scientific teams	√		√	√	
Partnership arrangements		√		√	√
Wide representation at meetings	√	√	√	√	√
Intersectoral/disciplinary brokering arrangements	√	√	√	√	
Intersectoral/disciplinary mentoring systems	√	√		√	
Role exchange/shadowing across sectors/disciplines	√	√	√	√	
Networks, forums, groups	√	√	√	√	√
Using "translators"		√	√	√	√
Using professional communicators			√	√	√
Face to face discussion	√	√	√	√	√
Use of World Wide Web	√	√		√	√

Two contexts, in which enhanced communication and use of knowledge must be considered, are:

- within the scientific community,
- between scientists and shoreline managers.

The first is not without problems because scientists need to begin the process of agreeing before they take their messages outside (see section on Impediments to Effective Communication). The second will operate at the levels of the ICM "project" (a bay, a town, an estuary, a coastal cell) and, also, at higher scales, both geographically and politically (regional scale analysis; projects and plans within government; between government and the private sector).

Mechanisms to enhance communication at ICM project levels can involve several types of activity:

- working consistently in integrated teams;
- talking together and using other means to promote mutual understanding;
- sharing experiences and resources (e.g., data);

- mentoring, or similar arrangements, whereby individuals are assisted in their work by others from different "sides" (i.e., shoreline managers are mentored by scientists, or vice versa);
- "brokering" arrangements between groups, similar to the above but using professional intermediaries (e.g., communication or science policy specialists);
- role exchange/shadowing across groups so that all involved understand the tasks of others, and the constraints under which they are working;
- newsletters, networks and other information, dissemination and exchange mechanisms.

The list of mechanisms can be elaborated by reference to case studies cited in the boxes of this paper and by other contributions in this volume. Mechanisms selected should be those that provide the best fit to the needs of a particular ICM or shoreline development context. This will vary between different types of shorelines and different shoreline resources at stake (e.g., fisheries, raw minerals, recreation, nature conservation, wetlands). Research is required to establish the optimal approach and mix of communication mechanisms in different contexts.

Communication must be undertaken in ways that will be understood by the receivers, rather than in the most convenient and easy way for the givers. The difficulties to nonscientists of understanding scientific information, and to scientists in communicating in "plain language," suggest that intermediaries may be required, particularly to convey scientific knowledge to managers and the wider community. Any communication strategy should identify the means of communication. This should make the most effective possible use of "bridge" personnel (Taussik and Singleton, in prep.; see Box 11.11), i.e., people already skilled in interorganizational, intersectoral, or interdisciplinary communication. These may be scientists with particular skills in communicating to a wider audience or individuals who operate at the interface between different groups, i.e., who act as "bridges" between these groups and who become particularly adept at communicating between the different groups. It may also be necessary to involve professionals (e.g., journalists and other media specialists), who have particular faculties in "translating" from scientific to generally understood language, in improving the wider communication of scientific knowledge. Advantage should be taken of any channels and methods that they have developed to facilitate communication between science and management, within the scientific community, or with the wider community.

Another way of understanding the operation of a particular group or discipline is to work with that group or in that discipline. In many spheres of activity, this has been successfully

Box 11.11 Importance of Bridges in Communicating between Different Groups

A network may be defined as a group or system resulting from the interlinkage and interaction of its members. In coastal management, many networks, groups, and forums exist, comprising people from different organizations, sectors, and/or disciplines. In this context, the interorganization, sector or discipline linkers, known as "bridges," have a particularly important role to play. They should be responsible for transmitting information between, and disseminating information from, organizations, sectors, and/or disciplines. Such information may be scientific or managerial; however, to be effective, "bridges" need to "translate" information from one group so that it can be understood by another group.

achieved by the exchange of individual roles, or by work shadowing (accompanying a person in all aspects of their position in order to develop understanding of a role or an organization), for limited periods of time. This can have substantial long-term value particularly if the exchanger/shadower is a good communicator in his/her own sphere. In science, this shadow strategy is the basis for transdisciplinary cooperation.

Structuring the processes in a different way, we have the following mechanisms that can be used to facilitate communication:

- promoting shared values,
- articulating and ranking goals and objectives,
- raising complicating issues and contrary views,
- option appraisal and ranking,
- making choices (and identifying conflicts between choices),
- identifying "winners" and "losers,"
- establishing how to make decisions and how to say "no," and
- monitoring and policy/project evaluation.

Successful communication between different interest groups activities should result in some transformation of ideas and convergence of mindsets. It may promote synergy of policies and resources between groups. It should, also, allow for conflicts to be identified so that resources can be focused on their resolution. This will allow the use of resources to be maximized to increase the value to society of the shorelines of the world. Nevertheless, we must recognize that some groups (e.g., aboriginal groups in the Great Barrier Reef region) may not accept this type of rationality. Additional communication will be required to resolve these conflicts. However, we must not forget that, ultimately, shoreline development involves choice. It may not always be possible to reach solutions that satisfy all groups.

CONCLUSION: THE GOLDEN RULES

Improved communication between science and management contributes to the more effective use of science in the management process associated with shoreline development. Additionally, greater availability of better scientific information increases opportunities for the development of bottom-up approaches to the management of shoreline development. We believe that such approaches are more effective, more sustainable, and are less resource intensive.

Clear and effective communication can result in mutual understanding and the changing of mind-sets, which is often achieved after appropriate levels of trust have been established. Where this occurs, positive feedback loops will be created by the generation of further trust, which may provide the context for wider cooperation and collaboration. This may develop through, for example, partnership (Crossland and Kenchington, this volume; Box 11.2). This will not be a speedy process. While building the framework of trust for optimally effective communication will take time, the rewards include:

- high levels of learning by multiple players,
- improved decision making because of its justification through objective, explicit, shared scientific information rather than through instinct or political pressure,

- demands for further improvements in, and more effective use of, science (social and natural) in decision making, and
- improvements in the multiple use and development of the shoreline area.

Some rules, which we named "Golden Rules," to strengthen the effective use and communication of science knowledge are listed in Box 11.12.

Box 11.12 The "Golden Rules"

To strengthen the effective use of science knowledge:

- Address shoreline issues at the appropriate spatial, temporal, and institutional scales, following the concept of adaptive management.
- Create awareness of time-scale discrepancies between the science policy process and the shoreline development response, hard-proven evidence may come too late.
- Implement Environmental Impact Assessment as a continuous process.
- Reward transdisciplinary scientific excellence on the same scale as mono-disciplinary scientific excellence.

To strengthen communication of science knowledge:

- Involve science as an equal partner from the start of the shoreline management process and develop science–management partnerships.
- Utilize clear language and invest in learning the language of partners/stakeholders
- Develop a communications strategy as an integral and effective part of research work proposals.
- Create a broker/facilitator function (institutions and people) in the information transfer process.

REFERENCES

Bailey, N., A. Barker, and K. MacDonald. 1995. Partnership Agencies in British Urban Policy. London: UCL Press.

Beck, U. 1992. Risk Society. London: Sage Publications.

Bijlsma, L., ed. 1993. Preparing to meet the coastal challenges of the 21st century. Report World Coast Conference. The Hague: Ministry of Transport, Public Works and Water Management/RIKZ.

Boudreau, P. 1999. A wisdom/willingness/wealth framework for successful ICZM — Some North American examples (Abstract). In: Report of the LOICZ Open Science Meeting 1999, Regimes of Coastal Change, ed. C.J. Crossland, H. Kremer, and J.I. Marshall, pp. 147. Texel, The Netherlands: LOICZ International Project Office, Netherlands Institute for Sea Research (NIOZ).

Cicin-Sain, B., and R.W. Knecht. 1998. Integrated Coastal and Ocean Management. Concepts and Practices. Washington, D.C., Covelo: Island Press.

Cunningham, W.J. 1954. A nonlinear differential difference equation of growth. *Proc. Natl. Acad. Sci. USA* **40**:709–713.

Dronkers, J., and I. de Vries. 1999. Integrated coastal management: The challenge of transdisciplinarity. *J. Coast. Conserv.* **5**:97–102.

Edwards, V., and N. Steins. 1998. Developing an analytical framework for multiple-use commons. *J. Theor. Pol.* **10(3)**:347–383.

Ehrlich, P.R., G.C. Daly, A.H. Ehrlich, P. Matson, and P. Vitousek. 1989. Global change and carrying capacity: Implications for life on earth. In: Global Change and Our Common Future, ed. R.S. DeFries and T.F. Malone, pp. 19–27. Washington, D.C.: National Academy Press.

European Spatial Development Perspective 1999. Luxemburg: Office for Official Publications of the European Communities.

Green, C., and E. Penning-Rowsell. 1999. Inherent conflicts at the coast. *J. Coast. Conserv.* **5**:153–162.

Hutchinson, C.E. 1948. Circular causal systems in ecology. *Ann. N.Y. Acad. Sci.* **50**:221–246.

JICA (Japan International Cooperation Agency). 1994. The Study on Recuperation of the Guanabara Bay Ecosystem, vol. 2. Tokyo: Kokusai Kogioco Lta.

Kiser, L.L., and E. Ostrom. 1982. The three worlds in action: A metatheoretical synthesis of institutional approaches. In: Strategies of Political Inquiry, ed. E. Ostrom. London: Sage Publications.

Millard, K., and P. Sayers. 1999. Maximising the use and exchange of coastal data. London: CIRIA.

Nonaka, I., and H. Takeuchi. 1995. The Knowledge-Creating Company. New York: Oxford Univ. Press.

Ostrom, E. 1990. Governing the Commons. Cambridge: Cambridge Univ. Press.

Pearce, D.W., and R.K. Turner. 1990. Economics of Natural Environmental Resources and the Environment, pp. 61–69. London: Harvester Wheatsheaf.

Rebello, A., C. Ponciano, and L.H. Melges. 1988. Evaluation of primary production and nutrient availability in Guanabara Bay. *An. Acad. Bras. de Cien.* **60**(4):419–430.

Taussik, J. 1999. The contribution of spoilt land to the sustainable development of the coast. *Mar. Poll. Bull.* **38. 9**:752–759.

Turner, R.K., and W.N. Adger. 1996. Coastal Zone Resources Assessment Guidelines. LOICZ Reports and Studies No. 4. Texel, NL: LOICZ.

van den Berg, C., and A. de L. Rebello. 1986. Organic copper interaction in Guanabara Bay, Brazil. *Sci. Total Env.* **58**:37–45.

12

Integrated Coastal Management in Tanzania and Eastern Africa

Addressing Diminishing Resources and a Forgotten People

M.A.K. NGOILE[1], J. DAFFA[1], and J. KULEKANA[2]

[1]National Environmental Management Council, P.O. Box 63154, Dar es Salaam, Tanzania
[2]Tanzania Fisheries Research Institute, P.O. Box 9750, Dar es Salaam, Tanzania

ABSTRACT

More than 30% of the 100 million inhabitants of the Western Indian Ocean maritime states reside in the coastal strip and these are heavily dependent on goods and services provided by marine and coastal resources. Coastal and marine areas house a bounty of biodiversity most of which is unknown to us. Marine and coastal ecosystems are also paramount for providing critical life support functions and play a significant role in balancing the extremes of climatic conditions. However, the resources we depend on are declining at an unprecedented rate, and our environment is degrading faster than we could ever have imagined.

Integrated coastal management (ICM) has been accepted globally as an approach that is appropriate for conserving and sustaining the use of marine and coastal resources. ICM aims at integrating government and stakeholders, science and management as well as sectoral and public interests; it is a truly ecosystem management approach. This chapter examines the challenges to managing coastal resources, constraints, and opportunities in developing countries. The coupling of ecosystem management and ecosystem science will ensure the sustainability of marine and coastal resources.

INTRODUCTION

Global Challenge

The world's oceans and their coastal zone traditionally have provided ample food for humanity. Particularly in the lower latitudes they are the sole protein resource for the coastal

Science and Integrated Coastal Management
Edited by B. von Bodungen and R.K. Turner

societies. The coastal zone supports the economies of many maritime nations through fishing and shipping industries, petroleum exploitation, seabed mining, energy, tourism, etc. It supports a wide range of biodiversity, most of which is unknown to us, especially that which is found in the deep oceans. Coastal and marine ecosystems perform critical life support functions and play a significant role in balancing the extremes of climatic conditions.

In a region that encompasses less than 20% of the world's inhabitable space, the coastal zone is home to more than 50% of the world's population. Seventy of the world's cities, each with populations of 2.2 million or more, are currently located near tidal estuaries; 12 of the world's 15 largest cities (megacities) are found in the world's coastal zone. Increasing population growth and further migration to coastal areas are estimated to result in a doubling of coastal population by the year 2020 (Olsen et al. 1999).

However, over the last decade, the burden of population migration and increasing population has resulted in an unprecedented decline in the very resources necessary for human existence. Currently, more than half of humanity's infrastructure and manufacturing, transportation, energy, processing, tourism, communication, and other services is centered along the coast (Olsen et al. 1999). This, combined with unplanned and uncoordinated urbanization, industrial development, and rapid population growth, especially in developing countries, has resulted in large-scale deterioration of coastal environments: declining quality of water, degradation or destruction of coastal habitats, decline and collapse of fisheries (Box 12.1), and biodiversity loss.

This decline denies humanity the goods and services necessary for life and puts the lives of millions of people at risk. Diminishing fisheries resources, for example, are threatening malnutrition to nearly a billion people, as well as the source of livelihood for an estimated 200 million people. It is therefore imperative for necessary measures to be taken to reverse this negative trend. Conserving and sustaining the use of marine and coastal resources are the means by which we can ensure our own survival.

Success in our efforts to manage coastal and marine resources are minimal compared to the forces causing degradation of coastal environments. For example, in the tropics, where the pace of coastal change is most rapid, coastal management is currently conceived and implemented as a scattering of pilot projects (Olsen et al. 1999).

Traditionally, less attention has been given to effective management of oceans and coastal resources than terrestrial resources. Until recently, oceans were perceived as expansive areas with inexhaustible resources. For centuries, coastal populations did not cause permanent harm to the coastal and marine environment. Uses of marine and coastal resources centered

Box 12.1 Vital and Critical Statistics from Ocean Fisheries (source: Hinrichsen 1998)

- Seventy percent of the world's commercially important marine stocks are fully fished, overexploited, depleted, or only slowly recovering.
- Some staple species, such as the northern cod and Atlantic halibut, have been excessively fished to commercial extinction.
- Worldwide governments pay an estimated US$54 billion per year in fisheries subsidies to an industry that catches only $70 billion worth of fish.
- Contemporary fishing practices kill and waste 18–40 million tons of unwanted fish, seabirds, turtles, marine mammals and other ocean life annually. This represents approximately one-third of the total world catch.

primarily on fisheries, causing some local environment changes but without threatening system sustainability. Today, however, use of marine and coastal resources has increased, as has beach erosion, and expected sea-level changes will have an amplifying effect. Unless we link good science and proper management, we can expect the resultant competition and conflict of interests to inflict degeneration and ultimate destruction of resources.

Challenges in the Western Indian Ocean Region

The pressures and declining trends in coastal resources experienced globally are manifested in the Western Indian Ocean. In terms of population pressure, it is estimated that more than 30% of the 100 million inhabitants in the region live in the coastal zone (Hinrichsen 1998).

Poverty is a major problem, particularly in the East African mainland states and Madagascar. In 1992–1993, the estimated gross national product (GNP) per capita ranged from $80 (Mozambique) to $330 (Kenya), making it one of the world's poorest coastal regions.

Poverty combined with rapid population growth and poor understanding and management of coastal resources has resulted in a number of environmental and resource use problems, including habitat destruction, overfishing, human-induced coastal erosion and flooding, and pollution due to intensified land use. The severity of environmental degradation tends to be limited to areas near urban settlements; however, increasing population and development pressures will increase zones of degradation to non-urban areas.

On a global scale, the Western Indian Ocean is not particularly productive in terms of fisheries. The Food and Agriculture Organisation of the United Nations (FAO) estimates that approximately 3 million tons of fish, crustaceans, and mollusks are harvested from this region every year, representing 4% of the world's total catch. The total harvest within the region increased by 72% from 1980 to 1990; however, annual fish consumption per capita decreased from 3.7 to 1.9 kg per year over the same period.

Several coastal marine ecosystems throughout the region have been destroyed or degraded through human activities, such as destructive fishing practices, sand and coral mining, land reclamation, and dredging. Land reclamation projects have filled extensive intertidal areas, particularly on islands, where flat coastal land is in high demand. In Victoria, Seychelles, for example, the airport, two ports, and relatively large residential areas and roads have been built on reclaimed land. This has also been the case in the main port and coastal industrial zones in Mauritius. Beaches and sand dunes in Mozambique and Tanzania are mined for construction materials and black sand minerals, causing direct habitat destruction and subsequent coastal erosion. Coastal vegetation, including the mangrove forests, are being cut and degraded in many areas throughout the region. Many coastal ecosystems are also damaged indirectly by pollution, particularly from land-based sources and hydrologic alterations to rivers.

The absence of environmental regulations in relation to tourism has resulted in growing environmental problems in all countries of the region related to coastal tourism development.

The industrial sector is not strongly developed in most of the region (with the exception of South Africa), and focuses primarily on the processing of agricultural products, petroleum, and other goods (e.g., textiles, fertilizers) for domestic consumption. In the coastal zone, these industries are clustered around the larger cities and ports, such as Mombasa, Dar es Salaam, and Maputo. In Mauritius, the zone set aside for the promotion of exporting industries has attracted considerable foreign investment, and manufacturing is now the largest component of the island's gross domestic product (GDP) (24%). Wastes from these industries tend to be organic and nutrient-rich, and are commonly discharged to sewers or directly to rivers and coastal waters without pretreatment.

There is little collection and no treatment of domestic wastes in the coastal cities of the region.

Development Opportunities in the Coastal Zone of the Western Indian Ocean Region

There are indeed several very strong arguments for improving the management of coastal areas in the Western Indian Ocean region; one, in particular, is the dependence of the economies of the region on the functional integrity of the coastal ecosystem. Compared to most other coastal areas around the Indian Ocean, population pressures along much of the extensive coastline of Eastern Africa are not that high. As a consequence, there are still considerable natural areas that are relatively undisturbed by human activities. This is particularly true for large parts of the coasts of Somalia and Mozambique, southern Tanzania and northern Kenya, as these areas have very little economic activity as well as few human settlements. These less-inhabited coastal zones have significant areas of pristine coral reefs, less exploited fisheries resources, white sand beaches, and undisturbed mangrove forests: resources that are important assets for development. However, with urban centers growing at rates of 5% to 10% per annum, it is likely that much of these unaffected habitats will be destroyed within a few decades, unless management measures are more successful than in the past.

Mariculture has been introduced only recently to the region, with the exception of the traditional fish farms (barachois) of Mauritius. This sector seems to have good development potential, as indicated by the recent success of seaweed culture in Zanzibar and prawn fishing in Madagascar.

In spite of the economic importance, most national fisheries in the region are still traditionally based and have, as such, not succumbed to the massive overcapitalization being experienced elsewhere in the world. Commercial coastal fisheries are a source of employment and foreign capital in the region, particularly in Mozambique and Madagascar, where the relatively wide continental shelves support large shrimp fisheries. Madagascar has the highest annual marine landings in the region, followed by Tanzania and Mozambique. Large open-ocean tuna fishing fleets are based in Seychelles and Mauritius, providing employment and foreign capital to these islands. Present concerns focus on pirate fishing in the exclusive economic zones (EEZs) of the Western Indian Ocean states. A more regional approach for the management of transboundary tuna resources in the Western Indian Ocean could optimize economic returns to the countries of the region.

Tourism in Eastern Africa has not developed to the same level as in many parts of the Mediterranean, Southeast Asia, and Caribbean. The amenities sought by tourists are still in pristine condition in the region. Opportunities do exist for better planning and management in order to ensure a sustainable tourism industry.

The Western Indian Ocean region is, therefore, at a crossroads. On one hand, pressures on the coastal environment do exist, in so-called "hot spots," located mainly in coastal urban centers. On the other, the opportunity is still there to effect sustainable coastal development.

This calls for a delicate balance between development drives and management. Since coastal and marine ecosystems are complex, there is an urgent need to move as rapidly as possible to ecosystem-level thinking and management.

The ecosystem management approach addresses people and environment as one system: the ICM approach.

MARINE SCIENCE AND COASTAL MANAGEMENT

Scientific knowledge on productive capacities and functioning of ecosystems (ecosystem health) and the involvement of stakeholders (communities and the private sectors) are the two key considerations for the successful implementation of sound policies, development plans, and management strategies for marine and coastal areas. Marine scientists must respond to coastal resources management needs, and coastal resource managers must take the advice of marine scientists (both natural and social) during their decision-making processes.

The link between science and management for the successful implementation of coastal management programs has been highlighted and emphasized at all levels and in numerous fora, policy statements, agreements, and conventions. Scientific information is critical in defining the scale of coastal management programs, elaborating selection of management options and types of management measures, as well as for informing the public through awareness and educational programs (GESAMP 1996b).

Natural sciences are vital to our understanding of the functioning of ecosystems (*state variables*) and the social sciences are essential to comprehending patterns of human behavior that cause ecological damage (*pressure variables*) and to finding effective solutions (*respective measures — management actions*) (Box 12.2). Scientists and managers must work together as a team through all stages of policy formulation and implementation. As a team, they must define the scientific work needed to address priorities and guide policy development (see also von Bodungen and Turner, this volume).

However, many factors impede the effective integration of science and management, resulting in poor performance (including delays) in the implementation of ICM programs:

1. The diversity and complexity of the marine and coastal ecosystems as well as the continuous interaction among the components have consequences that disturbances in any one of the components are experienced broadly, both in terms of time and geographic scales. In its endeavors, science has generated long lists of variables for collecting information and this has happened especially in situations where scientists are not collaborating with managers in setting/selecting the issues most relevant for management. This has resulted in a protracted scientific process, which frustrates the implementation of ICM programs. There is, in this respect, an urgent need to develop an objective and scientific approach/protocol to facilitate prioritization of issues which science should address.
2. The scientific approach employed during the coastal management issues identification stage has concentrated primarily on the state of the natural ecosystems — identifying the damage and disturbances of the ecosystem. In most cases, programs have entered into the planning stage without taking proper stock of the pressure on these

Box 12.2 Components of Ecosystem-based Research for Supporting Integrated Coastal and Marine Management

The following research components are important in ensuring ecosystem approaches to gathering information in support of integrated management of marine and coastal resources.

1. Assessment of Resources and the Environment
   a) Assessment of ecosystem and the environment:
      - marine biodiversity and biogeography,
      - populations, species, habitats, ecosystems,
      - fragile and sensitive ecosystems,
      - coral reefs, mangroves, coastal wetlands, coastal lagoons, estuaries, seagrass beds, small islands,
      - threatened and endangered species.

   b) Resources:
      - natural resources (fisheries, mangroves, seaweed),
      - nonliving resources,
      - services, e.g., transport, recreational.

   c) Threats:
      - land-based activities: coastal development, coastal and river basin agriculture and livestock, tourism, river impoundment, deforestation,
      - ocean-based activities: impacts associated with marine transport, seabed mining, hydrocarbon exploitation (ocean-based oil drilling).
2. Assessment of Socioeconomic Coastal Stakeholders
   a) Assessment of socioeconomic and demographic patterns and recommend mechanisms for:
      - safeguarding livelihoods,
      - equitable sharing of proceeds from resource uses,
      - indigenous knowledge is captured and used in policy, planning and management process,
      - ensuring transparent tenure systems,
      - assessment of the development alternatives including cost-benefit analysis.
3. Assessment of the Effectiveness of the Governance
   a) Assess existing policies and management mechanisms for their effectiveness.
   b) Choice of appropriate approaches to improve the management (e.g., ICM, MPAS, LMES).
   c) Assess the level of inclusion of the following characteristics:
      - community/stakeholder/private sector involvement,
      - cooperative management,
      - collaborative management,
      - cross-sectoral and multidisciplinary management,
      - actions based on scientific information.

   d) Development of policy, management plans, guidelines and legislation.
   e) Synthesize and document successful ICM experiences and facilitate their replication so as to avoid "reinventing the wheel." This will save time and money as well as enhancing the sharing of experiences.
   f) Promote the referencing of successful ICM experiences during regional and global debates and negotiations for the best code of conduct leading to the conservation and sustainable use of marine and coastal resources (GESAMP 1996a; Olsen et al. 1999).

ecosystems imposed by the various resource exploitations or at most using only deduction from the state of the natural resources.

3. Scientific information is presented (a) in a manner that is strange to advising management processes and (b) normally in scientific journals, which are not accessible to

many that deal with coastal management. It is important that procedures are developed which can facilitate and ensure a balanced synthesis of scientific information into coastal policy/management options and advice.

Fortunately, there are emerging methodologies that are able to facilitate the integration of marine science and coastal management. For example, ecological risk assessment, the analytical framework suggested in the GESAMP report for ICM (GESAMP 1996a), and the recently developed "Self-assessment Manual" (Olsen et al. 1999).

Coastal Science in the Developing World: Most Tools Not Available and Divided We Stand

In developing countries, the problems caused by the insufficient link between science and management are further compounded by several factors. First, the prevailing approach to scientific research on coastal and marine ecosystems is strongly discipline-driven and inadequate communication exists between scientists of different disciplines. Also, the philosophy of "schools of thought" that lingers in many people's minds usually reflects an uncompromising attitude based on a narrow scope of knowledge. Therefore, existing information is segregated rather than integrated.

Currently, the existing information base is inadequate and cannot provide the expected level of advice to public policy and management processes. The state of information in developing countries is further aggravated by the fact that most of it exists as "gray" literature under the custody of sectoral ministries and is rarely available in universities or public libraries.

There is inadequate communication between resource users, planners, and policy makers. The lack of coordination between different users together with an inadequate understanding of the dynamics of the coastal resources has resulted in highly sectoralized policies and management strategies which give little consideration to issues and the natural linkages between different resource uses and the pressures they impose.

ECOSYSTEM-LEVEL MANAGEMENT IN THE WESTERN INDIAN OCEAN

Past Management Experiences

An historic milestone in addressing the state of coastal and marine resources of the region was the Indian Ocean Expedition, which was implemented in the early 1960s. The expedition was able to increase the knowledge base on the physico–chemico–biological processes, especially in terms of how these are influenced by the monsoons. However, most of the countries bordering the Indian Ocean, and especially those on the Western Indian Ocean, were unable to participate effectively in the expedition because of lack of capacity and resources.

In the years following the expedition, the emerging independent states of Eastern Africa prioritized the development of fisheries resources to meet protein requirements of coastal populations as well as to provide foreign exchange earnings. National economic needs tipped the balance towards developing fisheries resources for economic benefits. As a consequence, marine resources development has been synonymous with commercial fisheries development to satisfy hard currency earning. The nations of the region have sought assistance for fisheries development, mostly through the FAO and bilateral mechanisms.

In addition, several countries (e.g., Norway, Canada, France, United Kingdom) assisted in the development of fisheries in Eastern Africa. During 1982–1984, the Norwegian research vessel, "Dr. Fridtjof Nansen," conducted fisheries surveys in Mozambique, Tanzania, Madagascar, and Kenya (Bianchi 1992). The survey results identified fisheries resources in offshore waters beyond the artisanal and prawn fishing through the improvement of fishing and preservation techniques.

Present Regional Initiatives

The processes leading to the United Nations Conferences on Environment and Development (UNCED) revealed the complexity of managing coastal and marine resources. This prompted the states of the Western Indian Ocean to make a concerted effort in addressing the sustainable management of coastal resources.

The concept of integrated coastal zone management was reviewed in the region during the Arusha Workshop and Conference on "Integrated Coastal Zone Management in Eastern Africa, Including the Island States" in 1993 (Linden 1995; Ngoile and Linden 1997). The workshop emphasized the need for integration of all issues, involvement of all players in the planning process, coordination between sectoral agencies, and application of cooperative management during the implementation of ICM.

The participating ministers discussed the recommendations presented by the scientists and they formulated and signed the "Arusha Resolution"; this policy statement calls upon the states of Eastern Africa to give emphasis to sustainable development and integrated management of coastal areas for the primary benefit of coastal communities. The resolution was signed by all the participating ministers from the region: Mauritius, Mozambique, Madagascar, Seychelles, Tanzania, and later by the minister from Kenya.

The Arusha Resolution issued three main instructions for follow-up. At the regional level, the participating governments recognized the importance of the Nairobi Convention and related regional environmental agreements, and called upon those countries that had not yet signed to do so. Second, it urged the implementation of ICM at the national level, including the establishment of a dialogue between the various stakeholders involved in coastal and marine resource use. Third, directed to the scientific community, it called for an inter- and multidisciplinary approach to research that will provide the required knowledge for ICM. The governments agreed to meet periodically in order to assess the progress made in the implementation of the resolution.

The Seychelles Conference, held in October 1996, took stock of the progress made during the intersessional period (1993–1996). The general conclusion was that the countries had just begun to implement ICM at different levels. A major achievement was the increased number of demonstration projects on ICM at local level.

The Tanzania Coastal Management Partnership Experience

Tanzania Coastal Management Partnership (TCMP) is a cooperative project between the Government of Tanzania through the National Environmental Management Council

(NEMC), the United States Agency for International Development (USAID), and the University of Rhode Island's Coastal Resources Center (URI–CRC). The initiative was established in 1997. TCMP works with the existing network of ICM programs and practitioners to facilitate a participatory, transparent process, in an effort to unite government and the community, science and management, sectoral and public interests to conserve and develop wisely coastal ecosystems and resources. The goal of the partnership is to "establish the foundation for effective coastal governance." TCMP is working towards achieving the following objectives:

- to develop an integrated coastal management policy that can be effectively applied to coastal problems at both the national and local levels,
- to demonstrate intersectoral mechanisms for addressing emerging coastal economic opportunities,
- to improve enabling conditions for integrated coastal management,
- to build human and institutional capacity for integrated coastal management, and
- to utilize Tanzania's coastal management experience by contributing to ICM regionally and globally.

The in-country partnership includes representation from all key national sectors that have a role to play in coastal management. A constituency for coastal management is growing within the highest levels of government. The relationship between national and local coastal management initiatives, including the private sector and nongovernmental organizations (NGOs), continues to strengthen. Major achievements of the partnership include:

- The initial issues, goals, and strategies for Tanzania's coastal policy/program were drafted and approved by the directors of all key government departments. This will guide the ICM policy/program for Tanzania. The completion of this ICM policy/program framework is noteworthy, as it is only the second such framework to be accepted by a nation in the Western Indian Ocean region.
- A maritime issue profile was developed and approved by government directors; this subsequently produced a national agenda for the development of mariculture guidelines.
- A network was built of coastal practitioners working through Tanzania; each is linked with ongoing coastal management programs, both on local and national levels.
- A partnership was built between local, regional, and global institutions for the purpose of building both human and institutional capacity for ICM in the country. This partnership completed a national needs assessment for formalized coastal management training in Tanzania.
- A range of documentation was created on important coastal management issues and problems, including reports on socioeconomics of the coast, a legal and institutional framework for coastal management, a profile of critical coastal management issues, and a profile of mariculture-specific issues.
- Awareness for environmental and coastal conservation has been raised through a series of activities, including a newsletter, video voices from the field, press articles, and a community awards scheme.
- In-country and external training opportunities have been provided for over 30 participants from 11 different government institutions. Training introduced the

candidates to a range of coastal management experience from around the world and broadened their view about how to address difficult intersectoral management issues.
- It supports the representation of the Government of Tanzania at important meetings and workshops throughout the Western Indian Ocean region.

These results have been achieved through a highly participatory process. TCMP has depended heavily on existing coastal management programs in Tanzania as well as the existing national-level technical expertise of sectors key to coastal management.

CONCLUSION

If we are to succeed in managing marine and coastal resources sustainably, we must begin orienting our scientific methodology to ecosystem-level science and we must apply ecosystem-level management. A regional approach to development of ecosystem-level marine science and management offers the possibility of optimizing capacities and financial resources.

I conclude by quoting E. Odum (1998, p. 116) "The real world, however, consists of open, far-from-equilibrium, thermodynamic systems that cannot be enclosed in glass test tubes or within laboratory walls. They are also much influenced by economic and political considerations that are rarely included in the scientists' model."

REFERENCES

Bianchi, G. 1992. Demersal assemblages of tropical continental shelves. A study based on the data collected through the surveys of R/V "Dr. Fridtjof Nansen," pp. 191–217. Ph.D. diss., Dept. of Fisheries and Marine Biology, Univ. of Bergen, Norway.

Bryceson, I., et al. 1990. State of the Marine Environmental in the East African Region. Regional Seas Reports and Studies No. 113. UNEP (United Nations Environment Programme).

Chua, T.E., and L.F. Scura, eds. 1992. Integrative Framework and Methods for Coastal Area Management. Conf. Proc. 37. Manila: ICLARM (International Center for Living Aquatic Resources Management).

Cicin-Sain, B., R.W. Knecht, and G.W. Fisk 1996. Growth in capacity for integrated coastal management since UNCED: An international perspective. *Ocean Coast. Manag.* **29(1–3)**:1–11.

Clark, J.R. 1992. Integrated Management of Coastal Zones. Fisheries Technical Paper No. 327. Rome: FAO (Food and Agriculture Organisation of the United Nations).

Dogse, P. 1995. Coastal Tourism Management Guidelines. Paper presented at the SREC/World Bank Integrated Coastal Zone Workshops in East Africa, February 1995.

Fagoonee, I., and D. Daby. 1995. Coastal zone management in Mauritius. In: Proc. Arusha Workshop and Policy Conf. on Integrated Coastal Zone Management in Eastern Africa including the Island States, Conf. Proc. 1, ed. O. Linden, pp. 217–249. Metro Manila: CMC (Coastal Management Center).

GESAMP (Joint Group of Experts on the Scientific Aspects of Marine Environmental Protection). 1996a. The Contributions of Science to Integrated Coastal Management. Reports and Studies No. 61. Rome: FAO (Food and Agriculture Organisation of the United Nations).

GESAMP. 1996b. Report of the Twenty-sixth Session. Reports and Studies No. 60, p. 6. Paris: UNESCO–IOC.

Hinrichsen, D. 1998. Coastal Waters of the World: Trends, Threats, and Strategies. Washington, D.C.: Island Press.

Hota, K., and I.M. Dutton, eds. 1995. Coastal Management in the Asia Pacific Region: Issues and Approaches. Japan: JIMSTEF (Japan Intl. Marine Science Institute and Technology Federation).

Insul, A.D., U.C. Barg, and P. Martosubroto. 1995. Coastal Fisheries and Aquaculture within Integrated Coastal Area Management in East Africa. In: Proc. Arusha Workshop and Policy Conf. on Integrated Coastal Zone Management in Eastern Africa including the Island States, Conf. Proc. 1, ed. O. Linden, pp. 19–36. Metro Manila: CMC.

IUCN/UNEP/WWF. 1991. Caring for the Earth. A Strategy for Sustainable Living. Gland: IUCN (The World Conservation Union)/UNEP (United Nations Environment Programme)/WWF (World Wildlife Fund).

Iversen, S.A., and S. Myklevoll. 1984. Proc. of the NORAD-Tanzania Seminar to Review the Marine Fish Stocks and Fisheries of Tanzania, Mbegani, March 6–8, 1984. Bergen: Institute of Marine Research.

Linden, O., ed. 1995. Workshop and Policy Conf. on Integrated Coastal Zone Management in Eastern Africa including the Island States. Conf. Proc. 1. Metro Manila: CMC.

Linden, O., and C.G. Lundin, eds. 1996. Integrated Coastal Management in Tanzania. Stockholm: SIDA (Swedish International Development Authority) in cooperation with the World Bank and the Government of Tanzania.

Ngoile, M.A.K., and O. Linden, 1997. Lessons learned from Eastern Africa: The development of policy on ICZM at national and regional levels. *Ocean Coast. Manag.* **37(3)**:295–318.

Odum, E. 1998. Ecological Vignettes: Ecological Approaches to Dealing with Human Predicaments. Amsterdam: Overseas Publishers Assoc.

Olsen, S., K. Lowry, and J. Tobey. 1999. A Manual for Assessing Progress in Coastal Management. Coastal Resources Center Report No. 2211. Narragansett: Univ. of Rhode Island.

Shah, N.J. 1995. The coastal zone of the Seychelles. In: Proc. Arusha Workshop and Policy Conf. on Integrated Coastal Zone Management in Eastern Africa including the Island States, Conf. Proc. 1, ed. O. Linden, pp. 275–290. Metro Manila: CMC.

Sorenson, J.C., and S.T. McCreary. 1990. Institutional arrangements for management of coastal resources. Coastal Management Publications No. 1. Washington, D.C.: U.S. Natl. Park Service/USAID (U.S. Agency for Intl. Development).

Winsemius, P. 1995. Commentary — Integrated policies: A requirement for coastal zone management. *Ocean Coast. Manag.* **26(2)**: 151–162.

13

River Basin Activities, Impact, and Management of Anthropogenic Trace Metal and Sediment Fluxes to Sepetiba Bay, Southeastern Brazil

L.D. LACERDA[1], R.V. MARINS[1], C. BARCELLOS[2], and B.A. KNOPPERS[1]

[1]Depto. de Geoquímica, Universidade Federal Fluminense, Outeiro de São João Batista s/n, 24020–007 Niterói, RJ, Brazil
[2]Fundação Oswaldo Cruz, Rio de Janeiro, RJ, Brazil

ABSTRACT

This case study quantifies the trace metal and suspended matter fluxes to Sepetiba Bay, Brazil, which is subject to impact by industrial activities, uncontrolled demographic expansion, agricultural practices, and engineering works in its watershed. The watershed retains the largest fraction of trace metal loads from industrial activities. Major loads of cadmium (Cd), zinc (Zn), lead (Pb), and mercury (Hg) are largely delivered to the soils. The atmosphere and the bay's waters receive loads by an order of magnitude lower. Increased erosion in the watershed has enhanced suspended matter loads to the bay and helped to wash out more efficiently the trace metals stocked in soils. The trace metal loads to the bay over the last 30 years have imposed significant contamination of the bay's sediments, which are frequently remobilized by dredging activities and harbor engineering works. Fisheries yields have also declined since industrialization began and several components of the biota are contaminated by trace metals. Part of the trace metals are exported to adjacent less impacted waters, where aquaculture and tourism are important economic activities. Although Brazil has developed a management policy for its coastal waters, there is still lack of an operative integrated coastal management (ICM) concept linking river basin activities and the recipient system and legislation is inadequate to cope with the fast pace of development and anthropogenic impacts. Partnerships between federal, state and environmental protection agencies, industry, scientific community, and nongovernmental organizations are now being sought.

Science and Integrated Coastal Management
Edited by B. von Bodungen and R.K. Turner

INTRODUCTION

Until the 1990s, most environmental impact studies on coastal ecosystems dealt with the activities occurring within coastal sites. Less attention has been given to the activities within river basins with high potential for creating impacts on the coastal areas. Most legislation regarding environmental conservation and sustainable utilization of natural resources of coastal areas has failed to consider basin activities as a key factor affecting these areas. Multidisciplinary approaches to understand and cope with the coastal impacts of river basin activities, including the legal, socioeconomic, historic, and transboundary issues involved, are now being given priority within the concept of integrated coastal management (ICM).

Over the last three decades, Brazil has expended a great deal of effort in assessing the environmental characteristics, anthropogenic impacts, and living and nonliving resources of its coastal and marine waters, as it has sought to develop a management policy for its coastal zone (GERCO–PNGC–PNMA 1996). Awareness — both by financial institutions and the public at large — of the need for environmental management and cleanup has increased at a fast pace. Some reasonable management plans have been proposed on a local scale for some systems, and the inflexible Brazilian environmental legislation is being applied more effectively. However, there is still a lack of quantitative information linking river basin activities and coastal impact for Brazil's estuaries, bays, and coastal waters, including those of Rio de Janeiro state.

The state of Rio de Janeiro has one of the most urbanized and industrialized coasts of Brazil. Immense anthropogenic pressures from river basin activities impact the renewable and nonrenewable goods and services. Examples of the most affected systems are the Paraíba do Sul River estuary and the Bays of Guanabara and Sepetiba.

We discuss trace element and sediment fluxes to Sepetiba Bay linked to the socioeconomic activities of its river basin. The concentration of industries, urban centers, and agriculture along the northeastern portion of Sepetiba Bay basin make this area of Brazil the most critical in terms of environmental contamination by trace metals (Barcellos 1995; Lacerda et al. 1987; Marins 1998; Amado Filho et al. 1999). In addition, engineering efforts carried out during the twentieth century, to improve water supply, energy generation, and land reclamation, have resulted in significant changes in sediment load to the bay. At present, industrial and harbor developments are pushed to relieve pressure from other areas, such as Guanabara Bay, and attend to the trade demands of the Mercosul economic agreement. Some of the environmental problems of Sepetiba Bay are thus intrinsically linked to transboundary issues.

PHYSICAL SETTING OF SEPETIBA BAY

Sepetiba Bay (Figure 13.1) is a semi-enclosed water body, connected to the sea by a small, shallow inlet in the east and a large natural channel, between the islands of Jaguanaum and Itacurussá, in the west. The latter has depths of up to 30 m and is responsible for the largest water exchange with the sea. The bay's area at high tide is 447 km^2 and at low tide 419 km^2. Mean water volume is 2.56×10^9 m^3, ranging from a maximum of 3.06×10^9 m^3 to a minimum of 2.38×10^9 m^3. Average depth is about 6 m. The tidal prism volume is 3.4×10^8 m^3 and the ratio between tidal prism and fluvial inputs is circa 0.03, characterizing the bay as a

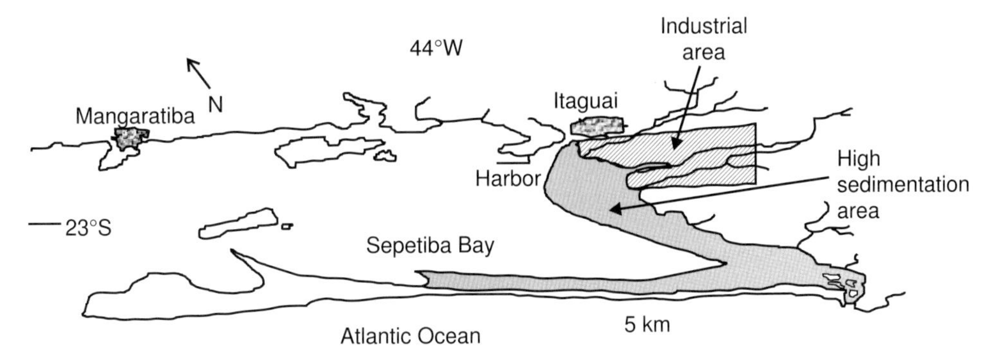

Figure 13.1 Map showing major physical aspects and anthropogenic activities of Sepetiba Bay, Rio de Janeiro State, southeastern Brazil.

well-mixed estuary. The mean turnover time of the water mass is estimated at around six days and the maximum current velocities at peak of tides range from 50 to 75 cm s^{-1} (Barcellos et al. 1997).

The climate is tropical humid with a mean annual precipitation of 1400 mm and an evaporation of 960 mm. The nine rivers draining the quaternary plain at the northeastern coast of the bay attain an annual flow of 7.6×10^6 m^3. The artificial Canal de São Francisco, with an annual flow of 6.5×10^6 m^3, accounts for 86% of the total fluvial inputs.

Circulation in the bay is driven by winds and tides. Dominant winds are from the southwest (250°), bringing seawater from the South Atlantic through the western channel. This water warms at the inner portion of the bay close to river mouths, creating a clockwise current pattern, driving freshwater and fluvial sediments southwards, and keeping the water around 30 PSU measurement units. During strong northeast (70°) winds, this clockwise pattern is disrupted, and most fluvial inputs move directly through the main channel to the Atlantic, dropping surface salinity to about 25 (Barcellos 1995; Marins 1998).

The islands and the rocky northern shore are covered by tropical rain forests. The bay supports 40 km^2 of mangrove forests, which are most dense at the inner eastern portion. They play an important role in providing nursery and feeding areas for the bay's fisheries (Lacerda 1998).

SOCIOECONOMIC SETTING

The lowlands of the eastern coast of Sepetiba Bay — with its good transport facilities, cheap and ample land, good freshwater supply, and low population density — became of interest to industrial development in the 1970s, after a large harbor was built. Over the last two decades, 400 industries, mostly metallurgical, have been installed in the region. At present, a petrochemical plant and two other large (>10,000 t y^{-1} production) pyrometallurgical factories are being constructed. The concentration of industry, urban centers, and agriculture along the northeastern portion of Sepetiba Bay basin, make this area of southeastern Brazil the most critical in terms of environmental contamination by trace metals (Lacerda et al. 1987). Table 13.1 summarizes the major economic activities of Sepetiba Bay basin. Apart from the industrial, urban, and agriculture developments, fishing and tourism are two other main activities

Table 13.1 Major economic activities which contribute with trace metals loads to Sepetiba Bay. Source: Barcellos and Lacerda (1994).

	Number of Plants	Production ($t\ y^{-1}$)	Number of Employees
Metal smelting			
Fe	3	1,102,000	4,053
Al	2	98,500	925
Zn	1	36,000	438
Manufactures			
Paper	5	534,000	4,041
Chemicals	16	176,900	3,887
Metallurgy	19	33,370	4,372
Plastics & Rubber	3	30,900	1,722
Food processing	8	16,700	1,645
Other	13	7,200	1,996
Thermoelectric plant	1	160 MW	250
Fishery & Agriculture	—	10,500	14,000
Harbor & Navigation	1	19,000,000	300
Total	72	—	37,879

in Sepetiba Bay. Fishing employs about 3000 people directly, with main catches of shrimp, flatfish, and mullet. Total commercial catches were about 800 tons in 1987 (270 tons of which was shrimp), after a peak in the early 1980s of about 1000 tons (550 tons of shrimp). The reduction in fish catch, in particular of shrimp, is probably related to overfishing, and consequently a fishing reserve has been established in the area. During the last decade, commercial annual fish catches remained around 1000 tons, with 30% to 50% of it being of shrimp.

Artisanal fisheries, mostly dependent on mangrove-dwelling species, were capable of production levels of the same order of magnitude as commercial fisheries (~1000 tons) during the 1980s. However, overfishing from commercial trawlers strongly affected nursery grounds, resulting in an 80% productivity reduction in artisanal fisheries. Unemployment was avoided by transferring the work force to tourism-related activities.

Tourism is growing rapidly in the region. More than 30 hotels and dozens of restaurants operate along the bay's shore; sightseeing boats attract anywhere from 5000 to 10,000 tourists on summer weekends. As a result of the unplanned development, environmental contamination of the bay is now in direct conflict with the many economic options for the region's development (Lacerda et al. 1987; Amado Filho et al. 1999).

SEDIMENT LOADS AND SEDIMENTATION RATES DURING THE 20th CENTURY

One of the most striking aspects of coastal zone change as a result of anthropogenic activities in Sepetiba Bay basin is related to sediment transport and sedimentation rates in the bay. Figure 13.2 shows the evolution of sedimentation rates during the past 100 years, based on

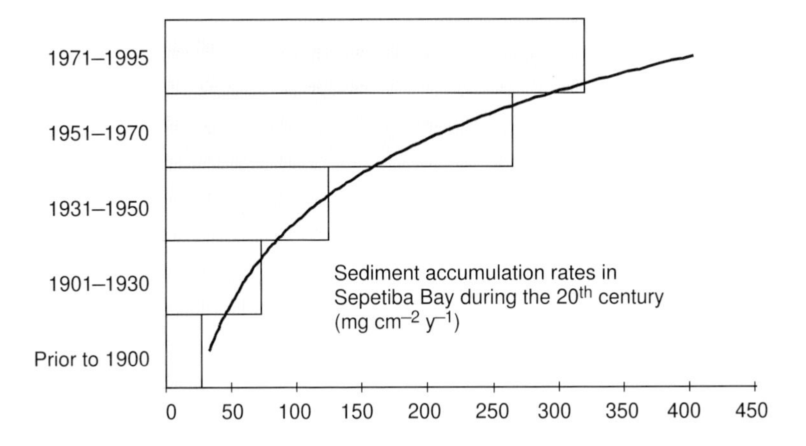

Figure 13.2 Sediment accumulation rates in Sepetiba Bay, SE Brazil, based on original measurements from Forte (1996).

^{210}Pb-dated sediment cores collected at the northeastern shore of the bay (Forte 1996). Sedimentation rates over the last century were about 30 mg $cm^{-2}y^{-1}$. At the beginning of the century, civil engineering was initiated in the bay's basin; river channels were dug and straightened, primarily to control malaria, and artificial canals were built. Although of relatively small scale, these engineering works more than doubled the sediment accumulation rates in the bay.

Civil engineering continued until the 1950s, when a large hydroelectric dam was built to collect waters from the adjacent Paraíba do Sul River basin to be diverted to the Sepetiba Bay basin. A large artificial canal, the Canal de São Francisco, with an almost constant flow volume of about 180 $m^3 s^{-1}$ throughout the year, was built to carry the waters to Sepetiba Bay. The amount of freshwater reaching the bay increased by a factor of 10 and sedimentation rates exceeded 250 mg $cm^{-2}y^{-1}$. After the 1970s, the creation of an industrial district and a subsequent population increase resulted in more extensive deforestation of the basin's former forest vegetation cover. This led to another increase in sedimentation rates to the present level of about 320 mg $cm^{-2}y^{-1}$.

TRACE METALS EMISSIONS TO SEPETIBA BAY AND BASIN

Several studies have characterized the industrial park and emissions of trace metals to Sepetiba Bay and quantified their loads (Barcellos 1995; Barcellos and Lacerda 1994). These studies identified Cd, Zn, and Pb as the major contaminants of the bay. Recently, estimates of Hg inputs from diffuse sources to Sepetiba Bay basin also highlighted the importance of this element to the overall contamination of the bay. Table 13.2 shows emission estimates for the four metals to Sepetiba Bay and basin (Barcellos and Lacerda 1994; Marins 1998).

The major sources of Cd and Zn are from smelters, a typical point source of trace metals. Estimated emissions for Zn and Cd reach 29 and 4170 t y^{-1}, respectively. Major emissions of Pb are due to iron and steel production, reaching about 350 t y^{-1}. Manufacturing, particularly

Table 13.2 Estimated Cd, Zn, Pb, and Hg loads (t y^{-1}) to Sepetiba Bay and Basin from major economic activities. Sources: Barcellos and Lacerda (1994) and Marins et al. (1999).

Activity	To Soil				To Air				To Water				TOTAL			
Metal Smelters:	Cd	Zn	Pb	Hg	Cd	Zn	Pb	Hg	Cd	Zn	Pb	Hg	Cd	Zn	Pb	Hg
Fe	1.00	168	165	?	0.33	30	55	0.1	0.1	17	8.3	?	1.34	215	228	0.1
Al	0.03	1.6	0.4	?	0.01	1.1	0.1	?	0.01	0.03	0.01	?	0.05	2.71	0.55	?
Zn	14	3,000	29	?	9	600	3.6	?	1	60	0.58	?	24	3,360	33	?
Power plant	0.05	3.0	4.4	—	0.02	0.5	1.1	0.5	0.01	0.6	0.88	—	0.08	4.10	6.4	0.05
Sewage	—	—	—	—	—	—	—	—	0.05	12	3.07	0.01	0.05	12.0	3.1	0.01
Solid waste disposal	0.40	42	14	0.2	—	—	—	?	0.02	2.1	0.28	?	0.42	44.1	14.3	0.2
Agriculture	0.01	0.15	—	0.01	0.01	0.1	—	—	0.01	0.01	—	—	0.03	0.27	—	0.01
Urban runoff	0.30	6	2	—	—	—	—	—	0.03	0.6	0.1	—	0.33	6.60	2.1	—
Harbor and navigation	0.50	95	19	—	—	—	—	—	0.05	10	0.95	—	0.55	105	20	—
Manufacture:																
Paper	0.30	58	29	—	0.01	0.03	0.03	—	0.01	2.4	0.25	?	0.32	58.4	29.3	?
Chemicals	0.40	35	0.02	—	0.01	0.2	0.12	—	0.04	0.02	0.20	0.01	0.42	35.2	0.34	0.01
Plastic & rubber	0.04	3.9	7.7	—	0.9	8.3	1.24	—	0.01	0.26	0.01	0.01	0.95	12.5	9	0.03
Metallurgy	—	10	?	0.02	—	1.0	—	—	—	1.1	—	0.04	—	12.1	—	0.06
Total	17.30	3,423	270	0.22	10.3	641	61.2	0.15	1.3	106	14.6	0.07	28.5	3,868	346	0.47

of paper and plastics, is also a significant source of Zn and Pb. In the case of Hg, however, diffuse sources such as solid waste disposal and energy generation by a thermoelectric power plant provide the majority input, reaching a total annual emission of about 0.3 tons per year (Barcellos and Lacerda 1994; Marins et al. 1999).

Emissions of all metals are mostly to soils, associated with tailings, and in solid wastes from metal smelting and a landfill. Contribution to soils from the total emission estimates reach 82%, 78%, 58%, and 53% for Zn, Pb, Cd, and Hg, respectively. Surprisingly, direct contribution to the basin water bodies are of less importance for all metals, in particular for Zn and Pb. Emission to waters reach only 10%, 8%, 5%, and 3% for Hg, Cd, Pb, and Zn, respectively. Atmospheric emissions can contribute to significant amounts of trace metals, in particular for Hg and Cd, reaching up to 36%, 34%, 17%, and 15%, for Hg, Cd, Pb, and Zn, respectively. The larger emission to Sepetiba Bay basin soils highlights the significance of soil use changes as a key factor controlling the further mobilization of trace metals to Sepetiba Bay.

TRACE METALS INPUT TO SEPETIBA BAY PROPER

Measurements of river and atmospheric inputs of trace metals to Sepetiba Bay are shown in Table 13.3. Major inputs are from rivers, particularly the three major rivers draining the most industrialized and urbanized sectors of Sepetiba Bay basin (Guandú River, Guarda River, and São Francisco Canal). Significant inputs of Zn and Cd (about 38% and 11% of the total, respectively), however, reach the bay through the atmosphere. Direct atmospheric Hg and Pb input to the bay corresponds to only 5% and 2% of the total inputs, respectively (Marins et al. 1999).

When comparing measured inputs with estimated emissions, the four metals present contrasting results. Inputs of Cd, Zn, and Pb are lower than the estimated emissions by a factor of 20 for Zn, 15 for Cd, and 8 for Pb. This suggests that a significant portion of the emissions are retained in the basin. Basin retention, therefore, reaches up to 95%, 93%, and 88% for Zn, Cd, and Pb, respectively. This also reflects the larger emissions to soils compared to the atmosphere and waters. In the case of Hg, however, measured inputs to Sepetiba Bay are higher by a factor of 1.7 as compared to the estimated emissions. Marins et al. (1999) suggested that fluvial inputs from the Paraíba do Sul River basin, which are diverted to Sepetiba Bay basin for the water supply for Rio de Janeiro city, may bring Hg to Sepetiba Bay basin. The Paraíba do Sul River has been reported to present larger Hg concentrations than the rivers of the Sepetiba Bay basin (Marins et al. 1999).

Table 13.3 Heavy metals inputs to Sepetiba Bay ($t\ y^{-1}$). Sources: Pedlowski et al. (1991), Barcellos and Lacerda (1994), and Marins et al. (1999).

Input/metal	Zn	Cd	Pb	Hg
Atmospheric	56	0.2	3	0.03
Fluvial	144	1.8	43	0.65
Total	200	2.0	146	0.68

FATE OF TRACE METALS IN SEPETIBA BAY

Once reaching the bay, trace metals are transported in association with suspended particles, following distinct patterns of dispersion and deposition. During plume dispersion, a large fraction is initially deposited near the river mouths. The remaining bypassed material undergoes gradual sedimentation during transport by the surface currents, generally moving clockwise along the southwestern coast of the bay to, unfortunately, the location of a fishery reserve. Previous works on the trace metal contamination of Sepetiba Bay biota have shown moderate to high contamination, in particular for Cd and Zn, and to a lesser level of Hg (Pfeiffer et al. 1985; Carvalho et al. 1993; Kherig 1995). Surface sediment concentrations of, for example, Zn lie between 700 and 2000 $\mu g\ g^{-1}$ (Cd between 5.0 and 8.5 $\mu g\ g^{-1}$) along the SW coast close to the river sources. However, similarly high concentrations are also encountered in some sections of the central part of the bay, where clay sediments also predominate. Thus, not all materials initially deposited are retained or transported close to shore. It has been shown that wind-induced mixing and tidal pumping results in considerable resuspension of bottom materials in shallow areas. Part of these are transported to deeper waters by consecutive cycles of resuspension, advection, and redeposition along the bottom, also known as material cascading.

Mass balance studies suggest that most of the trace metal loads to Sepetiba Bay are not buried immediately into bottom sediments. Instead, due to easy resuspension and high biological production in the water column, large fractions of the trace metals loads are exchanged between bottom sediments and the water column, which explains the contamination level found in the local biota.

MANAGEMENT ISSUES

Land-based emissions of several trace metals and Zn and Cd concentrations in surface sediments off the river mouths and in the central part of Sepetiba Bay are beyond the levels permitted by legislation. However, the effective control of the impact of trace element emissions harbors some specific problems, including transboundary issues. Trace elements are ubiquitous byproducts of most anthropogenic activities and reach coastal waters through rivers and the atmosphere, even when sources are located far from the coastline. For example, Hg contamination on ecologically and economically important coastal areas of southeast Brazil is due to diffuse inputs from the atmosphere, which has been receiving Hg from a variety of sources located along river basins far from the coastal area (Marins et al. 1996, 1999). As shown for Sepetiba Bay, trace metals are derived from diffuse and point sources by numerous socioeconomic activities spread over the drainage basin and also from the adjacent inland Paraíba do Sul River basin, as well as by abstraction of the rivers' freshwater to fulfill the growing demand of the local industry and population. Other examples can be listed here, suggesting that whatever control policies are applied to metal-emitting activities located at the coastal area, activities much further upstream in the basin can still contribute a significant input of trace metals to coastal ecosystems and time-delayed effects can happen even many years after the trace metal source has been banned, including Sepetiba Bay.

There is no simple solution to the management of trace metal impacts and other issues in Sepetiba Bay, and the radical options that prevail are difficult to digest from political and socioeconomic points of view. Solid scientific information and environmental legislation exist to support remedial actions and to press for monitoring within the scope of strategic management. However, effective control and strategic management of Sepetiba Bay is hampered by the lack of an integrated management structure, even though all legal instruments for management of Sepetiba Bay exist under present legislation (IFIAS 1988).

No central organ directs management policy for the region, although the Bay falls under the jurisdiction of the State Secretary for the Environment, which is responsible for regulating all activities that potentially affect the Bay's environment. Land-based activities are regulated by the local municipalities, harbor facilities by the state of Rio de Janeiro and federal authorities. Sea-based activities, such as shipping and the shoreline, are in the legislative sphere of the Navy and Fisheries, which is specially regulated by the Ministry of the Interior via its regional offices. Aspects related to the environment (e.g., licensing industries, preservation of mangroves and other natural habitats, and water quality) are covered by the National Council for the Environment through the CONAMA legislation of 1986 and are enforced by the local councils which control and fund such activities. The CONAMA resolution has made the presentation of an Environmental Impact Report (RIMA) compulsory for any new land-based development in view of its impact upon the system. RIMA represents a first step in proactive management and is, apart from the basic rules to be fulfilled, open to inclusion of new concepts and adaptations to the local needs of the environmental problems to be addressed. Its main caveat remains, however, the lack of guidelines to assess the state of the management structure of the environment in question. This problem is being addressed by the GERCO–PNGC–PNMA (1996) program, which envisages the establishment of an ICM plan for Brazil in the near future.

At present, there is an urgent need to integrate the management of Sepetiba Bay, as environmental problems related to trace metals emerge faster than they can be solved. Apart from problems in compliance of emission levels by industry, the industrial port complex of Sepetiba Bay is being expanded to reduce pressure from other industrialized areas in the vicinity of Rio de Janeiro and to accommodate the largest recipient harbor of the countries involved in the Mercosul economic market agreement. In 1997, for example, one of the dredging activities for port expansion aimed to deepen the main access channel to the sea. Estimates have shown, that the top 50 cm of the sediments in the nearshore area contained about 7 t of Cd, 4 t of Cu, 0.6 t of Pb, and 0.5 t of Hg. In 1997 alone, about 21×10^6 m^3 of sediments were dredged and henceforth deposited within the Bay, albeit in an area still little affected by industrial effluents. Macroalgae were monitored for their trace metal contents in the vicinity of the dumping site and exhibited a four-fold increase of Cd in their tissues shortly after the works (Amado Filho et al. 1999). The effect of the dredging activities upon other components of the biota is under study.

Public opinion is now concerned that Sepetiba Bay is evolving into a "second Guanabara Bay." A recent oil spill in the upper mangrove-dominated reaches of Guanabara Bay and recurrent fish dieoff in many coastal lagoons have triggered manifold joint actions between the state government, environmental protection agencies, industry, scientific community, and nongovernmental organizations. It is the hope that many of the now officially established partnerships will also focus on Sepetiba Bay to develop a practically feasible ICM plan. There are indications that pollution of Sepetiba Bay is now affecting adjacent bays which sustain important mariculture and tourism activities (Amado Filho et al. 1999). Such activities have been acting as a buffer to severe social problems due to increasing unemployment. Therefore, environmental protection is running parallel with socioeconomic development, bringing important support from portions of society that were previously passive to the increasing degradation of the environment.

REFERENCES

Amado Filho, G.M., C.E. Rezende, and L.D. Lacerda. 1999. Poluição da baía de Sepetiba já ameaça outras áreas. *Ciência Hoje* **25 (149)**:46–48.

Barcellos, C. 1995. Geodinâmica de Cádmio e Zinco na Baía de Sepetiba. Ph.D. diss., Instituto de Química, Univ. Federal Fluminense, Niterói, Brazil.

Barcellos, C., and L.D. Lacerda. 1994. Cadmium and zinc source assessment in the Sepetiba Bay and basin. *Env. Monit. Assess.* **29**:183–199.

Barcellos, C., L.D. Lacerda, and S. Ceradini. 1997. Sediment origin and budget in Sepetiba Bay (Brazil) — An approach based on multi-element analysis. *Env. Geol.* **32**:203–209.

Carvalho, C.E., L.D. Lacerda, C.E. Rezende, and J.J. Abrão. 1993. Titanium and calcium as tracers for continental and oceanic materials in the Brazilian continental shelf. In: Proc. III Simpósio de Ecossistemas da Costa Brasileira, Rio de Janeiro, pp. 122–127. Rio de Janeiro: Univ. Estadual do Rio de Janeiro (UERJ).

Forte, C.M.S. 1996. Determinação de taxas de sedimentação na porção nordeste da Baía de Sepetiba utilizando datação com radioisótopo ^{210}Pb. M.Sc. thesis, Instituto de Química, Univ. Federal Fluminense, Niterói, Brazil.

GERCO–PNGC–PNMA. 1996. Macrodiagnóstico da zona costeira do Brasil na escala da União. Gerenciamento costeiro, Programa Nacional do Meio Ambiente. Brasília: Ministério do Meio Ambiente (MMA).

IFIAS (Intl. Federation of Institutes for Advanced Study). 1988. Sepetiba Bay Management Study. IFIAS Research Programme on Coastal Resources Management, Workplan Report No. 69. Toronto: IFIAS.

Kherig, H.A. 1995. Estudo da contaminação por mercúrio em corvinas (*Micropogonias furnierii*) em quatro áreas costeiras do Brazil. M.Sc. thesis, Pontifícia Univ. Católica, Rio de Janeiro.

Lacerda, L.D. 1998. Biogeochemistry of Trace Metal and Diffuse Pollution in Mangrove Ecosystems. Okinawa: Intl. Soc. for Mangrove Ecosystems.

Lacerda, L.D., W.C. Pfeiffer, and M. Fiszman. 1987. Heavy metal distribution, availability, and fate in Sepetiba Bay, S.E. Brazil. *Sci. Total Env.* **65**:163–173.

Marins, R.V. 1998. Biogeoquímica de Mercúrio na Baía de Sepetiba. Ph.D. diss., Dpto. de Geoquímica, Univ. Federal Fluminense, Niterói, Brazil.

Marins, R.V., L.D. Lacerda, and R.C. Villas Boas. 1999. Relative importance of non-point sources of mercury to an industrialized coastal system, Sepetiba Bay, S.E. Brazil. In: Mercury Contaminated Sites, ed. R. Ebinghaus et al., pp. 207–220. Berlin: Springer.

Marins, R.V., E.V. Silva Filho, and L.D. Lacerda. 1996. Atmospheric mercury deposition over Sepetiba Bay, S.E. Brazil. *J. Braz. Chem. Soc.* **9**:177–181.

Pedlowski, M.A., et al. 1991. Atmospheric inputs of Fe, Zn, and Mn to Sepetiba Bay, Rio de Janeiro. *Ciência e Cultura* **43**:380–382.

Pfeiffer, W.C., L.D. Lacerda, M. Fiszman, and N.R.W. Lima. 1985. Metais pesados no pescado da baía de Sepetiba. *Ciência e Cultura* **37**:297–302.

14

Integrated Coastal Management

The Philippine Experience

L. TALAUE-MCMANUS

Marine Science Institute, College of Science, University of the Philippines, Diliman, Quezon City 1101, Philippines and
Division of Marine Affairs, Rosenstiel School of Marine and Atmospheric Science, University of Miami, 4600 Rickenbacker Causeway, Miami, FL 33149 U.S.A.

ABSTRACT

Integrated coastal management in the Philippines evolved in response to an overexploited capture fisheries, a degraded coastal environment, and government policies which exacerbated this environmental damage. While major government programs were implemented to mitigate these conditions, the development of community-based management began as a result of moderate success achieved in the agroforestry sector and in the perceived inability of government to curb overexploitation effectively. In 1991, the passage of the Local Government Code legally recognized the significant role communities could play in resource management. The code stipulated for the management of coastal waters by municipal governments, such as through the formulation of coastal and land-use development plans. However, devolution could not, by itself, guarantee effective management when local government units were not appropriately prepared for such a weighty task. Local communities and their leaders need to be empowered with knowledge and skills that best science could provide. Within the context of community-based coastal management, such empowerment of community members and their leaders has been pivotal in driving collective action towards effective interventions.

INTRODUCTION

The Philippines is the world's second largest archipelago next to Indonesia. It consists of 7100 islands, with an aggregate land area of 300,000 km^2 and a total coastline of 36,300 km. Its territorial waters up to the exclusive economic zone (EEZ) extends to 2.2 × 10^6 km^2 (7 times its land area). The shelf area to 200 m extends over 184,600 km^2 and accommodates one of the richest marine biodiversity worldwide, with 421 species in 70 genera of hard corals

Science and Integrated Coastal Management
Edited by B. von Bodungen and R.K. Turner

(Aliño 1994), about 294 species of reef fishes unique to the Philippines (Aliño 1994), 30 mangrove tree species (Spalding et al. 1997), and 16 seagrass species (Fortes 1995).

Inhabited by 76 million Filipinos, 80% of the population are coastal dwellers and depend heavily on the coastal and marine environment. The fisheries sector contributes about 4% to the gross domestic product and provides income for 1 million people (5% of the country's labor force, a gross underestimation because sustenance fisheries have not been appropriately estimated in the national accounts system). Food security is defined by access to fish protein, which contributes 60% of dietary requirement at 28.5 kg capita^{-1} y^{-1}.

Thus, a major impetus in the evolution of coastal management in the Philippines has been a desire to optimize the harvest of fishery resources as well as the revenues generated from the trade of raw and processed fishery products. Since the 1970s, the concerns of fisheries management have widened to subsume those of habitat degradation and pollution. The choice and implementation of management interventions during this period indicated how scientific information and data were used, misused, or ignored in justifying these measures. These experiences also highlight the major gaps in knowledge that science should address to make management effective and meaningful.

In this chapter, I review the Philippine experience to (a) determine the underpinning assumptions (scientific or otherwise) of early management initiatives, (b) describe evolving management approaches under a policy regime of devolved governance, and (c) underscore the management needs that scientific research should address.

EARLY INTERVENTIONS: SUCCESSES AND FAILURES

Fishery Policies of the 1970s and 1980s

Major laws on the disposition and management of fishery resources and associated coastal and marine habitats were formulated in the 1970s. The fundamental policy orientation of these laws was to develop the fisheries sector into a major contributor to an agriculturally based economy. While also containing provisions on the conservation of fish resources, they focused mainly on disposition, use, and exploitation. Presidential Decree 43 (1972) created the Fishery Industry Development Council whose mandate was to accelerate the development of the fishing industry (Kalagayan 1990). Presidential Decree 704, also known as the Fisheries Decree of 1975, permitted for the exportation of fish and fishery products to allow the industry to contribute significantly to the national economy. In 1976, Presidential Decree 977 was promulgated, creating the Philippine Fish Marketing Authority, to provide an efficient marketing system for the disposition of fishery products. In addition, major loans were secured from the Asian Development Bank during this period (Asian Development Bank 1997). A US$5.5 million loan to build the Navotas Fisheries Port in Metro Manila was approved in March, 1971. To expand freshwater aquaculture, the government borrowed $9 million in December, 1978, to finance the development of fish pens in Laguna de Bay. In 1979, $18 million was obtained to develop the capture fisheries of Northern Palawan.

Using retrospective analysis of archived fishery records, Silvestre et al. (1986) and Dalzell et al. (1987) showed that peak catches for small pelagic and demersal fisheries were reached in the mid-1970s. By the mid-1980s, fishing effort had exceeded, by 150–300%, that needed to harvest at maximum economic yield, and by more than 30% that required to fish at maximum sustainable yield (Silvestre and Pauly 1997). Thus, in the mid-1970s, the policies assumed that the high fish harvests would be sustained and that increase in effort through more intensive capitalization would lead to greater production. One may argue that the studies above, which were published in the following decade, could have constrained the expansionist desire of the Marcos era. However, a look at subsequent policy formulations of this government obliterates this possibility.

The rapid increase in fisheries production during the 1970s and the perception that fishery resources were abundant enough to increase fishing effort led the government to pursue the Expanded Fish Production Program from 1983 to 1987. The objectives of this program included: (a) self-sufficiency in fish; (b) optimal production and use of fishery resources; (c) increase in income of fishers; (d) promotion of import substitution; and (e) increase in export of fish and fishery products (Ferrer 1993). To reach these goals, artisanal fishers were encouraged to take out loans to build bigger, motorized boats and to acquire bigger nets and more efficient fishing gear. Projects such as the *Biyayang Dagat* (Bounty of the Sea) exemplified this orientation towards increasing fisheries production. Aquaculture was greatly encouraged so that mangrove conversion into fishponds resulted in the loss of 70% of mangrove area from 1918 to 1988 (Calumpong 1994).

From 1985 to 1994, total fish production increased from 2.05 to 2.69 million tons. The share of municipal fishers (those using boats weighing 3 gross tons or less) decreased from 60% in 1980 to 43% in 1992, while commercial fisheries (using boats above 3 gross tons) and aquaculture increased (Lacanilao and Fernandez 1997). The aggregate harvest of small-scale fishers decreased despite the doubling in the number of municipal fishing boats during the period 1980 to 1985. The concurrent decrease in municipal catch and increase in commercial harvest was a definite indicator that the total fisheries production (i.e., sum of municipal and commercial harvest) had reached its maximum limit so that these trends reflected changes in catch distribution, not in total production values. Encroachment of commercial vessels into the near-shore fishing grounds of small-scale fishers resulted in catch reallocation, reflecting differential access to the remaining capture fisheries.

Economic imperatives drove the formulation of policies and the implementation of fisheries programs towards the expansion of the capture and culture fisheries sectors. Within such programs, there were no stipulations to monitor the environmental impacts and socioeconomic feedback these interventions would have. In the case of coral reefs, the percentage of surveyed stations with 75% to 100% live coral cover decreased from 15% in the 1970s to 5% by 1990 (Gomez et al. 1981; Gomez 1990). For mangroves, the rate of area reduction was about 3700 ha y^{-1}, which was similar to the rate of expansion of fishpond areas at 4000 ha y^{-1} (Calumpong 1994). The rapid loss occurred despite laws that provided for mangroves to be inalienable, beginning in 1975. The government could lease areas up to 50 ha to individuals and 500 ha to corporations (Alcala and Van de Vusse 1994). Because there were no set limits in total area to be converted and in how these should be geographically distributed, the rapid loss of mangroves went on unabated. In 1981, through Presidential Proclamations 2151 and 2152, about 79,000 ha were declared as mangrove wilderness (4500 ha) and forest reserve (74,500 ha). The proclamations were made without consulting traditional users who, in turn, essentially ignored the new laws and continued with illegal mangrove conversion.

In effect, the move to expand the fishing industry through legislation, the establishment of institutions, and the implementation of fishery development programs worsened overexploitation, caused widespread habitat degradation, and resulted in further impoverishment of the sustenance fishers.

Early Management Interventions

Along with fisheries expansion, government made its first attempt to coordinate management of coastal habitats in 1977 when the Philippine President directed the National Environmental Protection Council to organize a multiagency task force, the Coastal Zone Management Committee (CZMC), which started with eight institutions as members (Kalagayan 1990). By 1979, the CZMC had grown to include 22 agencies and, among its first tasks, set out to define and delineate the spatial bounds of the coastal zone. Despite the fact that the team was multiagency, the CZMC failed to harmonize its mandates in the context of coastal zone management.

What prompted the government to establish the CZMC? The Stockholm Summit of 1972 spurred a global environmental movement, and the Philippines responded to this. In addition, the negative impacts from fisheries development were evident by the latter half of the 1970s. Poverty and environmental degradation in the coastal zone had to be addressed, but within a framework where these were considered as separate issues, not as impacts of resource exploitation. Kalagayan (1990, p. 47) noted that the policy shifts from one of use orientation to that of resource management did not neutralize the underlying economic driving forces, but "merely tempered the tendency to overexploit"

While government pursued its schizoid scheme to exploit natural resources further and attempted to manage the resultant impacts as if they were unrelated, academic and nongovernmental organizations decided to use an alternative paradigm for resource management in the coastal zone. Alcala (1998) cited three reasons for the adoption of this paradigm:

1. The increasing amount of scientific data and information from the use of self-contained underwater breathing apparatus (SCUBA), which linked decreased fish catch to degraded coastal ecosystems such as coral reefs.
2. The perceived inability of central government to mitigate significantly the degraded state of the marine environment.
3. The growing experience that agroforestry programs which engaged community participation generated relative success over those that did not.

Hence, the incipient stage of community-based coastal resources management (CB-CRM) began. Three such projects described below show the evolution of CB-CRM as a program framework.

The three chosen examples of coastal resources management highlight the potency of grassroots participation in resource management as shown in the functional marine reserves of Sumilon and Apo Islands. For Lingayen Gulf, the confluence of scientific information, legislation, and the establishment of the Lingayen Gulf Coastal Area Management Commission provided the necessary but as yet insufficient machinery for effective bay-wide management.

Sumilon Island Reserve

In 1974, the first functional marine protected area was established with grassroots participation in central Philippines (White 1987). Known as the Sumilon Island Reserve, it was declared a municipal reserve in 1974 and a national fish sanctuary in 1980 with the technical assistance of the Silliman University Marine Laboratory under the leadership of Dr. Angel Alcala. Before 1974, the Philippines had already gazetted a number of paper marine parks and reserves as a result of the passage of Republic Act 3915 in 1932, which also provided for the establishment of the National Park System (Kalagayan 1990). In Sumilon, fish yields at 14–21 t km^{-2} y^{-1} for the period 1976–1979 increased to 36 t km^{-2} y^{-1} in 1984 (White 1987), indicating that marine reserves may be a viable strategy to allow growth and reproduction to occur within natural habitats with protection over sufficiently long periods.

Apo Island Reserve and Community-based Management

Based on the experience in Sumilon Island, Silliman University implemented the Marine Conservation and Development Program (MCDP) in 1984–1986 in collaboration with the Asia Foundation and the United States Agency for International Development (USAID). The program was designed to assist three small island communities in establishing an integrated resource management of their reef systems and implementing, by themselves, the necessary action programs in marine management, agroforestry, community development, and institutional development and linkages (Calumpong 1993). For one of the sites, Apo Island, the sustainability of two local organizations and their ability to carry out coastal resources management was assessed four and six years after project initiation. Calumpong (1993) reported that program activities were continuously implemented by the Marine Management Committee, including the maintenance of a marine reserve. This has resulted in an increase in tourism and a doubling of the fish harvest, both of which increased the income of the island population. However, at that time the Marine Management Committee needed additional skills to cope with the increasing pressures of tourism. Alcala (1998) reported that he found a significant positive correlation of both density and species richness of large carnivorous reef fish during the 11-year period when Apo Island was protected, especially in areas closest to the reserve (within 200–300 m distance).

Scientific Research and Planning for Lingayen Gulf, Northern Philippines

Moving from small to big islands, the goal of ASEAN-US Coastal Resources Management Project (1986–1992) was to empower six ASEAN countries to develop their coastal resources on a sustainable basis (Chua and Scura 1992). Using a grant from the USAID and implemented by the International Center for Living Aquatic Resources Management (ICLARM), the regional project had three basic components: (a) research and coastal planning, (b) capacity building of government-based coastal managers, and (c) policy formulation. The Philippine site was Lingayen Gulf, 2100 km^2 wide and a coastline 160 km long, with 18 municipalities in two provinces. Research conducted over a two-year period (1986–1988) covered capture fisheries, coral reef fisheries, water quality, aquaculture, socioeconomic, and institutional analysis.

The yield to biomass ratio of 5.2 in capture fisheries of Lingayen Gulf was indicative of excessive overfishing (Silvestre et al. 1991). The use of nets with 2 cm mesh size in the cod-end of bottom trawls resulted in growth overfishing and a loss of 20% of the yield and 40% of the value per recruit. Ochavillo and Silvestre (1991) suggested that a mesh size of 4 cm is more appropriate. Furthermore, Calud et al. (1991) showed that the municipal gillnet fisheries were competing for the same fish assemblage (size as well as species) that were captured by encroaching commercial trawlers, which ignored the 7 km and 7 fathom delineation reserved for municipal fishing operations.

In the coral reefs, unequivocal signs of overharvest included decreasing adult fish density and diversity as well as a similar trend in the size of reproductively mature fish from 1988 to 1991 (McManus et al. 1992). A highly valuable sea urchin fishery collapsed in 1992 because of unabated harvest to meet an insatiable export demand (Talaue-McManus and Kesner 1995). Live coral cover in the reefs was fair to good (30–51%); siltation, dynamite, and cyanide poison were the chronic sources of stress (Meñez et al. 1991). Protection of ecologically critical sites in the reef was a recommended strategy to mitigate habitat degradation and overexploitation.

In terms of water quality, open gulf waters were within limits for heavy metals as set by the National Pollution Control Commission (Maaliw et al. 1989). However, nearshore waters adjacent to river mouths showed excessive levels of nutrients, heavy metals, suspended solids, and coliform bacteria that could adversely affect the culture of milkfish and prawns. Heavy metals were released by mining operations in the watershed areas of the rivers in the Cordillera mountain range.

Economic studies indicated that long lines and illegal dynamite fishing (when environmental impacts were not considered in the valuation) were among the most economically efficient gear on a net return basis. Because of a monsoonal climate, fishers in the gulf were gainfully employed for only 11 days per month, half that of a regular employee (22 days per month). Furthermore, the social tolerance shown by village members to fishers who use illegal fishing methods because they were given a portion of the fish catch for free, indicated that environmental education would have to change value systems and perceptions of socially acceptable behavioral norms (Galvez et al. 1989).

The results discussed above were used as basis for a planning exercise with representatives from the subregional economic and planning authority, local governments, nongovernmental organizations, and academia. The gulf plan was adopted for the period 1990–2020. However, the plan was not implemented until after the Lingayen Gulf Coastal Area Management Commission (LGCAMC) was created in 1994.

Two legislative measures, passed in 1993 and in 1994, institutionalized the gulf plan into the management agenda for Region I of the Philippines. In 1993, Presidential Proclamation 156 declared the gulf an environmentally critical area, based on the scientific information generated by the research phase of the project. In 1994, President Ramos issued Executive Order 171, creating the LGCAMC to implement the gulf plan.

LGCAMC was most successful in implementing activities for environmental education and policy advocacy in its first year of operation (Talaue-McManus and Chua 1997). It identified key policy issues that would need to be resolved by legislation, including those on the banning of commercial fisheries in the gulf, gulf-wide law enforcement, coastal development, and safeguards against the establishment of potentially pollutive industries. However,

LGCAMC lacks the necessary technical expertise for resource management and habitat restoration.

EVOLVING ICM APPROACHES IN THE 1990S

Policy Shift from Centralization to Devolution

In 1991, the Local Government Code of 1991 was enacted to provide a new crucible for coastal zone management in the Philippines (Talaue-McManus and Chua 1997). It was a major move to decentralize governance and devolve powers and responsibilities from the central government to local government units (towns, cities, and provinces). The code has been a singular driving force for coastal towns and cities to formulate their land-use and coastal development plans, and for provinces to develop land-use and physical framework plans that subsume plans of their component towns. However, the provision of a legal mandate to effect coastal resource management has not been sufficient without an appropriate orientation of skills and resources at all hierarchies of governance. A similar empowerment, mainly through knowledge and environmental awareness among management partners, including citizens' groups, academe, nongovernmental organizations, business and religious sectors, and others, needs to take place to achieve an integrated and collective stewardship of natural resources.

At the national level, the government embarked on the Fishery Sector Program (1990–1995), which aimed at reversing the trend in overfishing in municipal waters, overexploitation of resources, and habitat degradation that resulted from the expansionist policies of the previous decades. Formulated in 1989, its goals included the rehabilitation of the coastal zone, the reduction of extensive poverty of coastal communities, and the improvement of the productivity of the fishery sector. To achieve these goals, the project subsumed three major components:

1. Fisheries resource management that included data management, coastal resource management, planning and implementation, fisheries legislation and regulations, and community-based law enforcement, among others.
2. Diversification through community organizing, development of micro-enterprises, and provision of support for mariculture.
3. Capacity building of public institutions responsible for fisheries management at all levels but with special attention to those at the local government units.

The project was funded at a cost of $89 million which was to be implemented in 100 coastal municipalities surrounding 18 of 26 priority bays.

To provide a scientific basis for management, resource and ecological assessment studies (REAs) as well as those on socioeconomic and investment opportunities (SEOs) were conducted for each of the twelve bays as stipulated by the Fishery Sector Program. The REAs covered assessments of fisheries stocks, habitats, water quality, plankton, and physical oceanography, while the SEOs looked at demography, income indicators, socioeconomic characteristics, economic evaluation, entrepreneurship, economic resource potential, and options for alternative livelihood. However, no attempt was made to integrate REAs and SEOs in order to formulate clear management recommendations. Independently, these studies made separate recommendations that were at best limited to either the socioeconomic or ecological perspectives. Inability or lack of desire for integration constrains the development of integrated planning in both developing and developed countries. Clearly, there is a scientific need to develop analytical tools that require integration of natural science and economic data, and which is requisite to formulating scientifically sound management guidelines. Finally, the effective implementation of the latter requires a thorough institutional analysis to show how stakeholders can translate guidelines into action.

The Fishery Sector Program initiated coastal resource management in 12 priority bays with 128 municipalities. These included Carigara, Panguil, Calauag, San Miguel, Manila, Ormoc, Tayabas, Sorsogon, San Pedro and Sogod Bays; and the gulfs of Ragay and Lagonoy. Bay Management Councils were formed to provide stewardship of fishery resources at an integrated basin level. Management blueprints for each of the 12 bays were designed with broad participation of communities (Asian Development Bank 1997). The viability of these management institutions over time needs to be evaluated. A number of them have ceased to function because of a lack of understanding of their role, weak leadership, and a dependence on external funding sources. In 98 of the 128 municipalities covered by the program, municipal fishery ordinances were legislated and a number have committed funds to continue coastal resource management activities, including resource regeneration, law enforcement, and income diversification. The capability of participating municipalities to apprehend increased with the mobilization of 77 enforcement teams composed of 6200 fish wardens, as shown by a significant reduction of destructive fish activities. However, conviction rates remained low because of weak prosecution cases and unfamiliarity with fisheries regulations that need to be harmonized across governance levels.

An evaluation of the Fishery Sector Program by its donor, the Asian Development Bank, indicated a number of insufficiencies which a prospective program would have to address (Asian Development Bank 1997). Most important was the obvious lack of capability of the local government units to embark on thorough assessments of capacity and training needs of local governments that would be the prerequisite for future bank projects. A second point was the lack of integrated planning that would subsume all sectoral activities in the coastal zone and include the significant influence of those done in the watershed (such as deforestation and mining). Third was the lack of institutional power among bay management councils to manage resources effectively at the bay-wide level, which subsumed local government units (municipalities) that had the mandate to manage resources at the municipal level. The assessment iterated the earlier evaluation of the Lingayen Gulf Commission and served to underscore the major bane of devolution, i.e., devolving governance powers to insufficiently trained and thus incompetent local governments. Furthermore, the creation of bay- or gulf-wide institutions that subsume a number of local government units without providing the appropriate stipulations on when these could overrule municipal decisions was exacerbating institutional conflicts.

Community-based Coastal Resource Management

The major role that community groups could play in natural resource management was recognized in the 1950s in the field of agroforestry (Ferrer 1993) and applied for the first time in the context of marine reserves in Sumilon Island (Alcala 1998). From the foregoing discussion, the paradigm of CB-CRM was widely adopted by projects in the 1980s (see Table 14.1). By

Table 14.1 Major projects on coastal zone management in the 1980s.

Project Title (duration; fund source and type)	Spatial Scale	Project Components	Reference
Central Visayas Regional Project (1984–1991; World Bank loan)	Three small island communities	• Marine management • Agroforestry • Community development • Institutional development	White and Savina (1987); White (1989); Alcala and Van de Vusse (1994)
Marine Conservation and Development Program (1984–1986; USAID grant)	182 villages in four provinces of Central Visayas (Region VII)	• Upland agriculture • Social forestry • Nearshore fisheries	Bojos (1994); de los Angeles and Pelayo (1995)
ASEAN-US Coastal Resources Management Project (1986– 1992; USAID grant)	Six ASEAN countries, each with a demonstration site; Lingayen Gulf, NW Philippines	• Research and coastal planning • Capacity building • Policy formulation	Chua and Scura (1992)
Marine Conservation Project (1988–1993; Dutch Embassy grant)	San Salvador Island, Zambales	• Resource management and planning • Community organizing and education • Livelihood • Networking	Dizon and Miranda (1996)
Fishery Integrated Resource Management for Economic Development Program (1988–1993, grant)	Daram, Samar	• Community • Human resource development • Socioeconomic and livelihood development • Sustainable fisheries	Tan (1993)

the 1990s, the role people's organizations could play was recognized by law with the passage of the Local Government Code. Notwithstanding the scientific principle of analyzing resource-related issues at different spatial and temporal scales, the local governance scale was clearly the level at which science and policy would have to be strategically integrated.

A thorough analysis of the CB-CRM experience in the Philippines is beyond the scope of this chapter. As a paradigm, it empowers community constituents and their local government to implement a coastal management agenda through transfer of appropriate knowledge and skills (McManus 1995). Depending on who formulates the project, the priority of empowerment ranges from purely organized people's or community organizations (Añonuevo 1994) to major stakeholders, including elected officials and the business sector (comanagement *sensu* Pomeroy and Williams 1994; Pomeroy 1995). In reality, coastal resource management at the community level is achieved through consensus building and compromises within the constraints of existing economic and political realities.

The major components of CB-CRM in the 1990s included community mobilization and environmental education towards the creation of local organizations (e.g., fishers, farmers, women, youth, aquaculture operators) that would have coastal management as their major mandate (McManus 1995; see Table 14.2). Once organized and trained, these local groups could then engage in the action elements of management, including coastal development planning, resource management, livelihood development, networking, and advocacy. The viability of CB-CRM has been tested in a number of locations, including the Municipality of Bolinao in northern Luzon, where a project was implemented in 1993 (Talaue-McManus et al. 1999). To date, this municipality along with a number of local organizations has legislated a coastal development plan and implemented village-based mangrove reforestation, a reef-based marine protected area, and water quality monitoring of Caquiputan Channel, where coastal milkfish fish pens and cages have been set up. The coastal development plan was formulated in a participatory manner by a multisectoral drafting committee that was assisted by the CB-CRM Project of the University of the Philippines Marine Science Institute. The plan is the first of its kind and has been accepted by public consultation and legislation at the municipal and provincial levels.

Scientific contributions to CB-CRM initiatives are also best exemplified by the project in Bolinao, where the University of the Philippines Marine Science Institute has conducted research over the last 25 years. In particular projects, collaboration with social scientists and economists have widened the scientific data base beyond marine science. The major goals of the project have been the ecological sustainability and equitable distribution of economic benefits derived from management interventions, which would be implemented by user groups in collaboration with the local government. The project staff would prepare technical studies using existing scientific data, which would be presented to the user groups for them to make informed decisions. Where conflicts over use existed, the technical studies were crucial in achieving consensus. Where political alliances overcame technical soundness of recommendations, the stakeholders were made aware that science and policy would have to be integrated on the basis of sound science to sustain their resources. With active participation of community groups, coastal waters were spatially allocated on the basis of geomorphological characteristics and resource use patterns. Within relevant use zones, areas for mangrove reforestation and marine protected areas were identified and established based on unequivocal data that the reef-based fisheries were highly unsustainable (McManus et al. 1992). Waters in the channel were initially classified according to their ability to support coastal aquaculture on the basis of water residence time, bathymetry, and flushing rates. The technical data was presented to users and the legislative council as they sought to decide how to allocate the limited sea space that was appropriate for aquaculture. The technical basis for decision making was not accepted until water quality deteriorated and aquaculture production declined as a result of overcrowded fish pens and fish cages because of unregulated issuance of permits. During the advocacy against the establishment of a cement plant complex by the coast of Bolinao, a massive information campaign was launched to discuss the possible ecological and economic impacts of the plant, essentially an independent cost-benefit analysis to the environmental impact study for the plant (Yambao and Talaue-McManus 1999).

To date, the impacts of coastal aquaculture, eutrophication, and erosion on the structure and function of overharvested tropical coastal ecosystems remain to be elucidated and should be the foci of management science. As in the case of Bolinao, existing scientific data was used

Table 14.2 Major projects initiated in the 1990s with local government and community participation. The Local Government Code passed in 1991 provided the legal and institutional guidelines for devolving coastal resource management to local governments at municipal, city, and provincial scales.

Project Title (duration; fund source and type)	Spatial Scale	Project Components	Reference
Fishery Sector Program (1990–1995; Asian Development Bank loan)	12 bays	• Fisheries resource and related ecological assessment • Coastal resources management • Research and extension • Law enforcement • Credit • Infrastructure and marketing support	Asian Development Bank (1997)
Community-based Coastal Resources Management Project in the Municipality of Bolinao (1993–2001; International Development Research Centre, Canada grant (1993–1998) and Royal Dutch Embassy, Philippines, grant (1998–2001))	Municipal-wide	• Community mobilization and formation of local CRM groups • Environmental education • Coastal and land development planning • Marine protected areas • Aquaculture management • Capture fisheries management	McManus (1995)
Coastal Environment Program (1993–present; Government of the Philippines)	61 municipalities as of 1997	• Community organization • Training of local government environment personnel and community participants	DENR et al. (1997)
Regional Programme for the Prevention and Management of Marine Pollution in the East Asian Seas – Batangas Bay Demonstration Project (1994–1998; Global Environment Facility/Dept. of Environment/Provincial Government of Batangas)	Bay-wide	• Mitigation of land-based pollution • Partnership between industries and government • Enforcement of local and national pollution laws • Collaboration with nongovernment organizations	DENR et al. (1997)
Community-based Coastal Resources Management Project (1996–2002; U.S. Agency for International Development, technical assistance)	6 bays in Palawan, Visayas, and Mindanao	• Community management • Local government capacity building • National agency policy implementation • Information, education, and communication • Special activities	DENR et al. (1997)

as basis for management decisions. The scientific data gaps were recognized at both the project and community levels so that participatory monitoring was a logical step in generating new knowledge and in forging consensus necessary for collective action.

The Fishery Sector Program and the initiatives of community-based coastal management programs have converged for Phase 2 of the program. The priority sites for the program are those where the commitment of the local government to coastal resources management is unequivocal in terms of embarking on coastal development planning, complete with budget allocation and institutionalization of an implementing coastal resource management multisectoral organization. Furthermore, Phase 2 of the Fishery Sector Program aims to harmonize the coastal development plans of municipalities at the provincial and subnational levels to pave the way for functional bay-wide management.

DESIGN OF DEVELOPMENT AND MANAGEMENT PROJECTS

In developing countries like the Philippines, community initiatives within the context of local governance are major imperatives in coastal resource management. As such, the aim of management science in this milieu is to provide guidance on the most effective management interventions that communities can take. How does one mitigate sedimentation conveyed by river systems which tend to stress coastal habitats, including seagrass beds and coral reefs? Reforestations in watershed and coastal areas and along riverbanks are logical steps to take according to plant cover and land use in these areas prior to perturbations. Coastal development, as well as those that alter land use in the watershed and which exacerbate the release of sediments, will have to be analyzed for their potential impacts and mitigating measures appropriately evaluated to include environmental costs.

Answers to other management issues are not as clear-cut. What are the impacts of tropical coastal aquaculture on adjacent ecosystems that have been stressed by overharvest? Is eutrophication from domestic sewage and aquaculture loading altering planktonic and benthic flora that can change the food webs? When top predators and even herbivores are harvested, what happens to the ecosystem and will it be able to sustain existing fisheries? Can marine reserves allow for replenishment of fast and slow-growing fish species, and those that have long planktonic stages? For migratory species like milkfish, where a life cycle is completed across distant habitats, can marine reserves be effective both for recruitment and reproduction?

If fishing effort has to be decreased by 60%, what economic policies will need to be instituted to absorb this labor force? Access to education is one way to break away from the vicious circle of poverty. For 1 million sustenance fishers in the Philippines, this will have to be provided for through a multiple pronged approach—markets for alternative income generating activities, development of value-added fishery products, development of environment-friendly mariculture that is coupled with reseeding efforts, skills and training for nonmarine-based livelihood, among others.

The ability to monitor impacts of management interventions should be a major focus of systematic research, and should be enhanced and integrated in any development or management program at local or national levels. New knowledge should be incorporated in the periodic and timely review of existing policies so that management becomes truly adaptive. In all cases, knowledge empowerment among community groups and their leaders should constantly be taking place so that values and norms can evolve to hasten the degree to which innovative measures become socially acceptable.

REFERENCES

Alcala, A.C. 1998. Community-based coastal resource management in the Philippines: A case study. *Ocean. Coast. Manag.* **38**:179–186.

Alcala, A.C., and F.J. Van de Vusse. 1994. The role of government in coastal resources management. In: Community Management and Common Property of Coastal Fisheries in Asia and the Pacific: Concepts, Methods and Experiences, ed. R.S. Pomeroy, pp. 12–19. International Center for Living Aquatic Resources Management (ICLARM) Conf. Proc. 45. Manila: ICLARM.

Aliño, P.M. 1994. Patterns in the distribution of reef-associated fish communities in the ASEAN region. In: Status Reviews, Proc. Third ASEAN-Australia Symp. on Living Coastal Resources, vol. 1, ed. C. Wilkinson, S. Sudara, and L.M. Chou, pp. 11–22. Townesville: Australian Agency for Intl. Development, Australian Institute of Marine Science.

Añonuevo, C.T. 1994. On autonomous capability and technofascism: The role of NGOs (nongovernmental organizations) and LGUs (local government units) in community-based coastal resource management. *Lundayan* **5(4)**:38–44 [publ. by Tambuyog Development Center, Quezon City, Philippines].

Asian Development Bank. 1997. Report and Recommendation of the President to the Board of Directors on Proposed Loans to the Republic of the Philippines for the Fisheries Resource Management Project. Metro Manila: Asian Development Bank.

Bojos, R.M., Jr. 1994. The Central Visayas Regional Project: Experience in community-based coastal resources management. In: Community Management and Common Property of Coastal Fisheries in Asia and the Pacific: Concepts, Methods and Experiences, ed. R.S. Pomeroy, pp. 161–164. ICLARM Conf. Proc. 45. Manila: ICLARM.

Calud, A., E. Cinco, and G. Silvestre. 1991. The gill net fishery of Lingayen Gulf. In: Towards an Integrated Management of Tropical Coastal Resources, ed. L.M. Chou et al., pp. 45–50. ICLARM Conf. Proc. 22. Manila: ICLARM.

Calumpong, H.P. 1993. The role of academe in community-based coastal resource management: The case of Apo Island. In: Our Sea, Our Life, ed. L. Polotan-de la Cruz, pp. 49–57. Proc. Seminar Workshop on Community-Based Coastal Resources Management. Quezon City: Voluntary Services Overseas.

Calumpong, H.P. 1994. Status of mangrove resources in the Philippines. In: Status Reviews, Proc. Third ASEAN-Australia Symp. on Living Coastal Resources, vol. 1, ed. C. Wilkinson, S. Sudara, and L.M. Chou, pp. 215–228. Townesville: Australian Agency for Intl. Development, Australian Institute of Marine Science.

Chua, T.-E., and L.F. Scura, eds. 1992. Integrative Framework and Methods for Coastal Area Management. ICLARM Conf. Proc. 37. Manila: ICLARM.

Dalzell, P.P., P. Corpuz, R. Ganaden, and D. Pauly. 1987. Estimation of maximum sustainable yield and maximum economic rent from the Philippine small pelagic fisheries. Tech. Paper Series No. 10 (3), pp. 1–23. Manila: Bureau of Fisheries and Aquatic Resources.

de los Angeles, M.S., and R. Pelayo. 1995. The nearshore fisheries in Central Visayas, Philippines: An impact evaluation report on the Central Visayas Regional project – I (CVRP-I). In: Philippine Coastal Resources under Stress, ed. A. Juinio-Meñez and G.F. Newkirk, pp. 131–150. Halifax: Coastal Resources Research Network, Dalhousie Univ. and Quezon City: Marine Science Institute, Univ. of the Philippines.

DENR (Department of Environment and Natural Resources), DILG (Department of Interior and Local Government), DA-BFAR (Department of Agriculture – Bureau of Fisheries and Aquatic Resources), and CRMP (Coastal Resource Management Project). 1997. Legal and Jurisdictional Guidebook for Coastal Resource Management in the Philippines. Manila: Coastal Resource Management Project.

Dizon, J.C.A.M., and G.C. Miranda. 1996. The coastal resource management experience in San Salvador Island. In: Seeds of Hope: A Collection of Case Studies on Community-based Coastal Resources Management in the Philippines, ed. E.M. Ferrer, L.P. de la Cruz, and M.A. Domingo, pp. 129–158. Diliman, Quezon City: College of Social Work and Community Development, Univ. of the Philippines.

Ferrer, E.M. 1993. Overview of community-based coastal resources management (CB-CRM) in the Philippines. In: Our Sea, Our Life, ed. L. Polotan-de la Cruz, pp. 13–19. Proc. Seminar Workshop on Community-based Coastal Resources Management. Quezon City:Voluntary Services Overseas.

Fortes, M.D. 1995. Seagrasses of East Asia: Environmental and Management Perspectives. RCU/EAS Tech. Report Series No. 6. Bangkok: UNEP (United Nations Environment Programme).

Galvez, R., T.G. Hingco, C. Bautista, and M.T. Tungpalan. 1989. Sociocultural dynamics of blast fishing and sodium cyanide fishing in two fishing villages in the Lingayen Gulf area. In: Towards Sustainable Development of the Coastal Resources of Lingayen Gulf, Philippines, ed. G. Silvestre, E. Miclat, and T.-E. Chua, pp. 43–62. ICLARM Conf. Proc. 17. Manila: ICLARM.

Gomez, E.D. 1990. Coral reef ecosystems and resources of the Philippines. *Canopy Intl. Newsl.* **16 (5)**:5–7, 10–12.

Gomez, E.D., A.C. Alcala, and A.C. San Diego. 1981. Status of Philippine coral reefs. In: Proc. Fourth Intl. Coral Reef Symp., vol. 1, pp. 275–282. Quezon City: Marine Sciences Center, Univ. of the Philippines.

Kalagayan, B.N.V. 1990. Institutional and legal framework. In: The Coastal Environmental Profile of Lingayen Gulf, Philippines, ed. L.T. McManus and T.-E. Chua, pp. 44–65. ICLARM Tech. Report 22. Manila: ICLARM.

Lacanilao, F., and P.M. Fernandez. 1997. Ignoring sustainability and social equity in fisheries development. *Otolith Newsl.* **4(1)**:6–7.

Maaliw, M.A.L.L., N.A. Bermas, R.M. Mercado, and F. Guarin. 1989. Preliminary results of a water quality baseline study of Lingayen gulf. In: Towards Sustainable Development of the Coastal Resources of Lingayen Gulf, Philippines, ed. G. Silvestre, E. Miclat, and T.-E. Chua, pp. 83–91. ICLARM Conf. Proc. 17. Manila: ICLARM.

McManus, J.E., C.L.Nañola, Jr., R.B. Reyes, Jr., and K.N. Kesner. 1992. Resource Ecology of the Bolinao Coral Reef System. ICLARM Stud. Rev. 22. Manila: ICLARM.

McManus, L.T. 1995. Community-based coastal resources management, Bolinao, Philippines: An evolving partnership among academe, NGOs, and local communities. In: Coastal Management in Tropical Asia, September 1995, pp. 6–8. Colombo: Coastal Resources Management Project of the Coastal Resources Center, Univ. of Rhode Island.

Meñez, L.A.B., et al. 1991. Survey of the coral reef resources of Western Lingayen Gulf, Philippines. In: Towards an Integrated Management of Tropical Coastal Resources, ed. L.M. Chou et al., pp. 77–72. ICLARM Conf. Proc. 22. Manila: ICLARM.

Ochavillo, D., and G. Silvestre. 1991. Optimum mesh size for the trawl fisheries of Lingayen gulf, Philippines. In: Towards an Integrated Management of Tropical Coastal Resources, ed. L.M. Chou et al., pp. 41–44. ICLARM Conf. Proc. 22. Manila: ICLARM.

Pomeroy, R.S. 1995. Community-based and co-management institutions for sustainable coastal fisheries management in Southeast Asia. *Ocean Coast. Manag.* **27(3)**:143–162.

Pomeroy, R., and M.J. Williams. 1994. Fisheries co-management and small-scale fisheries: A policy brief. ICLARM PB1. Manila: ICLARM.

Silvestre, G., N. Armada, and E. Cinco. 1991. Assessment of the capture fisheries of Lingayen Gulf, Philippines. In: Towards an Integrated Management of Tropical Coastal Resources, ed. L.M. Chou et al., pp. 25–36. ICLARM Conf. Proc. 22. Manila: ICLARM.

Silvestre, G.T., and D. Pauly. 1997. Management of tropical coastal fisheries in Asia: An overview of key challenges and opportunities. In: Status and Management of Tropical Coastal Fisheries in Asia, ed. G. Silvestre and D. Pauly, pp. 8–25. ICLARM Conf. Proc. 53. Manila: ICLARM.

Silvestre, G.T., R.B. Regalado, and D. Pauly. 1986. Status of Philippine demersal stocks — Inferences from underutilized catch rate data. In: Resources, Management and Socio-economics of Philippine Marine Fisheries, pp. 47–96. Tech. Report No. 10. Manila: Dept. of Marine Fisheries.

Spalding, M.D., F. Blasco, and C.D. Field. 1997. World Mangrove Atlas. Okinawa: Intl. Soc. for Mangrove Ecosystems.

Talaue-McManus, L., and T.-E. Chua. 1997. The Lingayen Gulf (Philippines) experience: If we have to do it again. *Ocean Coast. Manag.* **37(2)**:217–232.

Talaue-McManus, L., and K.P.N. Kesner. 1995. Valuation of a Philippine municipal sea urchin fishery and implications of its collapse. In: Philippine Coastal Resources under Stress. Selected Papers from the Fourth Annual Common Property Conf., Manila, Philippines, 16–19 June, 1993, ed. A. Juinio-Meñez and G.F. Newkirk, pp. 229–239. Halifax: Coastal Resources Research Network, Dalhousie Univ. and Quezon City: Marine Science Institute, Univ. of the Philippines.

Talaue-McManus, L., A.C. Yambao, S.G. Salmo, III, and P.M. Aliño. 1999. Participatory coastal development planning in Bolinao, northern Philippines: A potent tool for conflict resolution. In: Community-based Natural Resource Management, ed. D. Buckles, pp. 149–157. Ottawa: Intl. Development Resource Centre/World Bank.

Tan, J.G. 1993. Poverty alleviation through an integrated approach to coastal resources management: CERD-FIRMED experience in Daram, Samar. In: Our Sea, Our Life, ed. L. Polotan-de la Cruz, pp. 39–49. Proc. Seminar Workshop on Community-based Coastal Resources Management. Quezon City: Voluntary Services Overseas.

White, A.T. 1987. Philippine marine park pilot site: Benefits and management conflicts. *Env. Conserv.* **14**:355–359.

White, A.T. 1989. The marine conservation and development program of Silliman University as an example for Lingayen Gulf. In: Towards Sustainable Development of the Coastal Resources of Lingayen Gulf, Philippines, ed. G. Silvestre, E. Miclat, and T.-E. Chua, pp. 119–123. ICLARM Conf. Proc. 17. Manila: ICLARM.

White, A.T., and G.C. Savina. 1987. Community-based marine reserves, a Philippine first. In: Proc. Conf. on Coastal Zone '87, pp. 2022–2036. New York: Am. Soc. Civil Eng.

Yambao, A.C., and L. Talaue-McManus. 1999. The proposed cement plant complex in Bolinao, Pangasinan: A case of ill-conceived development. In: Proc. Conf. on Integrated Management of the Coastal Fringe. Quezon City: Voluntary Services Overseas and the Palawan Council for Sustainable Development.

15

Integrated Management of the Benguela Current Region

A Framework for Future Development

M.J. O'TOOLE,[1] L.V. SHANNON,[2] V. DE BARROS NETO,[3] and D.E. MALAN[4]

[1] Ministry of Fisheries and Marine Resources, Private Bag X13355, Windhoek, Namibia
[2] Oceanography Dept., University of Cape Town, Cape Town, South Africa
[3] Instituto de Investigacao Pesqueira, Ministerio das Pescas, Luanda, Angola
[4] Chief Directorate: Marine and Coastal Management, Dept. of Environmental Affairs and Tourism, Cape Town, South Africa

ABSTRACT

Recent initiatives have been developed jointly by Angola, Namibia, and South Africa, three countries bordering on the Benguela Current Large Marine Ecosystem (BCLME), to address the legacy of fragmented management — a consequence of the colonial and political past — and to ensure the integrated sustainable management of the marine and coastal regions in the Southeast Atlantic. Examples of some of these activities and the processes followed are provided. Science and technology are recognized as fundamental building blocks underpinning the management process and, at all levels, the development of capacity — both human and material — is an overarching objective. Some recent successes of a regional fisheries-environment science and technology program BENEFIT are highlighted and serve to demonstrate the commitment of the three governments to collaboration in this area. At a country level, brief details are provided about coastal policy development by way of showing how South Africa proposes to correct some of the wrongs of the past and sustainably utilize one of its most valuable resources, i.e., the coast itself. At the regional ecosystem management level, information is given about an embryonic initiative, the BCLME Programme, which will provide a sound basis for the integration of science, technology, socioeconomics, and management to ensure a sustainable future for the Benguela Current as an ecosystem and the utilization of its coastal and marine resources. Transboundary issues feature high on the agenda.

Science and Integrated Coastal Management
Edited by B. von Bodungen and R.K. Turner

It is the view of the authors that the actions taken jointly by Angola, Namibia, and South Africa can serve not only as a blueprint for the application of science and technology in the southern African context, but also for the integrated sustainable management of marine and coastal systems which are shared by two or more countries elsewhere in the developing world.

BACKGROUND AND INTRODUCTION

The Benguela: A Unique Environment

The Benguela Current region is situated along the coast of southwestern Africa, stretching from east of the Cape of Good Hope in the south northwards to Cabinda in Angola and encompassing the full extent of Namibia's marine environment (see Figure 15.1). It is one of the four major coastal upwelling ecosystems of the world which lie at the eastern boundaries of the oceans. Its distinctive bathymetry, hydrography, chemistry, and trophodynamics combine to make it one of the most productive ocean areas in the world, with a mean annual primary productivity of 1–2 g C m^{-2} d^{-1} (Brown et al. 1991) — about six times higher than the North Sea ecosystem. This high level of primary productivity of the Benguela supports an important global reservoir of biodiversity and biomass of zooplankton, fish, seabirds, and marine mammals, while near-shore and off-shore sediments hold rich deposits of precious minerals (particularly diamonds), as well as oil and gas reserves. The natural beauty of the coastal regions, many of which are still pristine by global standards, have also enabled the development of significant local tourism initiatives. Pollution from industries, poorly planned and managed coastal developments as well as near-shore activities are, however, causing a rapid degradation of vulnerable coastal habitats in some areas.

The Namib Desert, which forms the landward boundary of the greater part of the Benguela Current system, is one of the oldest deserts in the world, predating the commencement of persistent upwelling in the Benguela (12 million years before present) by at least 40 million years. The upwelling system in the form in which we know it today is about 2 million years old. The principal upwelling center in the Benguela, which is situated near Lüderitz in southern Namibia, is the most concentrated and intense found in any upwelling regime. What also makes the Benguela upwelling and adjacent coast system so unique in the global context is that it is bounded at both northern and southern ends by warm-water systems, i.e., the tropical/equatorial Western Atlantic and the Indian Ocean's Agulhas Current, respectively (Shannon and Nelson 1996). Sharp horizontal gradients (fronts) exist at these boundaries of the upwelling system, but these display substantial variability in time and in space — at times pulsating in phase and at others not. Interaction with the adjacent ocean systems occurs over thousands of kilometers. For example, much of the Benguela marine environment, in particular off Namibia and Angola, is naturally hypoxic — even anoxic — at depth as a consequence of subsurface flow southwards from the tropical Atlantic (cf. Bubnov 1972: Chapman and Shannon 1985; Hamukuaya et al. 1998). This is compounded by depletion of oxygen from more localized biological decay processes. There are also teleconnections between the Benguela and processes in the North Atlantic and Indo-Pacific Oceans (e.g., El Niño). Moreover, the southern Benguela lies at a major choke point in the "Global Climate Conveyor Belt," whereby on longer time scales, warm surface waters move from the Pacific via the

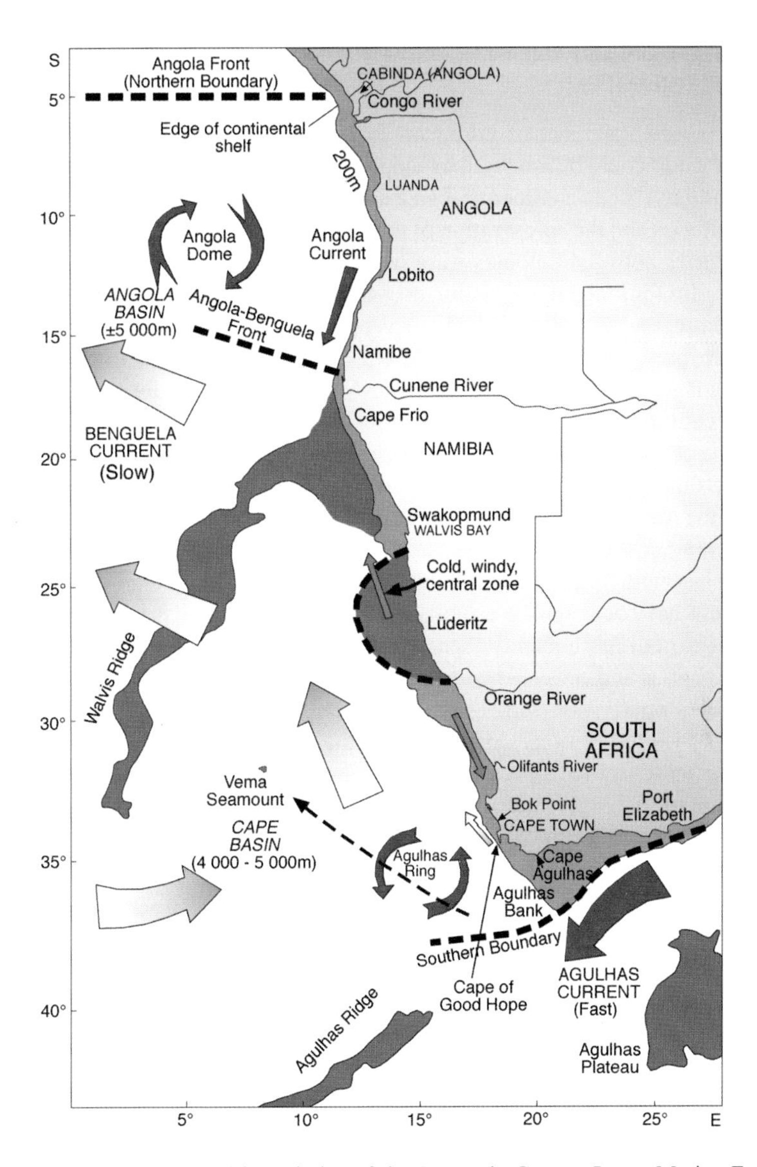

Figure 15.1 External and internal boundaries of the Benguela Current Large Marine Ecosystem, bathymetric features, and surface (upper layer) currents.

Indian Ocean through into the North Atlantic. (The South Atlantic is the only ocean in which there is a net transport of heat towards the equator!).

As a result, not only is the Benguela at a critical location in terms of the global climate system, but its marine and coastal environments are also potentially extremely vulnerable to any future climate change or increasing variability in climate — with obvious consequences for long-term sustainable management of the coast and marine resources.

Fragmented Coastal and Marine Resource Management: A Legacy of the Colonial and Political Past

Following the establishment of European settlements at strategic coastal locations where victuals and water could be procured to supply fleets trading with the East Indies, the potential wealth of the African continent became apparent. This subsequently resulted in the great rush for territories and the colonization of the continent — mostly during the nineteenth century. Boundaries between colonies were hastily established, often arbitrary and generally with little regard for indigenous inhabitants and natural habitats. Colonial land boundaries in the Benguela region were established at rivers (e.g., Cunene, Orange). The languages and cultures of the foreign occupiers were different (Portuguese, German, English, Dutch) and so were the management systems and laws which evolved in the three now independent and democratic countries of the region — Angola, Namibia, and South Africa. Moreover, not only were the governance frameworks very different, but a further consequence of European influence was the relative absence of interagency (or interministerial) frameworks for management of the marine environment and its resources and scant regard for sustainability. To this day, mining concessions, oil/gas exploration, fishing rights, and coastal development have taken place with little or no proper integration or regard for other users. For example, exploratory wells have been sunk in established fishing grounds and the wellheads (which stand proud of the sea bed) subsequently abandoned. Likewise, the impact of habitat alterations due to mining activities and ecosystem alteration (including biodiversity impacts) due to fishing have not been properly assessed.

Prior to the United Nations Convention on the Law of the Sea in 1982 (United Nations 1983) and declaration and respecting of sovereign rights within individual countries' exclusive economic (or fishing) zones (EEZs), there was an explosion of foreign fleets fishing off Angola, Namibia, and South Africa during the 1960s and 1970s — an effective imperialism and colonization by mainly First World countries of the Benguela Current Large Marine Ecosystem (BCLME) and the rape of its resources. This period also coincided with liberation struggles in all three countries and associated civil wars. In the case of Namibia, over whom the mandate by South Africa was not internationally recognized, there was an added problem in that prior to independence in 1990, an EEZ could not be proclaimed. In an attempt to control the foreign exploitation of Namibia's fish resources, the International Commission for the South-east Atlantic Fisheries (ICSEAF) was established, but this proved to be relatively ineffectual at husbanding the fish stocks. In South Africa prior to 1994, there was generally a scant regard for environmental issues or sustainable environmental management. Moreover, colonialism, civil wars, and the apartheid legacy have resulted in a marked gradient in capacity from south to north in the region. Another consequence of the civil wars has been the population migration to the coast and localized pressure on marine and coastal resources (e.g., destruction of coastal forests and mangroves), severe pollution of some embayments, and *de facto* impossibility of any form of integrated coastal zone management along large stretches of the Benguela coast.

While mineral exploration and extraction and developments in the coastal zones obviously occur within the geographic boundaries of the three countries, i.e., within the EEZs, and can to a large degree be independently managed by each of the countries, mobile living marine resources do not respect the arbitrary geographic borders. This has obvious implications for the sustainable use of these resources, particularly so in the case of straddling and shared fish stocks.

Thus the legacy of the colonial and political past is that the management of resources in the greater Benguela area has not been integrated within countries or within the region. The real challenge will be to develop a viable joint and integrative mechanism for the sustainable management of the coast and marine resources of the Benguela as a whole.

Regional Self-help: Joint Action by Three Developing Countries for a Sustainable Future

This historical scenario poses almost insurmountable problems for the countries bordering on the Benguela. Notwithstanding this, Angola, Namibia, and South Africa have, over the past three years, made substantial progress to address the science and management issues in a pragmatic, cost-effective manner. In this chapter we provide details about some of the joint (regional) and individual country actions which have been taken to overcome the difficulties, highlight some recent successes, and outline plans which the three countries collectively have to ensure that the greater Benguela Current region is sustainably used and managed through the proper application of science and technology. This approach could well serve as a blueprint in other parts of the world for the integrated management of marine and coastal systems which are shared between two or more developing countries.

At the *regional* level the approach has been somewhat different, however, from that which would normally be taken in developed countries and coastal areas with "concave coastlines" for the following reasons: First, much of the coast in the Benguela region is relatively pristine and/or inaccessible — except for small pockets of urban development. Second, many of the "coastal" issues concern the marine rather than the terrestrial system. Third, the transboundary problems which are amenable to management action are those relating to marine systems, e.g., shared fish resources. (It is just not feasible to attempt to address transboundary issues associated with, for example, the Congo River and its drainage basin). Fourth, (natural) environmental variability and change are major factors influencing natural resources and the way in which these are managed in an open "convex" system, such as the Benguela. While this cannot be controlled, cost-effective environmental monitoring and appropriate science for better predictability can improve marine and coastal resource utilization and management. Finally, but perhaps most important, is the need throughout the region to develop human and infrastructure capacity and to share available knowledge and skills. What better way to proceed than through the application of the appropriate science and technology! This is in keeping with the philosophy so well articulated by Sherman (1994).

A REGIONAL MARINE SCIENCE SUCCESS STORY — "BENEFIT"

In April, 1997, a major regional cooperative initiative was launched jointly by Angola, Namibia, and South Africa together with foreign partners "to develop the enhanced science capacity required for the optimal and sustainable utilization of living resources of the Benguela ecosystem by (a) improving knowledge and understanding of the dynamics of important commercial stocks, their environment and linkages between the environmental processes and the stock dynamics, and (b) building appropriate human and material capacity for marine science and technology in the countries bordering the Benguela ecosystem" (BENEFIT 1997). The Benguela–Environment–Fisheries–Interaction–Training Programme (BENEFIT) evolved out of a workshop/seminar on "Fisheries Resource Dynamics in the Benguela Current Ecosystem" held in Swakopmund in mid-1995 and hosted by the Namibian Ministry of Fisheries and Marine Resources in partnership with the Norwegian Agency for Development Cooperation (NORAD), the German Organization for Technical Cooperation (GTZ), and the Intergovernmental Oceanographic Commission (IOC) of UNESCO. BENEFIT has attracted substantial incremental support from overseas countries and international donor agencies. It remains, however, essentially a regional "self-help" initiative and has been endorsed by the Southern African Development Community (SADC) and accepted as a SADC program. It is providing a unique opportunity for development of partnerships within and beyond the southern African region in science and technology to promote optimum utilization of natural resources and thereby greater food security in the region.

BENEFIT has been planned in two five-year phases (1997–2002, 2002–2007). The science and technology component of BENEFIT has three foci: resource dynamics, the environment (of the resources), and linkages between resources and the environment. These foci are increasing knowledge of resource dynamics through improved research on the resources and their variable environment. The capacity development component of the program is being addressed through a suite of task-oriented framework activities to (a) build human capacity, particularly in areas of greatest need and greatest historical disadvantage, (b) develop, enhance, and maintain regional infrastructure and cooperation, and (c) to make the countries in the region and the region as a whole more self-sufficient in science and technology. The linkages between the three science foci and the suite of framework activities are illustrated schematically in Figure 15.2. BENEFIT has a Secretariat based in Namibia, while management meetings are held on a rotating basis in Angola, Namibia, and South Africa.

The launch of BENEFIT in April 1997 coincided with two major research cruises/surveys of the Angola–Benguela Front focusing on fisheries and environmental issues. (This front is situated west of Angola and is thought to play an important role as a permeable "boundary" between the tropical Atlantic and the upwelling region of the Benguela). During the past two years, BENEFIT has increasingly gathered momentum: funding for priority projects has been allocated and real progress in human capacity development has been made. Some recent achievements are briefly:

- Several reports and scientific/technical papers have been published on the results of the 1997 Angola–Benguela Front surveys, and several regional scientists and technicians received hands-on training at sea, in the laboratory, and in data analysis.
- A German-sponsored BENEFIT training course was conducted in Namibia in 1997, and a number of regional scientists received further training subsequently in Germany and in Norway.
- Fifteen fisheries and fisheries-environment projects were approved for funding in 1999.
- Two training workshops have taken place (1998 and 1999) and a BENEFIT Training Plan to complement the Science Plan is under development.

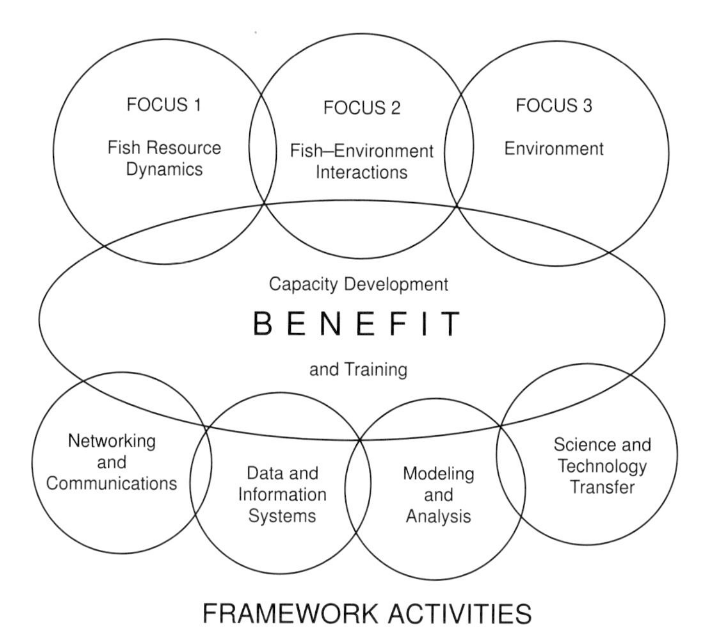

Figure 15.2 Schematic of the BENEFIT structure showing the interlinking science and technology foci and framework activities.

- In the first half of 1999, over 50 persons from the broad SADC region (i.e., including East African nations) have been trained during three BENEFIT cruises, including a 40-day survey of resources and the environment which extended between Cape Town and Luanda, primarily funded by the African Development Bank and the World Bank.

BENEFIT and related activities provide clear evidence of the desire and capability of Angola, Namibia, and South Africa to work together to solve common marine/fisheries science problems in the Benguela region in partnership with the international community.

A COUNTRY-BASED APPROACH TO ICM: SOUTH AFRICA AS AN EXAMPLE

The Need

In comparison with many other countries, South Africa's coastal areas have a low overall population density and large areas that are relatively underdeveloped, particularly in terms of opportunities for the poorer sections of the community. This is a result of a number of factors, including migrant labor, apartheid planning policies, and historical population movements. The coastal population is now growing as a result of recent political and economic changes in South Africa. These changes include the removal of barriers to movement, a decline in inland extractive industries, and increasing opportunities associated with coastal resources, such as tourism and port development. Future economic growth in South Africa is likely to concentrate along the coast and the pace of development in coastal areas is already accelerating. For example, five out of the eight Spatial Development Initiatives are linked to the coast and the direct contribution of coastal areas and resources to the gross domestic product of South Africa is estimated as 37%. This presents a unique management challenge for government, industry, and civil society.

Coastal Management Policy Programme in South Africa

The Department of Environmental Affairs and Tourism initiated the Coastal Management Policy Programme (CMPP) to promote integrated management of the coast as a system, in order to harness its resources for sustainable coastal development. An extensive process of public participation, supported by specialist studies, began in May, 1997, guided by a policy committee representing the interests of national and provincial government, business, and civil society. A Coastal Policy Green Paper was published in September, 1998, followed by a Draft White Paper for Sustainable Coastal Development in March, 1999 (Draft White Paper 1999).

The Draft White Paper advocates the following shifts in emphasis:

- In the past, the value of coastal ecosystems as a cornerstone for development was not sufficiently acknowledged in decision making in South Africa. The policy outlines the importance of *recognizing the value of the coast.*
- In the past, coastal management was resource-centered rather than people-centered and attempted to control the use of coastal resources. The policy stresses the powerful contribution that can be made to reconstruction and development in South Africa through *facilitating sustainable coastal development.* Maintaining diverse, healthy, and productive coastal ecosystems will be central to achieving this ideal.
- In the past, South African coastal management efforts were fragmented and uncoordinated, and were undertaken largely on a sectoral basis. The policy supports a holistic approach by *promoting coordinated and integrated coastal management*, which understands the coast as a system.
- In the past, a "top-down" control and regulation approach was imposed on coastal management efforts. The policy proposes *introducing a new facilitatory style of management*, which involves cooperation and shared responsibility with a range of stakeholders.

However, the institutional capacity to support the integrated approach required to manage coastal development is currently weak, and there is little awareness of the issues among key stakeholders. This both limits the opportunities associated with the coast and threatens the sustainability of development. An action plan to address these issues is presented in the Draft White Paper and centers on four key themes:

1. Developing and supporting an appropriate (integrated) institutional and legal framework across government.
2. Awareness, education, and training programs for government, the private sector, and civil society.

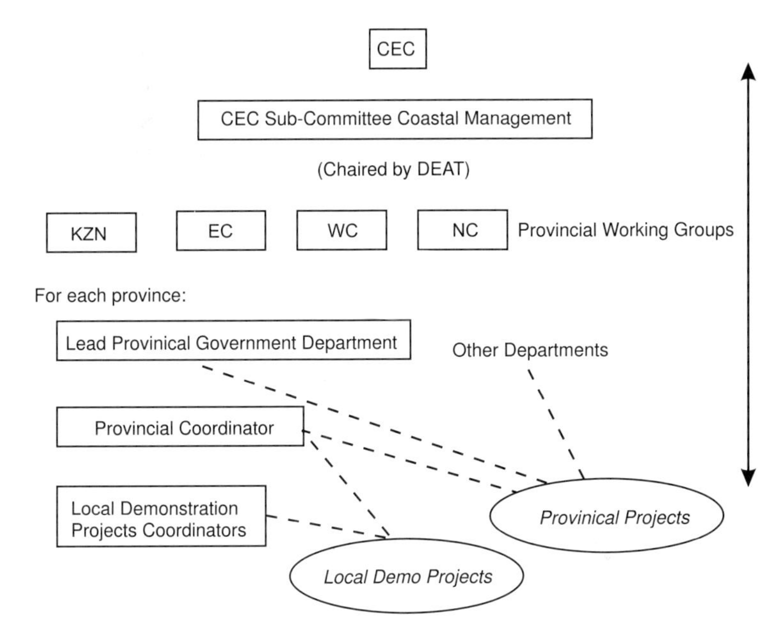

Figure 15.3 Proposed institutional structure for implementation of the Draft White Paper (1999) for sustainable coastal development in South Africa. CEC: Committee for Environmental Coordination (national body); DEAT: Dept. of Environmental Affairs and Tourism; KZN: Kwa-Zulu Natal; EC: Eastern Cape; WC: Western Cape; and NC: Northern Cape.

3. Information provision in the form of decision support for provincial and local government, monitoring programs, and applied research.
4. Local projects to demonstrate the benefits of effective coastal management and to address national and provincial priority issues.

A cyclical process of review and revision underpins the coastal policy. This allows successive implementation generations to reflect evolving priorities, visions, and institutional capacity. Figure 15.3 represents the institutional structures through which the coastal policy will be implemented.

Context of the Coastal Policy Within the Benguela Current Region of South Africa

The Draft White Paper provides a brief overview of South Africa's coast to sketch the context of the policy in relation to the thirteen coastal regions defined for the purposes of the policy formulation process. The two sections applicable to the Benguela Current region study are the Namaqualand and West Coast coastal divisions. These are bounded by the Orange and Olifants Rivers, and the Olifants River and Bok Point, respectively. The following extracts are lifted from the Draft White Paper.

Namaqualand

The Northern Cape province is comprised of only a single coastal region, the Namaqualand coastal region. Some 390 km long, the Namaqualand region is a sparsely inhabited area, much of which is semi-desert and is largely undeveloped. A lack of physical access to coastal resources and isolation from the center of provincial administration contribute to high poverty levels in the coastal communities.

Although dominated by large mining and fishing companies, the Namaqualand region has the second lowest economic growth rate in South Africa and unemployment has more than doubled since 1980. Challenges include declining fish stocks, poor road infrastructure, lack of sheltered bays for ports, and limited agricultural potential. The closure of many land-based diamond mining operations provides an opportunity for extensive rehabilitation programs to be carried out — to rehabilitate the natural environment and to create alternative livelihoods for people.

Potential exists for the harvesting of underutilized coastal resources, such as mussels and limpets, for small-scale industries that add value to fishing and agriculture, and for small-scale mining. Other natural assets, such as the annual wildflower display, a high diversity of succulent plant species, and the stark beauty of the area offer potential for nature-based tourism with community participation. More equitable distribution of mining and fishing concessions and the development of value-added activities could contribute to retaining revenue in local communities.

West Coast

The West Coast region has displayed significant growth and a relatively strong economy, although rural areas remain poor. Impetus for growth has come from the deep-water port of Saldanha and the proximity of the region to the Cape Metropolitan area. Despite the limited supply of freshwater, substantial investment has been attracted to the region for mariculture, shipping, industrial, manufacturing, tourism, and recreational activities. Much of the area is arid, which limits agricultural potential. The region is, however, well known for its strandveld and fynbos vegetation, which attracts many visitors to the region each spring. The region is at the center of South Africa's fishing industry, with rich fishing grounds supporting capital-intensive industries.

Economic development through industrialization, property development, and tourism has brought challenges for the management of the coast, including air and water pollution, salination of the coastal aquifer, restricted access to coastal resources, ribbon development, and inappropriate land use. Economic development is also attracting many job seekers to the region, increasing the need for infrastructure and government services. Potential exists in the region for the development of small-scale industries that add value to fishing, floriculture, and mariculture, and for tourism promotion initiatives, including the development of rail and air links.

Implementing the Coastal Policy in South Africa

Concurrent with the process of formal ratification of the Draft White Paper, the Department of Environmental Affairs and Tourism has translated the "Plan of Action" into a preliminary five-year implementation proposal. This is being used to seek implementation finance to ensure a reduced lead-in time between policy adoption and policy implementation. Once project funds are secured, detailed design will follow, involving provincial government, appropriate stakeholders, and specialist individuals and organizations. This process was completed in 2000, and it is hoped that the coastal policy will become government policy. Meanwhile, to maintain momentum of the CMPP, an interim phase program is underway to bridge the gap between the policy formulation and implementation phases. Activities include: specialist advice on an appropriate legislative framework, preparing for the appointment of National and Provincial Coordinators, initiating a needs assessment for public awareness and education programs, preparing and distributing a newsletter to coastal stakeholders, and developing a coastal management web site to better disseminate information.

AN INTEGRATED REGIONAL APPROACH TO THE SUSTAINABLE MANAGEMENT OF THE BENGUELA CURRENT AS A LARGE MARINE ECOSYSTEM: THE BCLME PROGRAMME

In a large marine ecosystem (LME), such as the Benguela Current System, sustainable management at the *ecosystem level* under conditions of environmental variability and uncertainty is a regional issue. The mobile components of the BCLME do not respect arbitrary geopolitical (country) boundaries. Several fish stocks straddle or are shared between the countries or otherwise migrate through the Benguela. Actions by one country, e.g., overexploitation or habitat destruction of their part of a migrating or shared resource, could in effect negatively impact on one or both neighboring countries. Joint management and protection of shared stocks is one of the few available options to the countries bordering the Benguela Current. In this manner, a better sense of ownership of the regions' resources can be attained, as "owners" tend to protect their property more than those enjoying a free service. There is thus a strong need for harmonizing legal and policy objectives and for developing common strategies for resource surveys, and investment in sustainable ecosystem management for the benefit of all the people in the Benguela region. Only concerted regional action with the enablement from the international community to develop regional agreements and legal frameworks and assessment/implementation strategies will in the longer term protect the living marine resources, biological diversity, and environment of the greater Benguela. While shared living resources present the most obvious case for comanagement, there are many examples of nonshared "resources" that can benefit from sharing of expertise and management structures developed and implemented in individual countries. These include *inter alia* mining, declining coastal water quality (pollution abatement and control, oil spill clean-up technology), oil/gas extraction, coastal zone development, tourism and eco-tourism development, mitigation of the effects of introduced species (exotics), and harmful algal blooms, which can also have system-wide impacts.

Whereas the governments of Angola, Namibia, and South Africa have made excellent progress in partnership with members of the international community in addressing the science and technology needs for *fisheries* in the region through the BENEFIT Programme, a viable regional *framework for management* for shared fish resources and *the ecosystem as a whole*, including the coastal zone, is lacking. Building on BENEFIT and on the success of LME initiatives elsewhere in the world (e.g., Black Sea; see Black Sea 1996), whereby incremental funding is made available by the Global Environmental Facility (GEF) of the World

Bank for the development of management structures that address transboundary problems (and which structures become self-funding after 3–5 years), Angola, Namibia, and South Africa with GEF assistance are in the process of developing a LME management initiative for the Benguela Current: the BCLME Programme. This program is a broad-based multisectoral initiative aimed at *sustainable integrated management* of the Benguela Current ecosystem as a whole. It focuses on a number of key sectors, including fisheries, impact of environmental variability, sea-bed mining, oil and gas exploration and production, coastal zone management, ecosystem health, and socioeconomics and governance. Transboundary management issues, environmental protection, and capacity building will be of primary concern to the program. It builds on existing regional capacity and goodwill, and could serve as a blueprint for the design and implementation of LME initiatives in other upwelling regions and elsewhere in the developing world. Moreover, the BCLME Programme will address key regional environmental variability issues that are expected to make a major contribution towards understanding global fluctuations in the marine environment, including climate change.

The BCLME Programme provides an ideal opportunity for the international community to assist the three countries in the region to develop appropriate mechanisms that will ensure the long-term sustainability of the ecosystem. In 1998 a small grant was made by the GEF via the United Nations Development Programme (UNDP) to facilitate the development of a comprehensive proposal. The process involved, which is lengthy and complex, follows prescribed procedures. In essence it is a participatory process involving all key stakeholders in the private and public sectors of the participating countries. Two regional workshops involving over 100 regional and international experts were held (Croll 1998; Croll and Njuguna 1999), consensus was built, a set of six comprehensive thematic reports or integrated overviews were commissioned (fisheries, environment, mining, coast, oil and gas, socioeconomics), an exhaustive Transboundary Diagnostic Analysis (TDA) was undertaken, a Strategic Action Programme (SAP) is being developed, and, finally, a Project Brief formulated. At first appearance the process appears overly bureaucratic and unwieldy, but having gone through the various stages it is clear that the process is rigorous, necessary, and logical. For example, the integrated overviews provided essential input into the subsequent TDA whereby the essential elements were formulated and prioritized through (consensus) group work as per the path issues → problems → causes → impacts → uncertainties → socioeconomic consequences → transboundary consequences → activities/solutions → priority → outputs → costs.

Key aspects of the TDA and SAP follow the next section, which considers BCLME external boundaries.

Geographic Scope and Ecosystem Boundaries

Conducting a comprehensive TDA is only possible if the entire LME, including all inputs to the system, is covered in the study. In the case of the Benguela, which is a very open system where the environmental variability is predominantly remotely forced, this should then include the tropical Atlantic *sensu latu*, the Agulhas Current (and its link with the Indo-Pacific), the Southern Ocean, and the drainage basins of all major rivers which discharge into the greater Benguela Current region, including the Congo River. Clearly, such an approach is impracticable, and more realistic and pragmatic system boundaries must be defined to develop and implement a viable ecosystem management framework. The principal external and internal system boundaries are shown in Figure 15.1.

Landward Boundary

With the exception of the Congo River, the main impact of discharges from rivers flowing into the South East Atlantic tends to be episodic in nature, i.e., in terms of significant transboundary concerns, these are limited to extreme flood events. (Their drainage basins nevertheless do include a major part of the southern African hinterland.) The Congo River, however, exerts an influence that can be detected over thousands of kilometers of the South Atlantic and drains much of Central Africa. From a practical point of view, it is quite beyond the scope of the BCLME to attempt to include the development of any management structures for a river such as the Congo. With respect to land sources of pollution in the BCLME (excluding the Congo River area), these are only really significant in the proximity of the principal port cities (e.g., Cape Town, Luanda, Walvis Bay), and the effects are generally very localized. Nevertheless, some of the problems experienced in these areas are common in nature and could be addressed through similar remedial actions. Like coastal development, their impacts generally do not have a transboundary character. (By contrast, pollution from ships, major oil spills, introduction of exotic species, and associated harmful algal blooms, for example, are transboundary concerns). From a BCLME perspective, the landward boundary can thus, for all practical purposes, be taken as the coast. Specific allowances can be made in some areas on a case-by-case basis (e.g., during episodic flooding from the Orange and Cunene Rivers, which are situated at the country boundaries of South Africa–Namibia and Namibia–Angola, respectively).

Western Boundary

The Benguela Current is generally defined as the integrated equatorward flow in the upper layers of the ocean in the South East Atlantic between the coast and the 0° meridian. The BCLME Programme will accordingly use 0° as the western boundary. For practical management purposes, however, the focus will be on the areas over which the three countries have some jurisdiction, i.e., their EEZs which extend 200 nautical miles seawards from the land.

Southern/Eastern Boundary

The upwelling area of the BCLME extends around the Cape of Good Hope, seasonally as far east as Port Elizabeth. This extreme southern part of the ecosystem is substantially influenced by the Agulhas Current, its Retroflection (turning back) and leakage of Indian Ocean water into the Atlantic south of the continent. As the variability of the BCLME is very much a function of the complex ocean processes occurring in the Agulhas Current–Retroflection area, this will be taken as the southern boundary with 27°E longitude (near Port Elizabeth), being at the extreme eastern end.

Northern Boundary

While the Angola–Benguela Front (Shannon et al. 1987) comprises the northern extent of the main coastal upwelling zone, upwelling can occur seasonally along the entire coast of Angola. There are, in any event, strong linkages between the behavior of the Angola–Benguela Front (and the oceanography of the area to the south of it) and processes occurring off Angola, especially the Angola Dome and the Angola Current. Unless these are considered as an integral part of the BCLME, it will not be feasible to evolve a sustainable integrated management approach for the Benguela. Moreover, there is a well-defined front at about 5°S, viz. the Angola Front (Yamagata and Iizuka 1995), which is apparent at subsurface depths. This front is the true boundary between the Benguela part of the South Atlantic and the tropical/equatorial Gulf of Guinea system. A northern boundary at 5°S would thus encompass the Angola Dome, the seasonal coastal Angola Current, and the area in which the main oxygen minimum forms, and the full extent of the upwelling system in the South East Atlantic. A pragmatic northern boundary is thus at 5°S latitude, which is close to the northern geopolitical boundary of Angola (Cabinda).

Issues and Perceived Main Transboundary Problems, Root Causes, and Areas Where Action Is Proposed: The TDA

Through the participatory TDA process involving regional stakeholders and international LME experts, seven major transboundary problems were identified, their root causes established, and suites of action formulated (BCLME TDA 1999). These are summarized conceptually in Figure 15.4 and expanded in the accompanying synthesis matrix (see Table 15.1). The latter is a "logistical map" which encapsulates the essence of the TDA.

Regional action is clearly required in three main areas: (a) sustainable management and utilization of resources, (b) assessment of environmental variability, ecosystem impacts and improvement of predictability, and (c) maintenance of ecosystem health and management of pollution. Within each of these areas is a suite of subactions. Each of these is examined more fully in the next level of the TDA to determine causes (of the relevant subproblem), likely impacts, risks and uncertainties, socioeconomic consequences, transboundary consequences, proposed activities/solutions, their priority and incremental costing (i.e., cost over and above costs presently spent by national governments), and anticipated outputs. By way of illustration we have extracted one of the several subtables from the BCLME TDA document (Table 15.2). This table is in reality only a summary of the comprehensive information and assessment which comprised the TDA process. It does, however, illustrate how transboundary concerns can be addressed through the application of a logical analysis framework, i.e., the TDA, which in turn provides essential input into the compilation of a SAP.

Strategic Action Programme for the BCLME

The Strategic Action Programme being developed (BCLME SAP 1999) is in essence a concise document that outlines regional policy for the integrated sustainable management of the BCLME as agreed by the governments of Angola, Namibia, and South Africa. The SAP spells out the challenge (regional problems), establishes principles fundamental to integrated

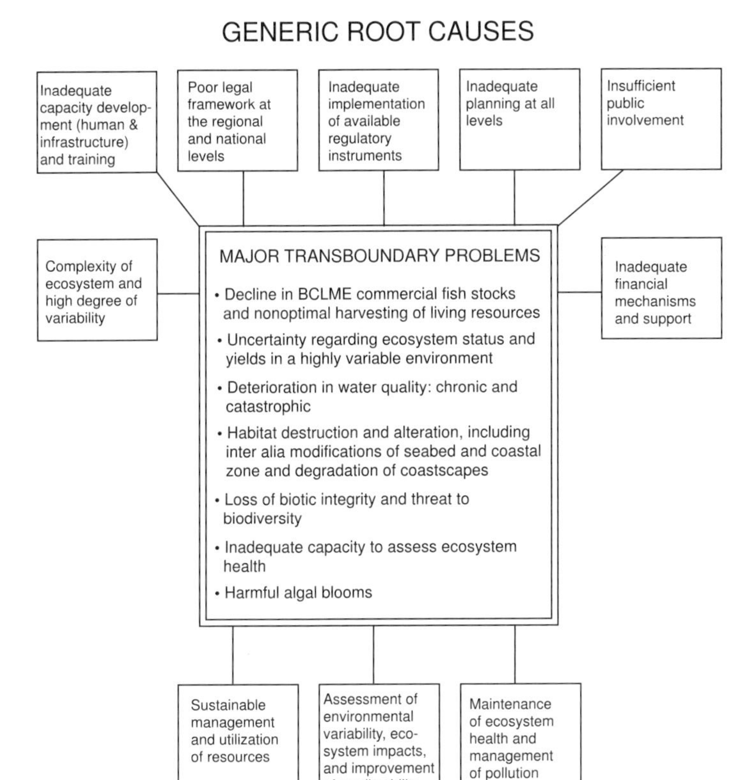

Figure 15.4 Results of the Transboundary Diagnostic Analyses (TDA): Overview of major transboundary problems, generic root causes, and areas requiring action in the BCLME.

management in the region, specifies the nature, scope, and timetable for deliverable management policy actions (based on TDA input), details the institutional arrangements (structures) necessary to ensure delivery, elaborates on wider cooperation (i.e., cooperation between the BCLME region and external institutions), specifies how the BCLME Programme will be financed during the start-up and implementation phase (five years), and outlines approaches to ensure the long-term self-funding of the integrated management of the BCLME.

Key details of the BCLME SAP are briefly as follows.

The Challenge

The legacy of fragmented management — inadequate planning and integration, poor legal frameworks and inadequate implementation of existing regulatory instruments, insufficient public involvement, inadequate capacity development, and inadequate financial support mechanisms — superimposed on a complex and highly variable environment have manifested themselves, for example, in the decline of fish stocks, nonoptimal utilization of

Table 15.1 Synthesis matrix.

Perceived Major Problems	Transboundary Elements	Major Root Causes*	Activity Areas**
Decline in BCLME commercial fish stocks and nonoptimal harvesting of living resources	Most of region's important harvested resources are shared between countries, or move across national boundaries at times, requiring joint management effort.	1, 2, 3, 4, 5, 6, 7	A, B, (C)
Uncertainty regarding ecosystem status and yields in a highly variable environment	Environmental variability and change impacts on ecosystem as a whole, and poor predictive ability limits effective management. The BCLME may also be important to global climate change.	1, 2, 3, 7	A, B, C
Deterioration in water quality: chronic and catastrophic	While most impacts are localized, the problems are common to all three countries and require collective action to address.	2, 3, 4, 5, 7	C
Habitat destruction and alteration, including *inter alia* modification of seabed and coastal zone and degradation of coastscapes	Uncertainties exist about the regional cumulative impact from mining on benthos and ecosystem effects of fishing. Degradation of coastscapes reduce regional value of tourism.	2, 3, 5, 6, 7	A, C, (B)
Loss of biotic integrity (e.g., changes in community composition, species diversity, introduction of alien species) and threat to biodiversity, endangered and vulnerable species	Fishing has altered the ecosystem, reduced the gene pool, and caused some species to become endangered/threatened. Introduced alien species are a global transboundary problem.	1, 3, 5, 6	A, C, (B)
Inadequate capacity to monitor/ assess ecosystem (resources, environment, and variability thereof)	There is inadequate capacity in the region to monitor the resources and the environmental variability, and unequal distribution of the capacity between countries.	1, 2, 5, 7	A, B, C
Harmful algal blooms (HABs)	HABs are a common problem in all three countries and require collective action to address.	1, 2, 3, 6, 7	A, B ,C

**Main Root Causes:*
1. Complexity of ecosystem and high degree of variability (resources and environment)
2. Inadequate capacity development (human and infrastructure) and training
3. Poor legal framework at the regional and national levels
4. Inadequate implementation of available regulatory instruments
5. Inadequate planning at all levels
6. Insufficient public involvement
7. Inadequate financial mechanisms and support

***Area Where Action Is Proposed:*
A. Sustainable management and utilization of resources
B. Assessment of environmental variability, ecosystem impacts, and improvement of predictability
C. Maintenance of ecosystem health and management of pollution

resources, increasing pollution, habitat destruction, threats to biodiversity, all of which have transboundary implications. The challenge is to halt the changing state of the BCLME and, where possible, reverse the process through the development and implementation of sustainable integrated management of the ecosystem as a whole. More specifically:

- Overexploitation of commercial fish stocks and the nonoptimal harvesting of living resources in the ecosystem are causes of concern, particularly as most of the important harvested resources are shared between countries and overharvesting in one country can lead to depletion of that species in another and changes to the ecosystem as a whole.
- Inherent high environmental variability in the marine system and associated uncertainty and poor predictability limits the capability to manage resources effectively. The challenge is to improve predictability of "events" and their consequences.
- Deterioration of water quality in the BCLME and associated problems are common to all three countries and require collective action to address. Catastrophic events (e.g., major oil spills, system-wide anoxia) can impact across geopolitical boundaries requiring sharing of expertise and technology.
- Habitat destruction, alteration and modification to the seabed and coastal zone, and degradation of coastal areas is accelerating. The regional cumulative impacts are unknown and need addressing to ensure sustainable resource utilization and tourism.
- Increased loss of biotic integrity and the introduction of alien species (e.g., ballast water discharges) threatens vulnerable and endangered species and biodiversity of the BCLME and impacts at all levels, system wide.
- There is inadequate institutional, infrastructural, and human capacity at all levels to monitor, assess, and manage the BCLME. Moreover, there is an unequal distribution of existing capacity.
- Harmful algal blooms occur in coastal waters of all three countries and all face similar problems in terms of impacts, monitoring, and management. Collective regional action is necessary.

Principles Fundamental to Cooperative Action

The following principles are being proposed for consideration by the three governments:

- Application of the precautionary principle.
- Promote anticipatory actions (e.g., contingency planning).
- Stimulate use of clean technologies.
- Promote use of economic and policy instruments that foster sustainable development (e.g., polluter-pays-principle).
- Include environmental and health considerations in all relevant policies and sectoral plans.
- Promote cooperation among states bordering the BCLME.
- Encourage the interests of other states in the southern African region.
- Foster transparency and public participation within the BCLME Programme.
- The three governments will actively pursue a policy of cofinancing with industry and donor agencies.

Institutional Arrangements (Structures)

It has been suggested that an Interim Benguela Current Commission (IBCC) be established to strengthen regional cooperation. Its Secretariat and subsidiary bodies could be fully

Table 15.2 Maintenance of ecosystem health and management of pollution: Improvement of water quality (taken from BCLME TDA 1999).

Problems	Causes	Impact	Risks/Uncertainties
C1. Deterioration in coastal water quality: Coastal developments and rapid expansion of coastal cities, much of which was unforeseen or unplanned, has created pollution "hotspots." Aging water treatment infrastrucure and inadequate policy/monitoring/enforcement aggravates the problem.	• Unplanned coastal development • Chronic oil pollution • Industrial pollution • Sewage pollution • Air pollution • Mariculture • Lack of policy on waste and oil recycling • Growth in coastal informal settlement	• Public health • Reduced yields • Unsafe edible organisms • Changes in species dominance	• Few or no baseline data • Performance standards and thresholds • National commitment to capacity building • Cause-effect relationship
C2. Major oil spills: A substantial volume of oil is transported through the BCLME region and within it, and this is a significant risk of contamination of large areas of fragile coastal environments from major accidents, damage to straddling stocks and coastal infrastructure.	• Sea worthiness of vessels/equipment • Military conflict • Sabotage • Human error	• Coastline degradation • Mortality of coastal fauna and flora	• Recovery period • Cost recovery mechanism • Return to peace in Angola
C3. Marine litter: There is a serious growing problem throughout the BCLME.	• Growth of coastal settlements • Poor waste management • Little public awareness and few incentives • Illegal disposal from vessels • Poverty of coastal communities • Ghost fishing • Fishing discards	• Faunal mortality • Negative aesthetic impacts • Damage to fishing equipment	• Accumulation zones • Illegal hazardous waste disposal

Table 15.2 continued

	Socioeconomic Consequences	Transboundary Consequences	Activities/Solutions	Priority	Anticipated Outputs
C1	• Loss of tourism • Higher health costs	• Transboundary pollutant transport	• Develop standard environmental quality indicators/criteria	1	• Shared solutions for water quality management
	• Altered yields • Reduced resource quality	• Migration of marine organisms, e.g., seals	• Establish regional working groups	1	• Regional protocols and agreements • Improved pollution control
	• Aesthetic impacts • Lowered quality of life	• Negative impacts on straddling stocks	• Training in marine pollution control	2	• Socioeconomic uplift
	• Loss of employment	• "Hot spots" common solutions	• Plan/adapt regional pollution monitoring framework	1	
			• Establish effective enforcement agencies	1	
			• Demo projects on pollution control and prevention	2	
			• Joint surveillance	2	
C2	• Opportunity costs (e.g. tourism, fisheries, salt production)	• Resource sharing for containment, surveillance, rehabilitation, etc.	• Port state control	3	• Regional contingency plan • Shared resources • Rehabilitation plans
			• Regional contingency plan development	1	
	• Altered yields • Reduced resource quality	• Ramsar site protection (border wetlands)	• Research/modeling of recovery periods	3	• Regional protocols and agreements
	• Aesthetic impacts	• Transboundary pollutant transport	• Public awareness of notification procedures	3	
C3	• Loss of fishing income	• Transboundary transport	• Litter recycling	2	• Cleaner beaches
	• Public health • Cleanup costs		• Harmonization of packaging legislation	3	• Education material/documents available regionally
	• Loss of tourism • Job creation in informal sector		• Public awareness	1	• Standardized policies and legislation on packaging/recycling incentives
			• Port reception facilities	1	
			• Regulatory enforcement	2	
			• Standardized policies	2	
			• Seafarer education	1	

functional by January, 2001. Meanwhile, the three governments have signed the SAP and the GEF Council recently approved the funds for the BCLME Programme (approximately US$ 15 million). As envisaged, the IBCC will be implementing the organization for the BCLME SAP and will be supported by advisory groups as necessary. The following initial advisory groups are likely to be:

- Advisory Group on Fisheries and Other Living Resources,
- Advisory Group on Environmental Variability and Ecosystems Health,
- Advisory Group on Marine Pollution,
- Advisory Group on Information and Data Exchange,
- Advisory Group on Legal Affairs and Maritime Law,
- Advisory Group on Industry and the Environment.

It is anticipated that the IBCC would regularly review the status and functions of the above advisory groups and also establish *ad hoc* groups to help implement the SAP. Within the IBCC, a Project Coordination Unit would play a key role in coordination, networking, communication and information exchange for the BCLME Programme. It has been proposed that three activity centers (one per country) be established to facilitate coordination within the partner countries and to serve as centers for specialist BCLME actions (e.g., resource assessment, methodology and calibration, regional environmental monitoring and networking, marine pollution, etc.).

Policy Actions

The policy actions by and large build on and give effect to (*with deadlines*) the actions specified in the TDA that are necessary to address the suite of identified priority transboundary problems and issues. As full coverage of these is beyond the scope of this paper, we present here a few examples which still need to be agreed and approved by the three governments:

- Joint surveys and assessment of shared stocks of key species will be undertaken cooperatively between 2001–2005 to demonstrate benefits of this approach. The three countries endeavor to harmonize the management of the shared stocks.
- A regional mariculture policy to be developed by December, 2002.
- The three governments commit themselves to compliance with the FAO (Food and Agriculture Organisation of the United Nations) Code of Conduct for Responsible Fisheries (FAO 1995).
- A regional framework for consultation to mitigate the negative impacts of mining be developed by December, 2002, and mining policies relating to shared resources and cumulative impacts to be harmonized.
- A regional network for reporting harmful algal blooms to be implemented in 2002.
- Wastewater quality criteria for receiving waters to be developed by June, 2002, for point source pollution.
- A strategy for the implementation of MARPOL 73/78 in the BCLME region be devised by December, 2000.
- Existing data series and material archives to be used to establish an environmental baseline for the BCLME.

- A regional biodiversity conservation management plan and framework to be developed by December, 2003.
- A comprehensive regional strategic plan for capacity development and maintenance for the BCLME to be finalized by June, 2001.

Wider Cooperation

The three countries, individually and jointly, would encourage enhanced cooperation with other regional bodies such as BENEFIT, SADC (Southern African Development Community), the future South East Atlantic Fisheries Organization (SEAFO), NGOs, UN Agencies, donors, and other states with an interest in the BCLME.

CONCLUDING REMARKS

We have attempted to illustrate the joint approach taken by Angola, Namibia, and South Africa in partnership, where appropriate, with the international community to manage the marine and coastal resources of the Benguela Current region sustainably through the application of science and technology. Examples have been provided of a fisheries science initiative, BENEFIT, country-based ICM, and holistic approach to regional marine and coastal management using the emerging BCLME Programme as the catalyst. These are some of the building blocks, but there are others. For example, at the science and technology level, strong links have been built with a number of parallel, but distinctly different, initiatives. These include (a) South Africa's established and internationally acclaimed Benguela Ecology Programme (BEP) (see Siegfried and Field 1982), which has resulted in the publication of thousands of publications on the Benguela ecosystem since 1982 (see, e.g., Payne et al. 1987, 1992; Pillar et al. 1998), (b) the ENVIFISH Programme (Environmental Conditions and Fluctuations in Distribution of Small Pelagic Fish Stocks), which is a three-year European Union funded project between seven EU states and Angola, Namibia, and South Africa focusing primarily on the application of satellite data in environment–fisheries research and management, and which commenced in October, 1998, and (c) VIBES (Variability of Exploited Pelagic Fish Resources in the Benguela Ecosystem in Relation to Environmental and Spatial Aspects), a bilateral French–South African initiative focusing on the variability of pelagic fish resources in the Benguela and the environmental and spatial aspects of the system, which also commenced in 1998. At the socioeconomic and management levels, bilateral arrangements between the three Benguela countries and various overseas states have materially assisted the development and application of sustainable management policies, while enhanced regional cooperation at all levels across disciplines is actively promoted by SADC. In the fisheries context, the future SEAFO is likely to play a pivotal role in the sustainable management of living marine resources.

In Figure 15.5 we have attempted to show how the various science and management initiatives fit together, both at the country level and regionally in the Benguela. Clearly, appropriate science and technology are the cornerstones of the integrated sustainable management. At all levels and in all disciplines and functions, strong emphasis has been placed on capacity development.

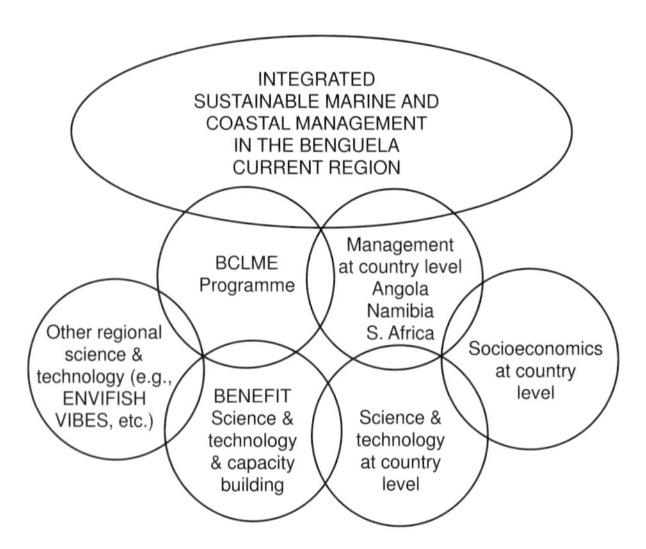

Figure 15.5 Schematic showing the interlinking of science, technology, and management in the BCLME at country and regional levels.

The collaborative approach by Angola, Namibia, and South Africa is highly relevant within a broader regional context, i.e., within SADC, as it provides an example how member states with very different resource bases (human, infrastructure, financial) can work together using science and technology as a unifying factor to underpin responsible management of a complex system. Taken one step further it can help convert the vision of an *African Renaissance* into reality. More than that, we suggest that the approach and action by the countries bordering on the Benguela Current could serve as a blueprint in other parts of the developing world for the integrated sustainable management of marine and coastal systems which are shared between two or more countries.

ACKNOWLEDGMENTS

What we have attempted to synthesize in this paper represents the collective wisdom and vision of a large number of local, regional, and international experts, and an example of the commitment by the governments of three southern African states to sustainable development and wise management of the region and its natural resources. In preparing this manuscript, we have drawn on published and unpublished documents as well as from the BCLME TDA and SAP and an article by L.V. Shannon and M.J. O'Toole entitled "The Benguela: Ex Africa Semper Aliquid Novi," which has been drafted for a book on large marine ecosystems edited by G. Hempel. We acknowledge permission given by the Windhoek Office of the United Nations Development Programme to use information from the TDA developed for the BCLME Programme and the invaluable input by Mr. C. Davis of the U.K. Department for International Development into the section dealing with ICM in South Africa.

REFERENCES

BCLME SAP (Benguela Current Large Marine Ecosystem Strategic Action Plan). 1999. Strategic Action Plan for the Integrated Management and Sustainable Development and Protection of the

Benguela Current Large Marine Ecosystem. Windhoek: UNDP (United Nations Development Programme).

BCLME TDA. 1999. Benguela Current Large Marine Ecosystem Programme (BCLME) Transboundary Diagnostic Analysis (TDA). Windhoek: UNDP.

BENEFIT (Benguela-Environment-Fisheries-Interaction-Training Programme). 1997. BENEFIT Science Plan. Swakopmund: BENEFIT Secretariat.

Black Sea. 1996. Strategic Action Plan for the Rehabilitation and Protection of the Black Sea. Istanbul: Global Environmental Facility/UNDP Black Sea Environmental Programme.

Brown, P.C., S.J. Painting, and K.L. Cochrane. 1991. Estimates of phytoplankton and bacterial biomass and production in the northern and southern Benguela ecosystems. *S. Afr. J. Mar. Sci.* **11**:537–564.

Bubnov, V.A. 1972. Structure and characteristics of the oxygen minimum layer in the South-eastern Atlantic. *Oceanology* **12**:193–201.

Chapman, P., and L.V. Shannon. 1985. The Benguela ecosystem. 2. Chemistry and related processes. In: Oceanography and Marine Biology: An Annual Review, vol. 23, ed. M. Barnes, pp. 183–251. Aberdeen: Aberdeen Univ. Press.

Croll, P. 1998. Benguela Current Large Marine Ecosystem (BCLME). First Regional Workshop, United Nations Development Programme, 22–24 July 1998, Cape Town, South Africa. Report prepared for Global Environmental Facility/UNDP Benguela Current Large Marine Ecosystem Programme. Windhoek: UNDP.

Croll, P., and J.T. Njuguna. 1999. Benguela Current Large Marine Ecosystem (BCLME). Second Regional Workshop, United Nations Development Programme, 12–16 April 1999, Okahanja, Namibia. Report prepared for Global Environmental Facility/UNDP Benguela Current Large Marine Ecosystem Programme: Windhoek: UNDP.

Draft White Paper. 1999. Draft White Paper for Sustainable Coastal Development in South Africa. Pretoria: Dept. of Environment Affairs and Tourism.

FAO (Food and Agriculture Organisation of the United Nations). 1995. Code of Conduct for Responsible Fisheries. Rome: FAO.

Hamukuaya, H., M.J. O'Toole, and P.M.J. Woodhead. 1998. Observations of severe hypoxia and offshore displacement of Cape hake over the Namibian shelf in 1994. *S. Afr. J. Mar. Sci* **19**:57–59.

Payne, A.I.L., K.H. Brink, K.H. Mann, and R.H. Hilborn, eds. 1992. Benguela Trophic Functioning. *S. Afr. J. Mar. Sci.* **12**:1–1108.

Payne, A.I.L., J.A. Gulland, and K.H. Brink, eds. 1987. The Benguela and Comparable Ecosystems. *S. Afr. J. Mar. Sci.* **5**:1–957.

Pillar, S.C., C.L. Moloney, A.I.L. Payne, and F.A. Shillington, eds. 1998. Benguela Dynamics: Impacts of Variability on Shelf-sea Environments and Their Living Resources. *S. Afr. J. Mar. Sci.* **19**:1–512.

Shannon, L.V., J.J. Agenbag, and M.E.L. Buys. 1987. Large- and meso-scale features of the Angolan Benguela Front. *S. Afri. J. Mar. Sci.* **5**:11–34.

Shannon, L.V., and G. Nelson. 1996. The Benguela: Large scale features and processes and system variability. In: The South Atlantic: Past and Present Circulation, ed. G. Wefer, W.H. Berger, G. Siedler, and D.J. Webb, pp. 163–210. Berlin: Springer.

Shannon, L.V., and M.J. O'Toole. 2000. The Benguela: Ex Africa semper aliquid nov, ed. G. Hempel, in press.

Sherman, K. 1994. Sustainability, biomass, yields, and health of coastal ecosystems: An ecological perspective. *Mar. Ecol. Prog. Ser.* **112**:277–301.

Siegfried, W.R., and J.G. Field, eds. 1982. A Description of the Benguela Ecology Programme 1982–1986. South African National Scientific Programmes Report No. 54. Pretoria: CSIR (Council for Scientific and Industrial Research).

United Nations. 1983. United Nations Convention on the Law of the Sea 1982. New York: United Nations.

Yamagata, T., and S. Iizuka. 1995. Simulation of tropical thermal domes in the Atlantic: A seasonal cycle. *J. Phys. Oceanogr.* **25**:2129–2140.

Standing, left to right: Mick O'Toole, Liana Talaue-McManus, Olasumbo Martins, Christiane Gätje, Claudio Richter, Bastiaan Knoppers
Seated, left to right: Magnus Ngoile, Wim Salomons, Sundararajan Ramachandran, Peter Burbridge

16

Group Report: Integrated Coastal Management in Developing Countries

C. RICHTER, Rapporteur

P.R. BURBRIDGE, C. GÄTJE, B.A. KNOPPERS, O. MARTINS, M.A.K. NGOILE, M.J. O'TOOLE, S. RAMACHANDRAN, W. SALOMONS, and L. TALAUE-MCMANUS

INTRODUCTION

Economic growth is a prerequisite for many countries as they seek to sustain an ever-growing population. At the same time, the resulting socioeconomic pressures cause the exploitation of natural resources at a level that often exceeds the carrying capacity of the ecosystem (WCED 1987). This is particularly true for tropical coastal areas, where a large proportion of the world's population is centered (IPCC 1994).

Proper management of the coastal area and its resources is imperative to ensure sustainable development, to maximize the benefits for the coastal community, and to minimize the negative effects on the environment. Coastal management must be implemented in the face of imminent problems caused by poverty, hunger, overpopulation, illiteracy, disease, etc. Lack of communication between institutions often hampers the exchange and integration of existing knowledge.

Clearly, many of the issues are not restricted to tropical countries; this permits the identification of common goals and the transfer of mutually applicable management strategies between developing and developed countries (Cicin-Sain and Knecht 1998). Others, for example, incorporation of traditional knowledge and sociocultural values in management schemes, will require innovative approaches.

Since changes in developing countries proceed at a faster pace than in western countries, the coastal manager in the former must seek short-term solutions to the problems that are at the same time cost-effective and easy to convey to local stakeholders and policy makers. To produce cost-effective strategies, it is necessary to identify and understand the hierarchies of the socioeconomic drivers, ranging from the local to the global scale, and how these interactions impinge on the use and trade of the tropical resources.

Science and Integrated Coastal Management
Edited by B. von Bodungen and R.K. Turner

A crucial issue concerns effective communication, which is essential to produce public awareness and to convince policy makers. For example, the economic valuation of goods and services of a coastal zone is a message that transcends all sectors of society, while a sophisticated ecological model is not.

Conceptually, there is a need to create a more robust scientific and philosophical foundation for integrated coastal management (ICM) in developing countries that encompasses natural and cultural aspects over a wide range of scales. Fundamental to this foundation is the development of holistic, multi-disciplinary, and integrative science that can be effectively communicated in forms appropriate to policy making, decision taking, and day-to-day management of coastal systems and activities they sustain.

We identified the following questions as guidelines for our discussions on ICM in developing countries:

- Why is ICM different from other management schemes?
- Are there discrete differences between developing and developed countries relevant to ICM? If so, what are they?
- Is there a recognizable hierarchy in the socioeconomic drivers?
- What tools do we need to establish linkages between socioeconomic and environmental issues relevant to ICM in developing countries?

THE INTEGRATED COASTAL MANAGEMENT PROCESS AND THE ROLE OF SCIENCE

What do we mean by ICM? Essentially, this refers to a *process* of developing a more effective use of human resources — including scientific expertise and knowledge, institutional arrangements, policies, laws and regulations, and other instruments to promote more efficient use of public and private capital and natural resources — to meet stated development objectives. This process starts with the *awareness* of issues of common concern; this facilitates a *dialogue* and exchange of views among interested and affected parties, which in turn supports *cooperation* among the parties as the basis for *coordination* of action which, over time, fosters *integration* of coastal management.

ICM also provides a conceptual framework within which individual strategies for resolving issues and promoting sustainable coastal development may be formulated. Strategies can range from highly centralized, national ICM initiatives which attempt to stimulate politically independent states or provinces to adopt coastal management (e.g., the Coastal Zone Management Act in the United States), to highly localized ICM based on adaptation of traditional management systems and practices (Cicin-Sain and Knecht 1998).

There is no "golden rule" or universal framework for promoting ICM. However, there are three major outcomes of ICM endeavors (adapted from Olsen et al. 1999), namely:

- development of a robust and well-informed ICM process;
- positive changes in societal behavior which enable economically rational, socially responsible, and environmentally sustainable forms of development;
- improvements in ecosystems and social systems.

The concepts and principles of ICM provide a sound basis for strengthening the development planning process. Some of the contributions that ICM can make are (Clark 1992):

- facilitation of sustainable economic development based on natural resources generated by coastal ecosystems;
- improvement of livelihoods of coastal communities;
- conservation of natural habitats and species;
- control of pollution and the alteration of shorelands and beach fronts;
- control of watershed activities that adversely affect coastal zones;
- control of excavation, mining, and other alteration of coral reefs, water catchments, and sea floors;
- rehabilitation of degraded resources;
- provision of a mechanism and tools for rational allocation of space and resources.

Box 16.1 lists the anticipated and actual benefits of ICM planning and implementation identified by practitioners in the East Africa Region. Achievement of each of these outcomes requires effective and timely contributions of expertise and knowledge from a diverse array of scientific disciplines.

DIFFERENCES BETWEEN ICM AND OTHER FORMS OF MANAGEMENT

Coastal zone management differs from other management forms in scale, complexity, and dynamics of the systems being managed. Compared to watershed management, coastal regions are subject to much higher inputs of energy than either upland or marine environments. Primary sources of energy include water draining from the land surface (hydrologic cycle), tidal and oceanic currents, as well as wind and wind-driven waves.

The high inputs of energy and materials, such as nutrients, derived from the hydrologic cycle, weathering, erosion, and upwellings stimulate the formation and maintenance of diverse and highly productive coastal ecosystems. The diversity and productivity of these ecosystems is generally higher than is found in purely terrestrial or offshore marine areas.

Linked to this diversity of ecosystems and high biological productivity are greater resource development opportunities than are common to other regions. This stimulates the development of economic and social activities, which are more diverse and enjoy more dynamic growth than is common in other land regions.

Coastal locations and economics of long-haul water-based transport lend themselves to trade and exchange of materials at "Break in Bulk" points along rivers, estuaries, and sheltered coasts. This in turn creates foci for industry, commerce, and human settlement—hence, the major industrial and commercial centers in coastal mega cities.

The resulting complex array of competing activities superimposed on dynamic and often fragile coastal ecosystems creates urban and rural management challenges that are more complex and risk laden than is found in other areas of human activity.

A broad knowledge of system processes is needed to understand the specific challenges, which may be quite different, depending on whether they occur, for example, in humid or arid regions. Generally, people tend to look at coasts as stable systems, while in fact they are

Box 16.1 The following benefits of ICM were identified by the participants in the "Experts and Practitioners Workshop on Integrated Coastal Area Management for Eastern Africa and the Island States" (Humphrey and Francis 1997).

Outcomes — Benefits to People and Environment

- Improved quality of life
- Reduction and easier resolution of conflicts
- Management for sustainable use of land and sea resources
- Sustainable/optimal utilization of resources (proactive versus reactive)
- Improved coastal and marine ecosystems
- Better distribution of available natural resources within the community
- Enhanced community welfare
- Improvement of infrastructure at the local level

ICM Process Benefits — Improving Efficiency and Effectiveness

- Enhanced transparency of policy making encourages questioning and involvement
- Improved cooperation between government departments/NGOs/education institutions
- Institutional cooperation and interlinkages
- Better coordination, cooperation, and communication
- Participatory approach (conflict, planning, education, multisectoral) to bring together politicians, business, general public, scientists
- Holistic problem-solving approach
- Rational use of funding and resources
- Better allocation of government resources
- Cross-sectoral collection, analysis, and transfer of information
- Dissemination of information assists decision making
- Facilitation of information exchange
- More active community involvement
- Integration of local knowledge of resources in reaching solutions

Sustainability — Making the Results Last

- Puts environment up front in the planning process
- Platform for conflict resolution
- Integrated legislation
- Enhanced capacity building
- Empowerment of stakeholders through participation
- Community involvement in management — ownership of solutions
- Enhanced community stability
- Sustainable utilization improves conservation awareness
- Institutionalization/integration of ICM in normal government
- Encourages appropriate use of components of the coastal zone
- Increases investment

transient systems that constantly adjust to the hydrodynamic energy regime (Pethick, this volume). Because the various ecosystems on a given coast are highly interlinked, human activities impacting any one of its integral parts impact the entire system. This is illustrated in Figure 16.1, where the various mutually supporting coastal ecosystems are represented as blocks of stone forming the arch of a Roman bridge. Degrading one or more of the ecosystem parts will weaken the functional integrity of the coastal system as a whole. Once the impacts exceed a critical load, the system breaks down. Maintaining ecosystem integrity and function for the sustained use of the coastal zone therefore requires a holistic approach.

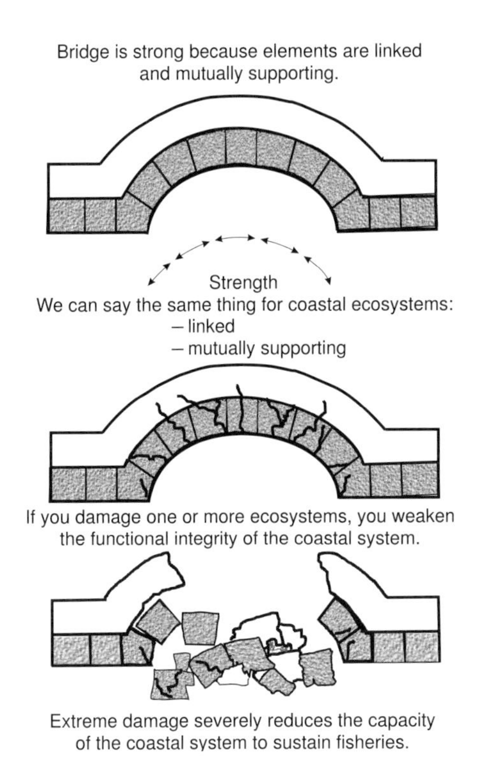

Figure 16.1 Interdependence of the various coastal ecosystems supporting the integrity of the coastal zone. Redrawn from Burbridge (1999).

The fundamental attitude of many nations and people is that there is still the paradigm that oceans are bottomless sinks and so you can dump as much as you want into the coastal sea and proceed without any check. Furthermore, the response time is so large, that respective effects remain unnoticed and policy makers see no reasons nor would have public support to initiate ICM synchronously with evolving economic planning.

At present, the need to integrate sound environmental management concomitant to rational economic planning is recognized by an increasing number of governments (Cicin-Sain and Knecht 1998). However, the complexity of the sectoral interests, lack of recognition of the economic benefits of integrated approaches to planning and managing the coastal development process, and lack of understanding of the risks inherent in disrupting natural coastal processes pose management challenges that are very different from other regions.

One of the challenges facing coastal management is to identify and assess the impact of external economic driving forces. Development of a robust ICM process, where relevant boundaries and linkages between coastal ecosystems and economic and social systems are assessed, will facilitate anticipation and prediction of system responses. Such "forecasts" can then be conveyed to the public and to decision makers, who may use them according to their specific requirements.

As a first step, it is important to identify all relevant environmental processes, ecosystem components, functions, resources, and users. Figure 16.2 illustrates the broad and diverse

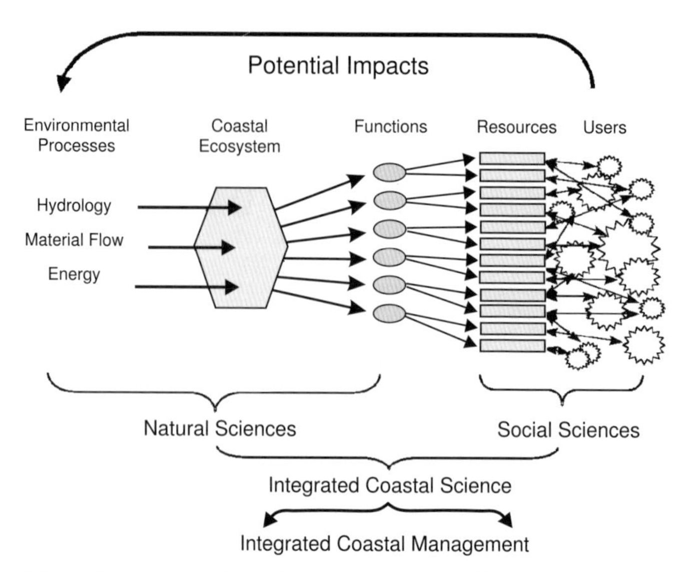

Figure 16.2 Schematic representation of a coastal system, which includes the environmental processes that maintain the ecosystem, the different functions the ecosystem hosts, the resources generated by those functions, and the different human activities that use those resources. Sectoral management generally focuses on maximizing the outputs of a limited range of products or on benefits to a subset of potential users of the ecosystem. While sustainability issues are increasingly considered, the implications of a particular activity for the full range of actual or potential users are all too often ignored. Integrated Coastal Management (ICM) provides a means to look at compatibility among different activities, as well as a means to resolve direct and indirect conflicts. Redrawn from Burbridge (1997).

array of human activities that are directly or indirectly supported by natural ecosystems. Points that should be noted from this figure are:

- the importance of environmental processes in maintaining the production of renewable resources;
- the potential for a particular resource to support more than one form of human activity;
- overexploitation of one resource can have a corresponding effect on other potential users of the resource,
- there is feedback from the users to the ecosystem.

The effects of coastal resource use usually extend beyond competition with other users, with direct and indirect impacts of resource exploitation felt on environmental processes and the coastal ecosystem, and consequently on its functions and services and on resource quantity and quality. Particular attention should be paid to environmental processes that may be affected by activities far from the ecosystems they maintain. Like a factory, the system will not continue to produce products unless attention is given to the amount and quality of inputs required to maintain the productivity and functional integrity of the system.

Figure 16.2 also illustrates that the management for sustainable use of coastal ecosystems requires a creative fusion of a wide range of disciplinary skills and knowledge. Towards the left-hand side of the figure, interdisciplinary ecosystem analysis has a major contribution to make to the development of planning and management practices that will maintain the health and productivity of the coastal systems. Towards the right-hand side, disciplines (such as economics, sociology, business management, or spatial planning) will make a major contribution to the development of resource utilization strategies to help ensure the sustainable use of the full range of renewable resources as well as functions and services produced by the coastal system (Burbridge 1999).

The vulnerability and carrying capacity of a coastal system will determine its response to natural or human impacts. Productive systems driven by external energy, such as coastal upwelling areas, are more resilient to environmental change than "closed" recycling systems such as coral reefs.

The resilience of upwelling ecosystems can be demonstrated by the widespread collapse of the Benguela Current ecosystem, which happened in 1993–1994. Here, changes in seasonal wind patterns, subsequent failure of upwelling, and a reversal in current direction caused system-wide eutrophication and anoxia, which resulted in catastrophic losses of pelagic and demersal fish, poor spawning and recruitment, and an unprecedented die-out of some 300,000 fur seals (Roux 1998). Two years later, in 1996, normal upwelling patterns returned which were followed by successive years of favorable oceanographic conditions suitable for fish spawning, eggs and larval survival, and recruitment. This resulted in the recovery of the ecosystem to levels where most fish stocks and seal populations now exceed pre-1993 levels. The unpredicted, catastrophic environmental perturbation, which occurred in 1993–1994, precipitated a change in the fisheries management policy in Namibia. Now, key marine environmental indicators (i.e., upwelling indices, thermocline depths, and the degree of mixing between warm Angolan and cool Benguela Current waters) are taken into account before fisheries management decisions are made (cf. O'Toole et al., this volume). A "State of the Marine Environment" report is now considered along with the "State of the Fish Stocks" report in reaching a consensus on fish quota allocations with a more conservative harvesting of resources likely in the event of poor marine environmental indicators.

By contrast, coral reefs, which feature high gross but only low net productivity may suffer irreversible damage from exploitation and eutrophication (Wilkinson 1998). Mangroves, which provide 70 direct products to multiple local users on a sustainable basis, have been converted to shrimp ponds sacrificing 69 products at the expense of only one product satisfying foreign demands on an unsustainable basis. Coastal areas that are allowed to adjust to a changing energy regime are more resilient than immobilized coastlines, calling for managed retreat options to restore the original ecosystem functions (Pethick, this volume). Box 16.2 provides a guideline to coastal management approaches. Adaptive management (Boesch, this volume) should be applied especially in cases where the scientific information is not available.

ICM-RELEVANT DIFFERENCES AND SIMILARITIES BETWEEN DEVELOPING AND DEVELOPED COUNTRIES

Coastal zones all over the world are experiencing anthropogenic impacts, such as high population pressure, exploitation of resources, degradation of habitats, resource-use conflicts. These are exacerbated by a general lack of understanding of the complexities of interactions between coastal system parts, the weak use of existing environmental information, and their biased application in support of sectoral interests.

Box 16.2 Summary of messages for coastal resources management

- Coastal ecosystems provide many different functions that help to sustain a very wide array of different forms of economic and environmental goods and services. This is reflected in both the concentration of population and diversity of economic activities found in coastal areas. To meet the needs and aspirations of current and future generations, it will be necessary to develop multiple use management approaches to the use of coastal ecosystems and resources that will allow different sectoral agencies to meet their economic objectives without adversely affecting the ecosystems that help sustain their economic activities. This will require an improvement in our scientific understanding of the functions performed by different coastal ecosystems and the resources they generate, and a move away from sectoral development planning to more integrated intersectoral plans and management strategies.
- In developing countries there is often a greater degree of dependence on renewable resources generated by coastal ecosystems with respect to food security and other basic human needs than is true of more developed countries. Traditional coastal resources management practices may not be sustainable as a result of rapid population growth and/or strong external economic demands for specific resources.
- There can be a high degree of mutual dependence upon the functions of individual coastal ecosystems by activities represented by different economic sectors. Therefore, it is in the common interest of these activities to promote policies, plans, and investment strategies that maintain the health and productivity of coastal ecosystems. It is important to shift emphasis away from coastal development based primarily upon controlling the end use of coastal ecosystems towards a more balanced approach. Emphasis should be given to maintaining the health and productivity of coastal ecosystems so that they can continue to supply flows of resources that sustain different forms of activity.
- Sectoral approaches to the development of resources normally seek to maximize the financial returns from a narrow range of renewable resources and effectively discount all other uses to zero. This often leads to nonsustainable and/or economically inefficient forms of development that disenfranchise other economic and social groups from the natural resources base. This is a major factor that contributes to the decline of the welfare of coastal communities and consequent rise in rural to urban migration.
- There is more to be gained in both economic and social terms by maintaining the health and productivity of coastal systems and the supply of their renewable resources than can be normally achieved through their allocation to single purpose and exclusive uses, or their conversion to alternative uses. Integrated management can enable coastal system functions (e.g., primary and secondary biological production, flood water retention, groundwater recharge, and maintenance of base water flows in river systems, etc.) to be maintained alongside such resource uses as controlled timber harvesting, fuelwood production, collection of secondary products, and conservation of breeding habitats for migratory wildfowl. Careful planning and timing of activities is required to prevent conflicts, for example, prohibiting logging during monsoonal wet seasons and within 2 km of nesting sites of bird species being protected (Humphrey and Burbridge 1997, 1999; Humphrey et al. 2000).

Differences in the ICM process between developed and developing countries arise from a combination of factors, which are characteristically (but not exclusively) found in tropical systems:

- The level of poverty provides fewer options for developing countries. Coastal communities, in particular, belong to the poorest of the poor. Because of a generally large degree of dependence on a narrow range of resources exported, resource users, such as fishermen, are left with only few options of occupational alternatives and are therefore resistant to change. Mitigation measures must therefore result in benefits palpable at the level of the resource user, to be perceived as effective.

- Informal groups are often pervasive. An example is the powerful influence of capital providers who control market prices and often exacerbate the exploitation of the natural resources at cheap costs to maximize profits. The lack of responsibility of these groups is a long-standing problem. Science should promote social and environmental awareness towards the adoption of globally accepted codes of conduct and environmental standards, such as the Eco-Management and Audit Scheme (EMAS), International Standards Organization (ISO) 14000, green tourism scheme, Social Accountability (SA) 8000. Once international recognition and acceptance is achieved, global players in the private sector, such as the tourism industry, may become promoters of market-leading environmental standards (Daily and Walker 2000).
- Lack of environmental awareness among the local users is a difficult issue. In pristine coastal areas, local communities know how to maintain the integrity of their system. In impacted regions, users may face great difficulties in bringing the system into balance.
- The institutional fabric is weaker in developing countries. The distribution of power is very often uneven. In addition, institutions and individuals involved in the ICM process change rapidly and often in an unpredictable manner, thus challenging the implementation of long-term management schemes.
- The pace of change is generally faster in tropical countries that have gained political independence only recently and which are in a stage of economic transition from a resource-based to a product- and service-based economy. In temperate countries, by contrast, this transition has already taken place, leaving fewer opportunities for development.
- Environmental databases are inadequate in most developing countries; long-term trend datasets, in particular, are almost nonexistent. This makes the formulation of predictive models very difficult, so that important decisions have to depend mainly on the subjective judgment of the coastal manager.

The scientific capacity, however, of developing countries to deal with complex ICM issues is often underestimated. A foreign "Chief Technical Advisor" is often perceived as the expert of choice. In many cases, this is not true and neglect of local capacity is disastrous.

Training for capacity building for effective science for ICM in developing countries has to happen on-site. Scientists have to do their fieldwork in their own countries or relevant environments. In this way, they contribute to the much-needed scientific information, build infrastructural capacity, and reduce brain drain. This has been demonstrated in the western Indian Ocean, where before 1989 most of the postgraduate training in marine science was taking place in Europe and North America. A majority of these experts have not returned to their home countries. However, the sandwich training programs that were initiated after 1999 have generated experts that are now contributing to science in the western Indian Ocean. The programs allow the postgraduate students to have short periods in universities in Europe or North America for purposes of literature review, technical expert advice, and writing of their thesis. All fieldwork and experimentation is conducted in home countries. In addition to individual training, international cooperation on an institutional level is needed to ensure the medium- and long-term success of the capacity-building measures. To be effective, both parties must benefit from the process. Because existing templates cannot be simply transferred between temperate and tropical systems, ICM is a challenging and mutually beneficial learning process for partner institutions from both developing and developed countries. Examples of joint research and management programs are the BENEFIT program in the Benguela Current region and the Red Sea Program in the Middle East (Hempel 1999; O'Toole et al., this volume). These multinational programs provide hands-on instruction to young scientists and technicians at sea, and promote training and cooperative research. This integrative approach is becoming increasingly recognized and has become the standard practice for, e.g., Brazil's governmental support institutions or Germany's Center for Tropical Marine Ecology in Bremen.

- The capacity of scientists and managers to engage in a meaningful dialogue is often weak. Communication of difficult environmental and management issues to the public presents another hurdle due to different levels of education, such as the low level of literacy in many tropical countries.

  As a consequence of communication impasses and mismanagement in many instances, local populations are often suspicious of scientific advice and government decisions. This may lead to the failure of well-intended programs because of the refusal of local users to follow government advice. Social barriers between ethnic and cultural groups are additional barriers to communication.

Similar needs are identified in developing and developed countries for improving the contribution of science to the ICM process:

- The ICM process is often hindered by the fragmentation of science and governance into various institutions and ill-defined constituencies dealing with the various ICM issues, particularly where government bureaucracies are diversified, divisions long-standing and entrenched, and based on divergent legal mandates (Cicin-Sain and Knecht 1998). For example, fishery issues are dealt in the EU in several directorates with little, if any cross-departmental links. A stronger integration of disciplines and governance bodies is therefore highly needed.
- The dialogue between managers, politicians, scientists, and stakeholders suffers from problems of conveying the relevant information in the appropriate manner at the right time. This impasse calls for joint efforts to establish consensus on the necessary mechanisms for interaction in order to improve ICM.

HIERARCHIC LEVELS OF SOCIOECONOMIC DRIVERS INFLUENCING ICM IN DEVELOPING COUNTRIES

In analyzing the nature of coastal management issues, it is necessary to understand that there are socioeconomic dimensions or causative factors that operate at various governance scales (local, national, international; Figure 16.3).

At the local level, income is generated; at the national level, the need for foreign revenues prevails; at the international level, the demand for coastal resources (fisheries and minerals) and services (tourism, waste disposal) is at issue. Recognition of these scales logically

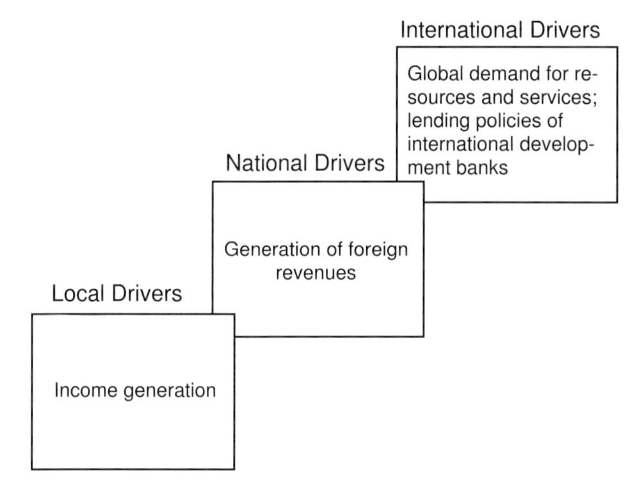

Figure 16.3 General scheme of the socioeconomic drivers operating at different hierarchic levels on tropical coasts.

dictates that science must address these dimensions at their appropriate scales to mitigate negative affects.

Examples of relevant issues include:

- At the drainage basin: deforestation caused by global wood trade and, to a lesser extent, slash-and-burn agriculture, driven by national and local factors.
- At the shoreline: exploitation of the coast through tourism and industrial facilities, usually by multinational companies which are extremely powerful, and environmental hazards by untreated sewage from coastal villages.
- In the water: shrimp trawling, shrimp mariculture, tuna fisheries and trade, trade of endangered species (e.g., marine turtles) but also artesanal fishery practices.

This hierarchical set of drivers will have to be the target of multi-scale integrated management. What are some of the major blocks to effective management at the various geopolitical scales?

- At the *international* level: Global trade and lending as well as development policies externalize environmental impacts and damages, and decisions are made on the basis of optimum profits and efficiency. Economic tools are needed that include environmental impacts and societal costs in, for example, the valuation of resources.
- At the *national* level: Inflow of foreign revenues supercedes domestic food (protein) and environmental security. Economic incentives (e.g., tax overrides) are used to encourage production for fisheries exports (prawns and tuna) and multinationally owned tourism facilities. A method is needed to address poverty and sustained national growth to strengthen a government's strategy of ensuring domestic food and environmental security.
- At the *local* level: Local communities need to sustain their income base. When environmental awareness is low, they can opt for development with short-term returns,

which would lead to environmental degradation. When made environmentally aware, they play a major role in implementing coastal management.

The concept of sustainability should apply at every scale. Conflicts, for example, in the case where the profit at the top level is at the expense of environmental degradation at the community level, should be resolved through a participatory process. Science should take the role in making society aware of the hierarchical nature of the problem and provide the knowledge to the people to assess its full scope. Information and transparency will allow for decisions according to the principle of self-determination of the people.

There seems to be a growing perception among governments of developing nations that the Western model of progress, i.e., a shift from an agro-based to an industrial-based economy, is the way to proceed. Do developing nations need to industrialize the way the West did?

Economies in transition should not develop in a similar manner to Western societies. Neither their resources nor environment could sustain such a development trajectory (Figure 16.4a).

Emerging economies exploiting their natural resources for economic growth have to manage pollution and overexploitation at the earliest possible stage to sustain further economic growth (Figure 16.4b). Examples of countries in different states of development are found in Table 16.1.

It is important to note that while postindustrial economies are focused on pollution abatement, emerging economies have responded to the increasing global demand for coastal living resources and services often at the expense of their environment. This development pattern is not sustainable. ICM must be proactive at local, national, and international scales to ensure a healthy living resource base which emerging economies must have for sustainable economic growth.

For many developing nations, there is direct dependence on the health of their living resources and natural systems. The extent of resource overexploitation and environmental degradation severely constrains the ability of these nations to ensure food and environmental security which underpin economic progress. Where population growth is high and poverty widespread, the key to economic progress may be to put the health of the environment at the top of the development agenda.

Table 16.1 Stages of development: Timing and pace of development in developing and developed countries from the preindustrial phase (I) through industrialization (II) and pollution management (III) to the postindustrial phase (IV).

	1900s	1960s	2000
Western Europe, eastern U.S.A.	II	III	III
Eastern Europe and Central Asia	I	II	II
South and Eastern Asia, South and Central America	I	II	(III)
Amazonia, parts of Africa, Irian Jaya	I	I	II
Singapore	I	I	IV

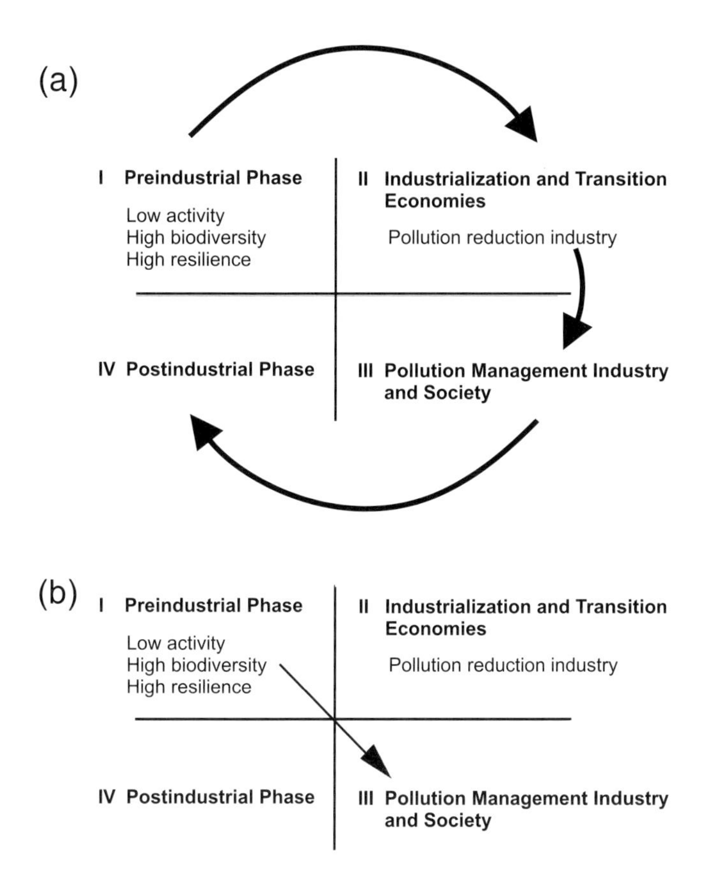

Figure 16.4 Changes of state and environmental management cycle, where (a) is the development cycle following the (usual) reactive management approach, and (b) the desirable development scenario following proactive management goals.

TOOLS LINKING SOCIOECONOMIC AND ENVIRONMENTAL ISSUES RELEVANT TO ICM IN DEVELOPING COUNTRIES

The limited success of ICM to date raises the question as to why the available scientific information has not been used effectively to inform the coastal policy, planning and management process. We believe that the main causes are rooted (1) at the international level, demonstrated by the lack of will to implement the resolutions of the Rio Conference (Agenda 21), (2) at the national level, in the lack of capacity to ensure continuation of ICM programs, (3) at the institutional and individual level, in the lack of engagement, (4) through inappropriate use and misinterpretation of scientific data, and (5) through inadequate tools to establish the necessary links between the relevant socioeconomic and environmental factors.

To tackle effectively the problems resulting from the pressures occurring at the different levels, a *Plan of Action* is needed:

1. Agenda 21

At the international level, it was recognized and declared at the Rio Conference that coastal regions play an important role in fulfilling international and national development objectives (Chapter 17, UN 1993). We believe that there has been a lack of strategic investment in science and management to fulfil the objectives of Agenda 21 (cf. also Boesch 1999). A review of progress and action plan is needed to foster the wide-scale implementation of the Rio resolutions.

2. Governance Capacity

Lack of governance capacity is an international problem. International funding agencies spend large sums without assessing with sufficient vigor the capacity of the governance institutions to implement the project. Management science provides one measure to assess managerial procedures and points to ways of improvement (Olsen et al. 1997). Science can also improve management capacity, if it finds the right way of communicating. Scientists are needed to provide the necessary scientific information to managers to frame the right questions.

3. Engagement

Scientists, like other professionals striving to reach their career goals, are often conditioned by their peer review systems. The reason for their lack of engagement in external services to the society lies in the failure to understand what their science should be, namely an act of responsibility to serve human society and their country. However, lack of engagement is caused by institutions that provide little, if any, incentives to engage in services to society. Success is measured by different standards: peer review sets the standard in science; achievement of stated objectives and outcomes sets standards in management. The reward system comes from different sources: from a scientific institution or from diffuse sources, such as the beneficiaries of ICM, which (e.g., societal recognition or consciousness) are not quantifiable. Consequently, external services are grossly undervalued in the scientific peer review system. They should, however, form an integral part of the recognition of the contribution of science to the achievement of societal objectives as well as support the career advancement of individual scientists.

There is a danger that the pursuit of scientific excellence without concomitant pursuit of social benefits will seriously hinder the development of more effective use of scientific expertise and communication of scientific information to support ICM. The peer review system must be broadened to strengthen the recognition of the value of interdisciplinary research and the contribution that science makes in informing the ICM process. Appropriate incentives are also required to alleviate the fear of scientists that engagement in interdisciplinary research and contribution to the effective application of science to help meet management objectives will carry negative consequences, such as loss of scientific prestige. We suggest that the relevant institutions concerned with the ICM process set aside funds for international awards to promote recognition of the important role of both fundamental and applied science. These awards should acknowledge both the contributions from scientists and their institutions.

4. Scientific Approach

The scientific approach has mainly been reactive to changes in the environment. However, scientific expertise and knowledge can be misdirected if it is locked into a reactive approach. Where such resources are scarce, as in many developing countries, they could be better utilized in the ICM process by adopting a proactive approach. This can be done in two ways. The first is by using environmental impact assessment (EIA) at the very early stage in planning development so that alternative sites for specific developments are evaluated (Clark 1992). The second is to use information from different sciences to form databases, which can be integrated to help opportunities for economic development and to assess options (Figure 16.5).

Options can then be tested using a more focused formal EIA, where EIA techniques are applied to assess the relative benefits and costs of alternative forms and locations of potential development. Once options have been chosen, spatial planning should be used to identify appropriate locations while monitoring would be used to ensure that development is carried out in an approved manner and that unexpected inputs can be identified and remedial action taken.

The proactive approach linked to the ICM process is adaptive, can be used to guide development to appropriate locations, and can be used to update information and coastal development plans. Fundamental to this approach is the recognition that ICM is not a given static frame but an iterative process. Usually it receives an environmental or socioeconomic stimulus (crisis that needs attention, an interest group that wants to set up an enterprise, etc.) which sets off the process. This is when dialogue begins, where the scientist has to be active, not only to provide information and point out hindrances, but also to identify opportunities that will ensure long-term enjoyment of the investment.

Science should be engaged at the *beginning* of an ICM process, be it policy or program. Scientific information provides the content and context for discussions, negotiations and decision making. Without scientific information, the ICM process succumbs to a political and/or pressure group process. Scientists themselves stand to benefit greatly through this

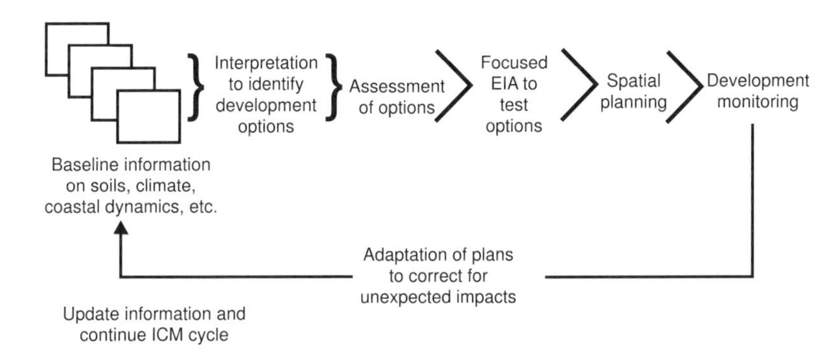

Figure 16.5 Stages involved in the iterative proactive ICM process (EIA: environmental impact assessment).

engagement because the identified gaps, inadequacies, and uncertainties can intrigue and motivate them into scientific investigations that have a societal purpose. A positive example for the early and regular engagement of science is the Arctic Council for the protection of the indigenous people of the Arctic, which seeks extensive briefing from scientists prior to each meeting (J.M. Pacyna, pers. comm.).

5. Tools

In the context of the proactive ICM process in developing countries, sophisticated diagnostic tools are a scientific luxury, where rapid-assessment techniques can give first-order estimates on the state of a system at a much lower cost (Pido and Chua 1992). Proactive science follows a holistic approach that does not require sophisticated tools, but rather builds on the integration and sound interpretation of scientific information gathered over a wide range of disciplines. This makes it particularly appealing for tropical countries, where expensive equipment is often not available.

Two types of toolboxes are available for the ICM process: one for assessing the status quo of the system, another for monitoring change. The latter is more important to management.

A prerequisite for both is that they should be simple, provide quick answers, and allow the incorporation of sound science into an easy-to-understand message. The following descriptors are not generally valid or particularly successful, but rather good examples of the interaction between socioeconomic and environmental aspects. Clearly, there are no prescriptive tools or templates that are universally transferable, and existing models must be customized to the particular socioeconomic and environmental setting in order to be effective.

Environmental and socioeconomic descriptors or indices (e.g., Human Development Index [HDI], life expectancy, gross national product [GNP]) are powerful tools, that can capture crucial ecological issues and confront people with trade-offs between development options as a basis for their decisions. They are also scientifically appealing in that they foster cross-learning and holistic thinking by drawing comparisons on a global scale. However, they may also be misunderstood or used inappropriately. For example, nutrient concentrations may not be useful indicators of nutrient stress in coastal waters, where rapid uptake by algae may lead to serious underestimates of nutrient inputs.

Performance indicators are used to assess, for example, the economic efficiency of development options. Several tools are available that can be used in the ICM process:

a) *Economic valuation of the environment.* Many development decisions are made on economic grounds. Economic valuation can be a powerful tool to not only aid and improve wise management policies but also to establish the linkages between socioeconomic and environmental issues relevant to ICM (Dixon and Hufschmidt 1986; Lipton et al. 1995; Daily 1997). The objective of economic valuation of the environment is the valuation of ecological functions. To achieve this, the importance of integrating ecological and economical approaches is critical. However, choosing a correct economic assessment approach is very important. To do this, it is imperative to define, identify, and prioritize components, functions, and attributes to relate these to use value. For instance, the coral reef economic value can be related to its coastal protection function, fisheries production function, and tourist attraction function. When it is correctly assessed, the coral reef ecosystem function becomes so vital, both ecologically and economically, that a decision to ban coral mining becomes very easy to make (e.g., in Sri Lanka; Cicin-Sain and Knecht 1998). However, economic valuation is not a panacea for all decisions. It is just one input; the political, social, and cultural considerations are also very important.

b) *Input–output analysis.* This has become increasingly important as a major tool to link natural and social sciences (Isard 1972; Constanza 1997). Its application has become a flexible and tractable mechanism in view of the advanced methodologies, such as Social Accounting Matrices (SAM) and Natural Resource Accounting Systems. There are several advantages of applying input–output analysis in ICM:
    - the inclusion of direct and indirect impacts into the pressure and response system,
    - the possibility of developing scenarios, and
    - the applicability to decision making and the estimation of a measure of sustainability.

c) *Geographical Information System (GIS) and Spatial Decision Supporting Systems.* In recent years GIS has been used very widely in decision-making processes (Sample 1994; Lyon and McCarthy 1995). The ability of GIS to integrate spatial and nonspatial variables and to present these variables in any desired combination makes it a powerful tool to educate policy makers and the public alike. Decision-supporting systems using GIS have also been used increasingly to convince policy makers in the decision-making process. GIS can serve the major functions of inventory, database, analytical tool, and also management tool.

d) *Computer Simulation Models.* Simple two-dimensional computer simulation models with "if" and "what" scenarios could also be used as a tool to establish the linkages between socioeconomic and environmental issues in ICM (see also Emeis et al., this volume). Powerful user-friendly software is now available to explore the nature and dynamics of complex human–environment interactions as a basis for informed decision making (Ruth and Hannon 1996; Hannon and Ruth 1997).

e) *Data and Information Systems and Networking.* The use of databases and information networks could also serve to link socioeconomic and environmental issues.

In addition to these tools, there are several theories of decision making, such as games theory, negotiation theory, "ecological footprint" (Wackernagel and Rees 1995; Clayton and Radcliffe 1996; Costanza 1997). However, these are very complex to use and difficult to communicate to user groups. Below we cite examples of model indices that can merge socioeconomic and natural processes:

- Terrados et al. (1998) compared changes in species richness and biomass of seagrass communities along siltation gradients in sites in the Philippines and Thailand. Results indicated that the species richness and community leaf biomass decreased with increasing percentage of silt and clay in the sediments. The lack of negative relationship between species richness and increasing community biomass suggested weak competition in multispecies seagrass beds. The positive relationship between seagrass richness and biomass, on the other hand, indicated that reduction in species diversity could lead to significant changes in biomass in areas that are heavily impacted by siltation. The gradient in sediment load can be taken as a proxy for anthropogenic influence that can be quantified (anthropogenic load vs. natural load).

- Hodgson and Dixon (1988) examined the potential conflicts between hillside logging and coastal activities, including coral reef fisheries and tourism as a result of sedimentation. Continuation of logging over a ten-year period in a hillside in Palawan, Philippines, was predicted to cause a decline in coral species richness, a decrease in live coral cover, and a reduction in fish catch. The forecast was based on an intensive analysis of sediment loading with decreasing forest vegetation, of fish production and harvest, and a study of the local tourism industry.

- Talaue-McManus et al. (1999) synthesized primary data and secondary information on biogeochemical processes in Lingayen Gulf (northern Philippines) and its associated catchments and the socioeconomic drivers that influence these. These were obtained over a four-year study (1996–1999). The analytical framework used was based on a watershed perspective and considered interactions and feedbacks among upstream and downstream economic activities and the cycling of nutrients, sediments, and freshwater within the Gulf. It linked the biogeochemical budget approach of the Land–Ocean Interactions in the Coastal Zone Core Project and an economic modeling tool (input–output model) to examine the impacts of anthropogenic drivers on the processing of materials at the coastal zone.

- Economic input–output models could be used to determine, for example, carbon/nitrogen/phosphorus (CNP) residuals from economic activities. One can compare these estimates with ambient concentrations of CNP and get first-order estimates of how much economic activities contribute to what is present in coastal waters. A variety of relatively simple and low-cost approaches may be adopted to infer, for example, the potential or degree of eutrophication. These include empirical estimates of nutrient loading from the watershed's populations (rural and urban) and fertilizers; statistical models from system comparisons linking the trophic state, nutrient loading, and the residence times of waters; the application of the LOICZ biogeochemical budget model; reconstruction of CNP depositional phases in conjunction with age determinations in surface sediment records; and, finally, observations of typical eutrophication symptoms (e.g., nuisance algal blooms and fish kills, bad odor from extensive plant decay and hydrogen sulfide, shifts in local fisheries species) (Rast and Holland 1988; Knoppers et al. 1991; Vollenweider 1992; Nixon 1995; Gordon et al. 1996).

These examples show the wide range of economic, environmental, and combined descriptors. However, the challenge for the scientific community is to customize the descriptors for the ICM process, with the ultimate goal of building up a "hierarchy of indices" that will take into account the hierarchy of socioeconomic drivers influencing tropical coastal areas.

FUTURE ACTIONS

To address the inadequacies confronting developing countries, the global community — especially scientists — have a responsibility to:

- improve the information base of ICM in developing countries;
- build the capacity for science and management to threshold levels to sustain the dialogue for effective ICM;
- generate appropriate methodologies and tools (indicators including socioeconomic indicators) that can give first-order estimates on the state of the marine and coastal ecosystems;
- improve the utility of scientific information so that it can support coastal policy, planning, and management and generate positive and equitable benefits to the diverse array of people whose welfare depends on coastal resources;
- improve the communication of available and new scientific information in forms appropriate to the needs of the wide array of potential users ranging from subsistence fishers to industrial concerns;
- develop the creative interface between the social and natural sciences to address gaps in information required to support ICM and to create new knowledge and expertise;
- increase investment in science to support the development and communication of information appropriate to the accelerating pace of economic transformation of all coastal societies;
- identify potential reward systems to facilitate effective scientific communication and linkage to integrated coastal management.

ACKNOWLEDGMENTS

We would like to thank the chairmen of this Dahlem Workshop, Bodo von Bodungen and Kerry Turner, the Dahlem Workshop participants, as well as the organizers, Klaus Roth, Julia Lupp, and their professional staff for an extraordinary and stimulating meeting. Thanks are due to Julia Lupp for drawings and careful revision of the manuscript.

REFERENCES

Boesch, D.F. 1999. The role of science in ocean governance. *J. Ecol. Econ.* **31**:189–198.

Burbridge, P.R. 1999. The challenge of demonstrating the socio-economic benefits of integrated coastal management. In: Perspectives on Integrated Coastal Zone Management, ed. W. Salomons, R.K. Turner, L.D. de Lacerda, and S. Ramachandran, pp. 35–52. Berlin: Springer.

Cicin-Sain, B., and R.W. Knecht. 1998. Integrated Coastal and Ocean Management: Concept and Practices. Washington, D.C.: Island Press.

Clark, J.R. 1992. Integrated Management of Coastal Zones. Fisheries Technical Paper No. 327. Rome: FAO (Food and Agriculture Organisation of the United Nations).

Clayton, A.M.H., and N.J. Radcliffe. 1996. Sustainability — A Systems Approach. London: Earthscan.

Costanza, R. 1997. Frontiers in Ecological Economics. Cheltenham: Edward Elgar.

Daily, G. 1997. Nature's Services: Societal Dependence on Natural Ecosystems. Washington, D.C.: Island Press.

Daily, G.C., and B.H. Walker. 2000. Seeking the great transition. *Nature* **403**:243–245.
Dixon, J., and M. Hufschmidt. 1986. Economic Valuation Techniques for the Environment: A Case Study Workbook. Baltimore: Johns Hopkins Univ. Press.
Gordon, D.C., P.R. Boudreau, K.H. Mann, J.-E. Ong, W. Silvert, S.V. Smith, G. Wattayakorn, F. Wulff, and T. Yanagi. 1996. LOICZ Biogeocehmical Modelling Guidelines. LOICZ Reports and Studies 5. Texel, The Netherlands: LOICZ.
Hannon, B., and M. Ruth. 1997. Modeling Dynamic Biological Systems. New York: Springer.
Hempel, G. 1999. Reflections on international cooperation in oceanography. *Deep Sea Res.* **II 46**:17–31.
Hodgson, G., and J.A. Dixon. 1988. Logging Versus Fisheries and Tourism in Palawan: An Environmental and Economic Analysis. Honolulu: East–West Environment and Policy Institute, Univ. of Hawaii.
Humphrey, S.L., and Burbridge, P.R. 1997. Survey of Existing Integrated Coastal Management Monitoring and Evaluation Methodologies, Indicators, and Instruments. New York: UN Development Programme.
Humphrey, S.L., and Burbridge, P.R. 1999. Thematic Study D: Planning and Management Processes: Sectoral and Territorial Cooperation. Brussels: Euopean Demonstration Programme on Integrated Coastal Zone Management.
Humphrey, S.L., P.R. Burbridge, and C. Blatch. 2000. US Lessons for European coastal management. *Marine Policy* **24**:275–286.
Humphrey, S., and J. Francis, eds. 1997. Sharing Coastal Management in the Western Indian Ocean. Dar es Salaam: Western Indian Marine Sci. Assn.
IPCC (Intergovernmental Panel on Climate Change). 1994. Preparing to Meet the Coastal Challenges of the 21st Century, Proc. World Coast Conference 1993. The Hague: Netherlands Ministry of Transport, Public Works and Water Management, National Institute for Coastal and Marine Management/RIKZ.
Isard, W. 1972. Ecologic-economic Analysis for Regional Development. New York: Free Press.
Knoppers, B., B. Kjerfve, and J.P. Carmouze. 1991. Trophic state and water turn-over time in six choked coastal lagoons in Brazil. *Biogeochemistry* **14**:149–166.
Lipton, D.W., K.F. Wellmann, I.C. Weiher, and R.F. Weiher. 1995. Economic Valuation of Natural Resources: A Handbook for Coastal Resource Policymakers. Silver Spring: NOAA (National Oceanic and Atmospheric Administration) Coastal Ocean Office.
Lyon, J., and J. McCarthy. 1995. Wetland and Environmental Applications of GIS. Boca Raton: CRC Press.
Nixon, S.W. 1995. Coastal marine eutrophication: A definition, social causes, and future concerns. *Ophelia* **41**:199–219.
Olsen, S., K. Lowry, and J. Tobey. 1999. A Manual for Assessing Progress in Coastal Management. Coastal Resources Center Report 2211. Narragansett: Univ. of Rhode Island.
Olsen, S., K. Tobey, and M. Kerr. 1997. A common framework for learning from ICM experience. *Ocean Coast. Manag.* **37**:155–174.
Pido, M.D., and T.E. Chua. 1992. A framework for rapid appraisal of coastal environments. In: Integrated Framework and Methods for Coastal Area Management, ed. T.E. Chua and L.F. Scura, pp. 144–147. ICLARM (International Center for Living Aquatic Resources Management) Conf. Proc. 37. Manila: ICLARM.
Rast, W., and M. Holland. 1988. Eutrophication in lakes and reservoirs: A framework for making management decisions. *Ambio* **17**:1–12.
Roux, J.P. 1998. The impact of environmental variability on the seal population. *Namib Brief* **20**:138–140.
Ruth, M., and B. Hannon. 1996. Modeling dynamic economic systems. New York: Springer.
Sample, V.A. 1994. Remote Sensing and GIS in Ecosystem Management. Washington, DC: Island Press.

Talaue-McManus, L., et al. 1999. The impact of economic activities on biogeochemical cycling in Lingayen Gulf, northern Philippines: A preliminary synthesis. *LOICZ Newsl.* **10**:1–2.
Terrados, J., et al. 1998. Changes in community structure and biomass of seagrass communities along gradients of siltation in Southeast Asia. *Estuar. Coast. Shelf Sci.* **46**:757–768.
United Nations. 1993. Agenda 21: A Blueprint for Action for Global Sustainable Development into the 21st Century. United Nations Conf. on Environment and Development (UNCED), Rio de Janeiro, June 3–14, 1992. New York: United Nations.
Vollenweider, R.A. 1992. Coastal marine eutrophication: Principles and control. In: Science of the Total Environment, ed. R.A. Vollenweider, R. Marchetti, and R. Viviani. Amsterdam: Elsevier.
Wackernagel, M., and W. Rees. 1995. Our Ecological Footprint: Reducing Human Impact on the Earth. Gabriola Island: New Society Publishers.
WCED (World Commission on Environment and Development). 1987. Our Common Future. Oxford: Oxford Univ. Press.
Wilkinson, C. 1998. Status of Coral Reefs of the World: 1998. Townsville: Australian Institute of Marine Science.

17

River Basins under Anthropocene Conditions

M. MEYBECK

SISYPHE, UPMC/CNRS, Université de Paris 6, 4, place Jussieu,
75232 Paris Cedex 05, France

ABSTRACT

Despite major progress over the last thirty years, our knowledge of riverine fluxes in human-impacted basins (pollutants, nutrients) is still limited to a few descriptors of water quality. River flux surveys need to be optimized, particularly in terms of the sampling frequency and particulate matter, to consider the residence time of many pollutants. Concentrations versus water discharge relationships can be set up for that purpose. In the absence of long-term flux, sediment profiles of elements and compounds can be used as environmental archives to determine the relative fluxes of metals, persistent organic pollutants, and some nutrients over time scale of decades to centuries. From the few well-documented flux records, several types of trends can be discerned:

1. steady state for many ions and particulate metals not affected by anthropogenic activities,
2. moderate and regular increase of some major ions,
3. fast increase, exceeding one order of magnitude, for descriptors affected heavily by anthropogenic activity, such as NO_3^- in many rivers and metals in some of them,
4. sudden occurrence of xenobiotic substances,
5. bell-shaped trends when pollution reduction has been successful (e.g., NH_4^+, PO_4^{-3}, BOD_5, COD, and most metals in the Rhine),
6. slow decrease in dissolved silica for biologial uptake by aquatic biota,
7. stepwise decrease of total suspended solids (TSS), particulate nutrients, and pollutants due to reservoir storage after damming.

Whole-basin flux models combining hydrological and biogeochemical processes are now being developed and used to explore future scenarios or to reconstruct past long-term patterns in evolution.

Although riverine fluxes of pollutants, nutrients, and carbon species are now considered in environmental studies, they have long been regarded as minor indicators compared to riverine concentrations, on which most water quality criteria are based (Chapman 1992). Geochemists were the first to compute river fluxes systematically as a measure of land erosion, on one hand, and of inputs to oceans, on the other (Livingstone 1963). In the 1960s, limnologists working on eutrophication started to

Science and Integrated Coastal Management
Edited by B. von Bodungen and R.K. Turner

link systematically the primary productivity to the total amount of phosphorus received per unit area of water body. This new approach was developed, in particular, by Richard Vollenweider in his 1968 report for the Organisation for Economic Co-operation and Development. This report, written by a hydrobiologist, focused on riverine fluxes, their sources, and sinks. These concepts, however, had already been developed by F.A. Forel, eighty years before, when he measured inputs from the Rhône River to the Léman in the founding stage of limnology, the "oceanography of lakes."

In this chapter, the evolution of riverine fluxes during the Anthropocene is discussed. Anthropocene is a term recently coined by Crutzen and Stoermer (2000). It is the current geological epoch when the growing impacts of human activities on the Earth's system are equal to the natural forcing. Although these authors assign Watts's invention of the steam engine (1784) as the starting point of the Anthropocene, I prefer to refer to 1950 as the key date for its full development, i.e., the point at which many indicators of human impacts (e.g., land use, dam constructions, urbanization, CO_2 increase, waste release) reached a global extension.

RIVER FLUX SURVEYS

Rivers contain naturally occurring compounds, such as major ions (e.g., Ca^{2+}, HCO_3^-), plant nutrients (SiO_2, NO_3^-, NH_4^+, orthophosphates), organic compounds (e.g., humic acids and hydrocarbons), and xenobiotic substances (synthesized by humans). In these latter substances, thousands of products or by-products can be found at very low concentrations (ng l^{-1} to μg l^{-1}) but possessing toxic properties. They are, therefore, termed organic micropollutants (e.g., pesticides, polyaromatic hydrocarbons [PAH], polychlorobiphenyls [PCBs], solvents). When they are very stable in the aquatic environment or in soils, they may not degrade for years and can eventually accumulate in either sediments, soils, or even higher organisms. These specific compounds are now termed persistent organic pollutants (POPs). Metals — particularly the heavier metals, such as lead (Pb), cadmium (Cd), mercury (Hg), zinc (Zn), chromium (Cr), and copper (Cu) — as well as some nonmetallic elements, such as arsenic (As), antimony (Sb), and tin (Sn), are naturally found in rock-forming minerals. Variability in weathering and erosion processes causes large natural variations in the background concentration of these elements. When concentrations of these elements are elevated by human activities to toxic levels (for biota and humans), they are termed inorganic micropollutants, often simplified as heavy metals. Major ions, metals, nutrients, and carbon species that reach harmful levels from the human perspective by natural processes (variability) cannot be considered pollutants. Natural variability in concentrations over several orders of magnitude are observed dependent on individual chemical indicators and compounds as well as on river basin size (Meybeck et al. 1989; Meybeck 1996; Kimstach et al. 1998). This variability is one to two orders more at finer spatial scales (<100 km^2) than found at the coarsest scale (> 10^6 km^2).

By considering both the current river quality surveys, for example, those collected at the global scale by the UNEP GEMS/WATER Programme (United Nations Environment Programme Global Environmental Monitoring System/Freshwater Quality Programme), and academic studies, we can obtain an overview of current knowledge of riverine geochemistry and water quality (Table 17.1). Despite considerable efforts to improve the quality of river surveys in some locations over the last 10 to 15 years, concentration of many elements are not or are only insufficiently known, particularly those for micronutrients (metals) and organic pollutants (Table 17.1).

Table 17.1 Present state of river survey for metals and organic compounds in rivers.

	Metals		Natural Compounds	Organic Pollutants
	Dissolved	Particulate		Xenobiotic
Very seldom analyzed	At, Bi, Ga, In, Os, Pd, Po, Re, Rh, Ru, Ta, Tl, Zr, W	At, Be, Bi, In, Os, Pd, Po, Re, Rh, Ru, Tl, W	Dissolved fatty acids	Chlorobenzene, chlorophenols, phthalates, halogenates, esters, acrolein
Seldom analyzed	Rare Earths: Ag, Au, Be, Br, **Cd***, Co, Cs, Ga, Ge, I, Hf, **Hg***, Ir, La, Li, Mo, Ni, Pt, Rb, Sb, Sc, Se, Sn, Ti, V, Te, Y, Zn*, Zr	Ag, Au, B, Ba, Br, Cl, Cs, Ga, Ge, Hf, I, Ir, La, Mo, Nb, Pt, Se, Nb, Rb, Ta, Te, Y, V, Zr	Fatty acids, amino acids, vitamins, nonaromatic hydrocarbons, adenosine triphosphate	Benzene, toluene, xylene, ethylbenzene, endrin, triazines, PCBs and isomers, solvents, PAHs
Some monitoring	Al, **As**, B, Ba, Cr, Cu, Fe, Mn, Sr, Th, U	**As**, Co, **Cd**, Cr, Cu, F, **Hg**, Li, **Pb***, S, Sb, Sc, Sn, Sr, Th, U, Zn	Total organic nitrogen, total glucids, lipids, total fulvic and humic acids, total hydrocarbons,	Total phenols, Σ DDT; PCBs, aldrin, dieldrin, lindane
Regular or sufficient monitoring	C, Ca, Cl, F, H, K, Mg, N, Na, O, P, S, Si	Al, C, Ca, Fe, K, Mg, N, Na, O, P, Si, Ti	Dissolved and particulate organic carbon, chlorophyll and phaeopigments	Detergents

Boldface text: most toxic elements
* Contamination problems

PCB = polychlorobiphenyls
PAH = polyaromatic hydrocarbons

There are many obstacles to adequate surveys. First, the cost of a single sample analysis with key organic micropollutants (a few dozen), heavy metals (a dozen), nutrients, and general descriptors of water quality exceeds US$ 1000. Considering the variability of concentrations, the frequency of optimal surveys is inversely proportional to the basin size, ranging from 12 samples per year in larger basins (10^6 km^2) to more than 52 samples per year for the smaller ones (< 10,000 km^2), where water discharge can be highly variable. In the 1970s and 1980s, national surveys in most industrialized countries in Europe and North America analyzed dissolved heavy metals in surface water. Lead concentration levels measured by Federal Agencies were found to be regularly decreasing; this trend, however, was primarily due the fact that there was less contamination of the samples. At the same time, very elaborate academic studies of archived records of particulate Pb from Mississippi delta sediments have shown a definite, major decline in Pb content after 1971 as a result of the Clean Air Act, which banned Pb from gasoline (Trefry et al. 1985).

RIVERINE FLUXES: DEFINITIONS AND VARIABILITY

Instantaneous flux (F) is the product of the concentration (C) (expressed in ng l^{-1} to g l^{-1}) and instantaneous discharge (Q) (expressed in $m^3 s^{-1}$). Since both C and Q vary over time, fluxes will combine these variabilities:

$$F(t) = C(t) \; Q(t). \tag{17.1}$$

For particulate compounds or elements, it is necessary to combine their contents (c) (expressed in percent to 10^{-9} g g^{-1}) and the total suspended solids (TSS) (in mg l^{-1}), to calculate the concentration of the particulate pollutant or nutrient per volume of water:

$$C(t) = c(t) \; \mathrm{TSS}(t). \tag{17.2}$$

In some monitoring and/or for some water quality descriptors, the analysis is performed on an unfiltered volume of water, thus giving a total (i.e., dissolved + particulate) concentration. This is the case for total Kjeldhal nitrogen (TKN), for total phosphorus (TP) or total organic carbon (TOC), and for many organic micropollutants. Such an approach has also been found for metals; however, its use should not be encouraged because a metal analysis cannot be complete unless a total digestion by acids is performed on a dry aliquot of particulates, and such total metal concentrations are always very dependent on TSS levels.

The concentration versus water discharges are highly variable; some are illustrated for the polluted Seine River in Figure 17.1. When considering dissolved substance concentrations (C), most decrease as a power function of discharge, $C = aQ^b$, where a and b are coefficients characteristic of each station and each substance. These dilution patterns (C, D, E types in Figure 17.1) are always limited, with $-1 < b < 0$. Other patterns of dissolved substances include leaching, as for DOC (type H) or NO_3^- (type I), and biological uptake at low flows by aquatic plants and phytoplankton, as is sometimes observed for SiO_2 (type B). Since the dilution is generally limited, C typically decreases, from 100% to 50% or 20%, when Q increases, from 100% to 10,000% or more; therefore, *all fluxes of dissolved substances increase with Q.*

Particulate pollutants present peculiar patterns. The contents of pollutants or nutrients per gram of suspended particulates are usually less variable than most dissolved concentrations by a factor of two to ten, while the TSS increase with water discharge can be enormous. Under the least variable conditions, the TSS range at a given river station is over one order of magnitude (Saint Lawrence) and can exceed two orders of magnitude, e.g., from 10 to more than 1000 mg l^{-1} (see type T in Figure 17.1, where $C = aQ^b$, where $b > 1$). Therefore, as a general rule, (a) *the concentration of particulate nutrient or pollutant per volume of river water is mostly related to the amount of TSS* and (b) *their instantaneous fluxes can range over three to five orders of magnitude, depending on water discharge range and on TSS variability.*

The flux during a period t_1 to t_2 results from the time integration of instantaneous fluxes:

$$F(t_1, t_2) = \int_{t1}^{t2} C(t) \; Q(t). \tag{17.3}$$

River discharge is available on a continual basis; however, river chemistry information is generally discontinuous and highly fragmentary. Sampling is generally realized in a few minutes, but most of the time it can be considered to represent the whole sampling day. A typical monthly sampling provides only partial information on flux variability. Very few countries, one of which is Switzerland, utilize a continuous sampling strategy: discharge-weighted

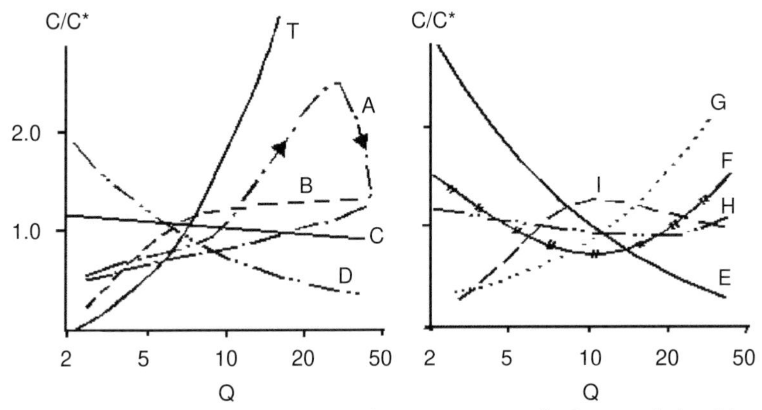

Figure 17.1 Schematic examples of concentrations versus water discharges relationships observed in the Seine River basin. Concentrations (C) are normalized to their yearly weighted average (C*); discharges (Q) are in log scale of specific runoff. A: total suspended solids (TSS); B: dissolved SiO_2, C: Ca^{2+}; D: Na^+; E: P–PO_4^{-3}; F: total P; G: particulate organic carbon (POC) in mg l^{-1}; H: dissolved organic carbon (DOC); I: N–NO^{3-}. T: TSS pattern for typical turbid rivers, not found in the Seine basin. The average yearly runoff is 7 l s^{-1} km^{-2}, the range is between 1.5 and 30 l s^{-1} km^{-2}.

water aliquots are sampled and analyzed every week. This is the most appropriate strategy to measure fluxes; however, it must be noted that short-term pollution peaks are minimized in such integrated surveys. For generalized discontinuous sampling, fluxes may be calculated in numerous ways and should be based on the relative variabilities of $c(t)$, $C(t)$, TSS(t) and $Q(t)$ (see Walling and Webb 1985; Meybeck et al. 1994).

TIME DURATION OF FLUXES

When long series of daily fluxes are available, either through direct measurement or reconstruction, the time distribution or duration curves of fluxes can be performed (Table 17.2 provides an example of this for the Tuy River, downstream from Caracas). The duration of pollutant fluxes depends on (a) the duration of water discharge and (b) the concentration–discharge relationship. For pollutants that are diluted (types C, D, E in Figure 17.1), as for P–PO_4^{-3} in the Tuy, the main fluxes occur at the low-water stage. For pollutants associated with riverine particulates, such as particulate organic carbon (POC) (type G in Figure 17.1), main fluxes occur at the highest flows: 65% of Tuy POC flux occurs in 13 days or 3.55% of the time, for only 14% of water volume. The establishment of such patterns may greatly help to set up monitoring strategies. In most rivers, 50% of the riverine flux is discharged in less than 25% of the time. In very variable conditions of TSS and Q, the 50% pollutant flux duration period can drop to 5% and even 1% of the time. Particulate pollutants are always discharged in less time than the water discharge, while for diluted pollutants, the contrary is observed (Figure 17.2) except for rare patterns (B and I in Figure 17.1).

Since regular monitoring of river water quality is usually geared to check the compliance of concentrations with water quality standards, it is seldom appropriate for flux determinations. For example, in November, 1994, the Rhône (France) reached a very high level of water

Table 17.2 Flux duration in the Tuy River (Venezuela) downstream from Caracas (Ramirez et al. 1988).

Discharge m^{-3} s^{-1}	Days	Duration %	Water Volume %	Pollutant Fluxes % DOC	POC	N–NH_4	P–PO_4
> 600	2	0.55	3.2	2.5	27.7	2.9	0
344–600	11	3.0	10.8	8.3	37.1	9.6	1.7
164–344	69	18.9	33.8	19.3	23.5	34.5	5.3
90–164	110	30.1	31.8	32.1	9.5	25.2	12.3
46–90	109	29.9	15.4	26.1	1.8	16.3	40.3
25–46	64	17.6	5.0	11.8	0.3	11.4	40.3

after 77 days of flooding (Q> 3000 m^3 s^{-1}), with a peak discharge at 9760 m^3 s^{-1}. As a result, the TSS annual load was 11.3 × 10^6 t, i.e., 81% of the 1994 TSS load was discharged during only 22% of the time. The maximum level of TSS measured was 5200 mg l^{-1}, and the water discharge-weighted average TSS was 683 mg l^{-1} (A. Thomas, pers. comm.). Note, however, that during the ten years of regular monthly monitoring carried out by French Water Authorities, the measured maximum never exceeded 549 mg l^{-1}, and the average calculated TSS was estimated at 41.5 mg l^{-1}.

In highly variable dry and semiarid conditions, flux variability can be even greater. In the Oued Medjerda (Tunisia, basin area: 23,300 km^2), an exceptional flood (return period 200 to 300 years) carried 80–100 × 10^6 t of suspended load in 8 days, with TSS reaching 35 g l^{-1} for a maximum Q of 3150 m^3 s^{-1} (Claude et al. 1977). This sediment load was nearly equivalent to the yearly Nile load (120 × 10^6 t y^{-1} prior to the Aswan High Dam construction). Also, in

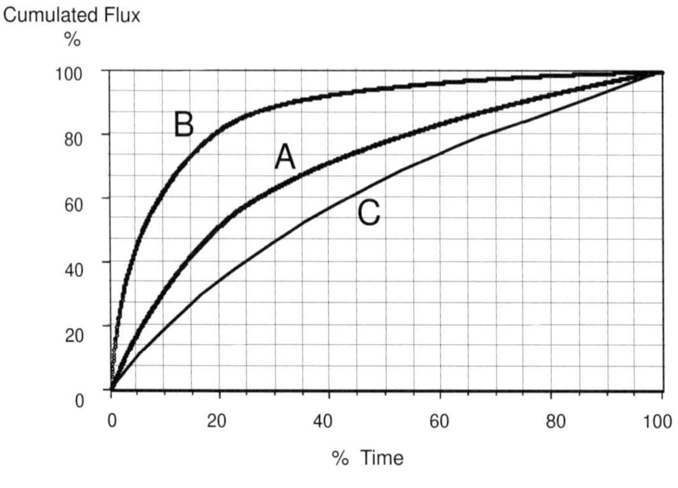

Figure 17.2 Distribution of cumulated fluxes in rivers as a function of elapsed time (duration curves). A: water volume; B: riverine fluxes increasing faster than river discharge; C: riverine fluxes increasing slower than river discharge. Example for the Marne River (France): B = TSS flux; C = Na^+ and Cl^- fluxes. For B, 63% of the flux is discharged in 10% of the time (Meybeck 1998b).

Tunisia, the Oued Zeroud (8950 km^2) does not normally reach the Mediterranean Sea. The 1969 flood event discharged 240×10^6 t of sediments and 2.4 km^3 of water in two months to the sea, with a maximum discharge estimated at 17,000 $m^3 s^{-1}$ (Colombani and Olivry 1984). Also, in the Mediterranean basin, the Ebro River (Spain) discharged about 0.72×10^6 t of TSS in one hour in October, 1907, compared to 3×10^6 t for the estimated long-term average (Ibanez et al. 1996) prior to the river's damming. During such events, the particulate matter that has accumulated in the riverbed and on the riverbanks over years or decades can be mobilized and expelled with its related particulate nutrients and pollutants. These are catastrophic events and their study is very rarely realized; however, they must be taken into account in flux estimations, particularly in the long term.

These Mediterranean examples can be regarded as the upper limit of flux variability. The lowest limit is observed for more regular water discharge regimes combined with the smoothing and trapping influence of lakes. For instance, in the middle Rhine at Maxau, about 150 km downstream from Basel and all the Swiss lakes, TSS fluxes are much less variable. The ratio of maximum/minimum annual TSS load is only 3.2 for 20 years of daily TSS record—an exceptional series (see TSS trends in Figure 17.3D). When considering the daily TSS fluxes, this ratio is up to 700. For the Seine, the corresponding figures are 16 and 500 over 35 years of reconstructed fluxes (Idlafkih 1998).

Once reliable annual average fluxes are determined at a given station, they can be compared within stations, for the same basin, or between river basins by computing the specific fluxes per unit area, usually as t $km^{-2} y^{-1}$ (equivalent to g $m^{-2} y^{-1}$, but ten times higher than kg $ha^{-1} y^{-1}$, a unit commonly used in agronomy).

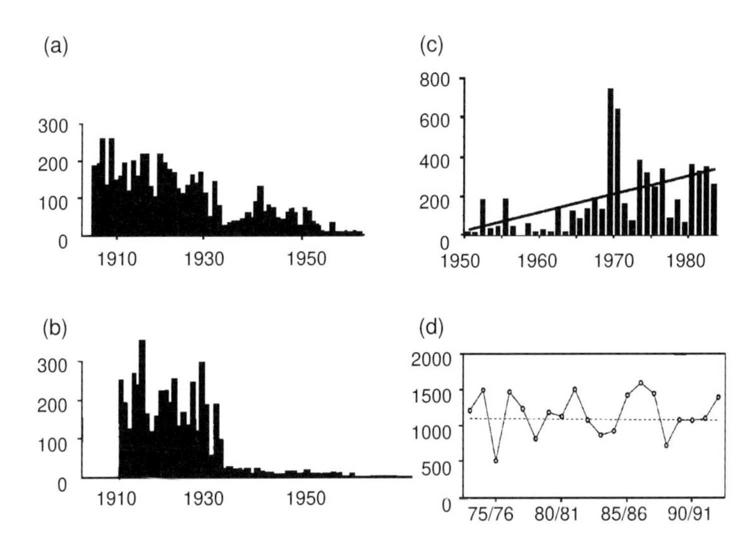

Figure 17.3 Long-term trends of water sediment discharge in major rivers. Colorado trend of (a) water discharges (in $km^3 y^{-1}$) and (b) sediment load (in $10^6 t y^{-1}$) at the U.S./Mexican border before and after the construction of the Hoover Dam (from Meade and Parker 1985). (c) Trend of suspended sediment yield (t $km^{-2} y^{-1}$) in the Dniester River (Bobrovitskaya, quoted by Walling 1999). (d) Trend of yearly sediment load (10^3 t y^{-1}) in the Rhine at Maxau (Idlafkih 1998).

LONG-TERM TRENDS OF RIVERINE FLUXES

Trends of riverine fluxes may result from numerous causes and can affect concentrations, particulate contents, and river runoff. If TSS load and water discharge (*Q*) vary significantly, these trends will affect the load even if the quality of water and/or particulates is stable. The most striking example is the evolution of TSS and *Q* related to water use and damming.

River Damming and Irrigation Affecting Water and Sediment Discharges

The golden age of reservoir construction began in 1935, with the construction of the Hoover Dam on the lower Colorado River (Arizona–Nevada). Before this, river dams were restricted to small rivers and/or tributaries, a water engineering activity that began 4,000–5,000 years ago. Major dams currently harness main rivers on all continents (Dynesius and Nilsson 1994; Vörösmarty et al. 1997b) and store 90–99% of the incoming sediments (Vörösmarty et al. 1997a). An example of this spectacular feat can be seen in the Colorado River (Figure 17.3B). Such trends in TSS loads are observed in all dammed rivers (Columbia, Mississippi, Rio Grande, Nile, Zambezi, and Indus) and will be observed for the Chiang Jiang (Yang Tse Kiang) after completion of the Three Gorges Dam.

Many of the reservoirs created behind the dams are used for irrigation and most of this diverted water is evaporated. In dry and semiarid river basins, this means that the amount of river water discharged to the ocean has been lowered considerably (as in the Indus, Ebro, and Rio Grande) or reduced to within only a few percent of its original level (e.g., the Nile or the Colorado) (Figure 17.3A). The largest river in Central Asia, the Amu Darya, has been completely diverted and used for irrigation; it no longer discharges to its natural endpoint, the Aral Sea. The Huang He, once a perennial river, is now drying up during the summer period.

Not all long-term trends of TSS are negative ones. Major changes in land cover and land use (e.g., deforestation, particularly clearcutting, wetland filling, and intensive agriculture) can enhance soil erosion and the corresponding river fluxes. This signal, which originates from the headwaters (first stream orders), is filtered by the landscape and waterscape, where sediment sinks are numerous (e.g., in the flood plain) (Walling 1999). In small and medium basins, such TSS flux increases are found (Figure 17.3C). The sediment routing is complex and can take from months to millennia to develop, depending on floodplain dynamics and on the grain size of riverborne particulates (Meade 1988).

Trends of Major Ions

Since most of the regular river monitoring did not start before the 1960s or 1970s, it is rare to find long-term records of riverine fluxes. In those river basins where the influence of reservoirs and irrigation can be neglected, the annual water discharge varies much less than concentrations. Concentration trends are good proxies of flux trends, as illustrated in Figures 17.4–17.7.

The well-studied lower Don is an excellent example of combined influences from river damming, mining operations, and industrialization (Tsirkunov et al. 1998). The Tsimlyansk reservoir generally had a smoothing effect on the seasonal pattern of concentrations. Since the April peak flood levels have been reduced, the corresponding fluxes have not changed

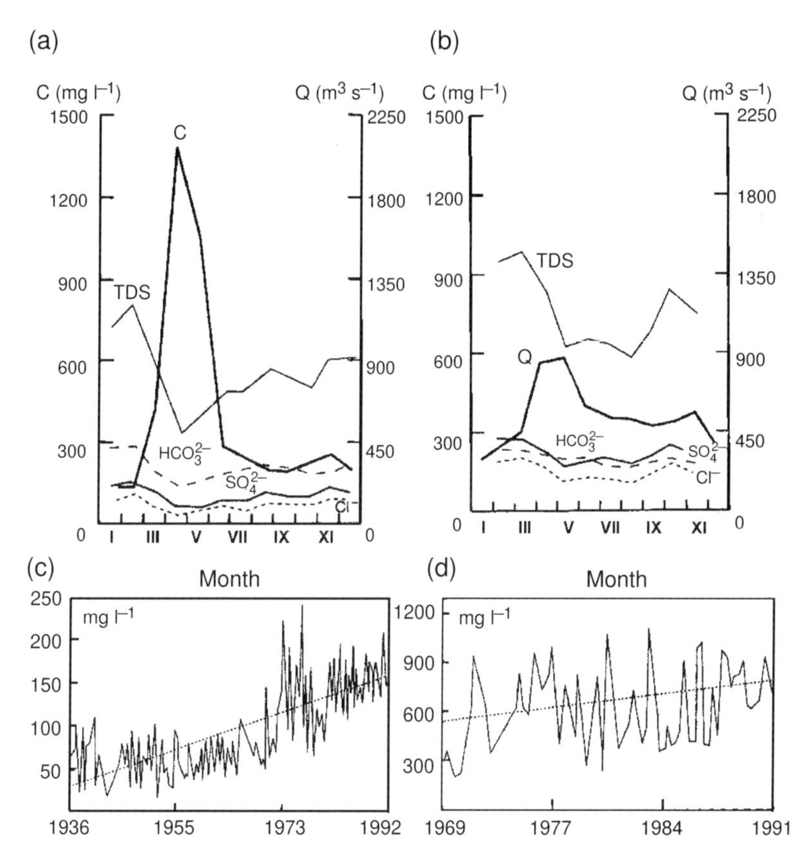

Figure 17.4 Long-term evolution of water quality in Russian rivers. (a) Don River before reservoir construction (1949–1952); (b) Don River after the Tsimlyansk reservoir construction (1970–1975); (c) Cl^- trend in Don River at Razdorskaya; (d) $SO4^{2-}$ trend in the Kundryuchya River (Don basin) (Tsirkunov et al. 1998).

much (Figure 17.4, A and B). The long-term trend of Cl^-, however, presents a marked increase in the Don (Figure 17.4 C). The same applies for SO_4^{2-} in the Kundryuchya River (Figure 17.4 D). Chloride levels are now reaching the World Health Organization recommended level (200 mg l^{-1}) while sulfate levels are exceeding those limits (250 mg l^{-1}).

Similar increases of Na^+, Cl^-, and SO_4^{2-} are observed in most rivers impacted by chemical industries, big cities, or where deicing salts are used, as in Switzerland. With additional ion sources from active mining operations or from leaching of old mine tailings (Cl^-, SO_4^{2-}, and even K^+), drinking water standards in rivers can be reached or even exceeded. In the Rhine, for instance, the Cl^- concentration downstream from the brine release from the Alsace potash mines has increased by about a factor of 10 between Basel and Strasbourg, where it is regulated at 200 mg l^{-1}, a level about 40 times the natural background. A careful look at daily records of Cl^- concentrations shows a rare weekly variation of Cl^- fluxes — a xenoperiodicity that does not exist naturally — with a maximum/minimum ratio of 2.0 between average Saturday fluxes (120 kg Cl^- s^{-1}) and Wednesday fluxes (60 kg Cl^- s^{-1}) (Idlafkih 1998).

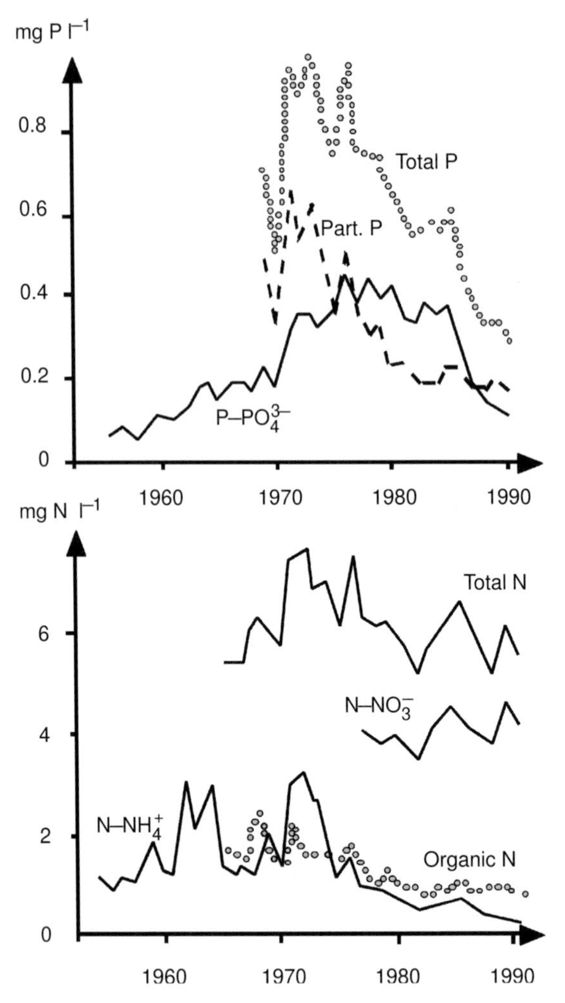

Figure 17.5 Schematic evolution of nutrients levels in the Rhine (from Van Dijk and Marteijn 1993).

Nutrients Trends

One of the best-studied water quality stations on a major river over a long period is the Rhine River at the Dutch–German border (Bimmen/Lobith). Nutrient trends (Figure 17.5) over 30 years show a marked improvement of BOD_5, COD, total P, and NH_4^+ concentrations and fluxes. This trend began in the 1970s, with the significant improvements in sewage and treatment over the whole basin. For orthophosphates, improvement came later and was less marked: the first generation of sewage treatment plants did not eliminate PO_4^{3-}, and phosphate-containing detergents were still used in most parts of the basins. The efficient ban on phosphates, implemented in 1985 in Switzerland, the upper Rhine basin, very efficiently curbed the PO_4^{3-} levels. Such phosphorus decrease is observed only in rivers where specific

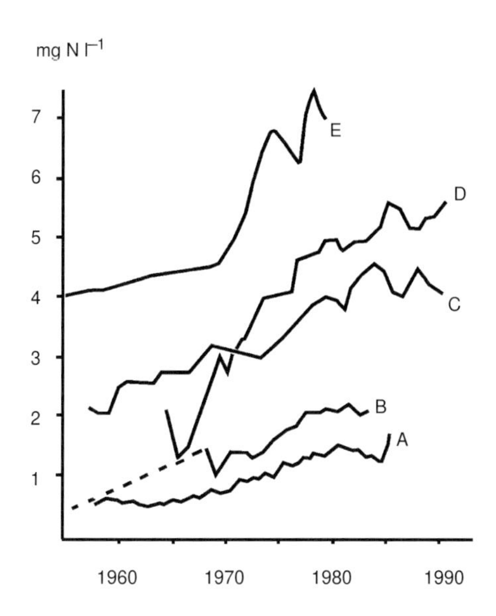

Figure 17.6 Recent trends of nitrate concentrations in some rivers. A: Mississippi at mouth; B: Danube at Budapest; C: Rhine at the Dutch/German border; D: Seine at mouth; E: Thames at mouth (data from A: Bryan et al. 1992; B: Hock 1986; C: ICPR 1987; D: Service Navigation de la Seine, Rouen (A. Ficht, pers. comm.); E: Toms 1987).

P-control measures have been set up, as in northern Europe. In southern Europe, phosphate levels are still increasing, for example in the Ebro and the Axios rivers in Spain and Greece, respectively. In the Seine basin, most water quality stations located upstream of Paris still demonstrate an increase of $P–PO_4^{3-}$ and of total P. However, as the result of recent P-control measures in the giant sewage treatment plant of Paris, which treats the waste of 8 million people, $P–PO_4^{3-}$ levels downstream have recently stabilized and are now slowly decreasing.

Nitrate trends observed in most rivers in Europe and North America reveal a marked increase since the 1950s, concomitant to strongly increasing use of nitrate fertilizers in agriculture (Figure 17.6). Contrary to BOD_5, NH_4^+, and PO_4^{-3}, which originate from point sources in most cases, nitrate is a typically diffuse source of pollution and can only be curbed by reducing fertilizer use and changing agricultural practice. This is a very slow and difficult process in a market-dominated economy. In regions where fertilizer use is limited, atmospheric pollution (e.g., from fuel combustion) can be a relatively important nitrogen source, as is the case in New England. So far, no incentive system (e.g., pollution charge, levy, or self-regulation from farmers) has been successful in decreasing fertilizer use. When considering the amount of NO_3^- still stored in the unsaturated zone, and in phreatic aquifers that feed most rivers at low stage (Garnier et al. 1998), a decrease of NO_3^- cannot be envisaged for the rivers presented in Figure 17.6 over the next 10 to 20 years, due to the very slow renewal of water in soil and groundwaters. In terms of fluxes, this picture is even more dramatic since NO_3^- maximum levels are observed in winter, at medium to high flows, not in summer during low flows (type I or Figure 17.1): river fluxes actually increase faster than arithmetic means of NO_3^- concentrations.

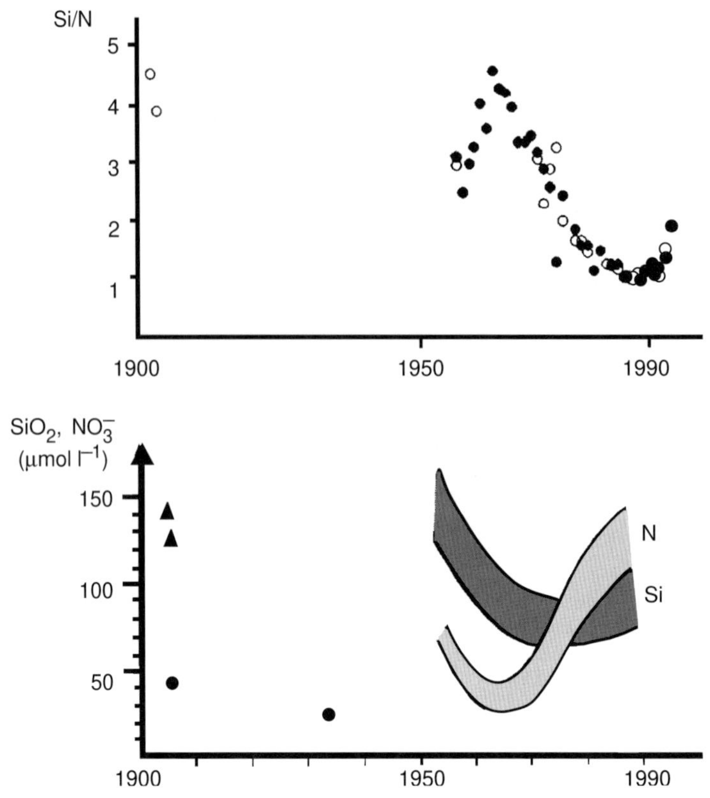

Figure 17.7 Long-term comparative evolution of dissolved SiO_2 (▲) and nitrate (●) and molar Si:N ratio at the mouth of the Mississippi River (from Turner and Rabalais 1991).

The proportions of riverine nutrients (Si, N, P) that reach the coastal oceans are not constant since the pattern of these species depends on each element and is also station dependent (Figure 17.7). These ratios ("Redfield" ratio) can control algal development (e.g., decreasing diatoms at lower Si:N ratios) that occurs in continental waters, lakes, reservoirs, and the lower parts of big rivers (Turner and Rabalais 1991; Justic et al. 1995; Garnier et al. 1998). Examples of changes in Redfield ratios in the Po and Mississippi rivers are presented in Table 17.3. Similar changes are likely to occur in other regions, even on a global scale, which can cause major shifts in coastal primary production and related bottom water anoxia (Justic et al. 1995).

Metal Trends

Difficulties in metal surveys have already been discussed. The current approach for metal flux determination is to perform metal analysis on filtered or centrifuged particulates and to couple these qualitative surveys with an appropriate high-frequency TSS survey (a minimum weekly sampling). Thereafter, each metal analysis is associated with a period for which the sediment flux is accurately estimated, and its metal content assumed to be constant. As most

Table 17.3 Changes in molar ratio of Si–SiO_2, total inorganic N, and total P in the Po and Mississippi rivers (Justic et al. 1995).

	Po River		Mississippi River	
	1968–1970	1981–1984	1960–1962	1981–1987
Si : N	1.06	0.70	4.25	0.95
N : P	62	37	9	15
Si : P	65	26	40	14

metals are carried by rivers with particulates — from 80% to 99%, depending on metals and on TSS levels — *metals fluxes are extremely dependent on hydrological conditions, not on pollution*, on any scale. In the Seine River, a wet year that follows a dry year may carry 4–10 times more metals and other particulate pollutants that have been stored on the riverbed and remobilized through increased water flow.

The long-term evolution of particulate pollutant contents is very useful in determining the overall degradation or improvement of a river basin. Prior to 1960, such direct analyses were rare and had been only performed on past riverine TSS samples found on occasion. In the Rhine River, Salomons and De Groot (1977) were among the first to reconstruct a river evolution for Hg, Cd, Zn, Pb, and As over an 80-year period, but only at a few incremental stages. For most metals, peak pollution levels were reached in the 1950s and 1960s; observed levels were 10–50 times higher than the estimated background levels. Thereafter, a general decrease was observed, albeit at different rates for the metals involved.

Trends in metal concentrations are more and more reconstructed from core analysis in lakes and in coastal areas where a continuous and regular deposition of fine-enriched particles — typically silts and clays — can be found. In rivers, such calm deposition zones are rare and found only in still water zones. Sediments from river floodplains can also be used although they are generally coarser and may be less contaminated. In the U.S., a national program of estuarine coring was implemented in the 1980s (Valette-Silver 1993). Profiles of heavy metals and POPs have hence been systematically obtained and dated using the ^{210}Pb method. The resultant environmental archives of past micropollutant contents show different patterns, depending on the local history of pollution (Figure 17.8):

1. A near-constant natural background of metals was observed at the bottom of the core prior to any human impact.
2. A slow or sudden increase of natural substances, such as metals or PAHs, due to contamination, was found.
3. A sudden increase in xenobiotic substances was detected after their first use or leak into the environment (e.g., PCBs).

Such profiles can give valuable information on relative flux trends although they cannot be used to represent riverine flux unless a detailed reconstruction of riverine particulate loads is performed. For instance, in the Rio Tinto basin (Spain), gold and mercury mines have been in operation for the past 4000 years. Metal pollution peaks have been effectively found in estuarine deposits at different periods between which a partial environmental recovery is observed (F. Elbaz-Poulichet, pers. comm.).

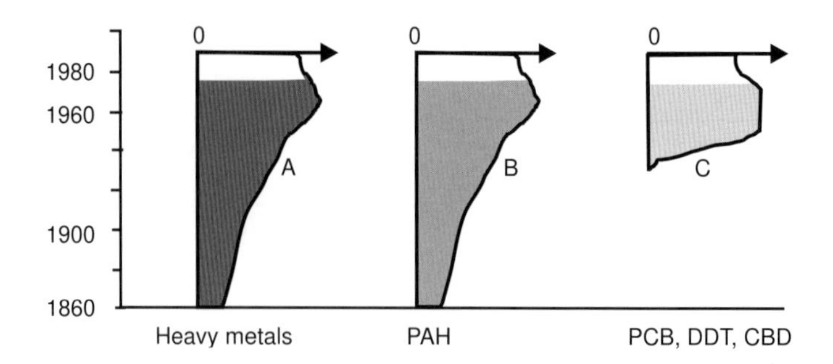

Figure 17.8 Schematic evolution of estuarine sediment contamination in industrialized country (modified from Valette-Silver 1993). (A) Heavy metals such as Hg, Cd, Zn; (B) polyaromatic hydrocarbons; (C) xenobiotic pollutants such as PCB, DDT.

In the absence of any measurement of river water quality, of particulate analyses, or of available sediment archives, qualitative indications of pollutant flux can still be obtained. The impact of mines on riverine environments has already been stressed. There is no doubt that Welsh tin mines discharged enormous amounts of heavy metals into the Bristol Channel during Roman times. In Brittany there is a historical record of massive contamination from lead mines in the 18th century (M. Thibault, pers. comm.) that has nearly disappeared nowadays, though this remains to be confirmed through estuarine and coastal deposits.

Several types of riverine flux, normalized to pre-anthropogenic figures, can be schematically depicted from the many examples we have considered (Figure 17.9). Type A concerns those elements that are barely affected by anthropogenic activities (e.g., Ca^{2+}, Mg^{2+}, HCO_3^-, particulate Al, Si, Fe, and Co). Type B represents a gradual increase (Na^+, Cl^-, K^+, SO_4^{2-}, and so far, NO_3^-) in most basins exposed to increasing human pressure (agriculture, urbanization, industry). Bell-shaped evolution (type D) characterizes a successful control (NH_4^+ and PO_4^{-3}) in some basins, while type C relates to the steady decrease of dissolved silica in eutrophied rivers. The negative stepwise trends (type E) are typical of reservoir impact on all particulates, while type G is the gradual decrease of all fluxes (water runoff, dissolved and particulate matter) as a result of the consumption or transfer of water resources. Actually, in some basins, or for some issues, more complex trends have been described: multiple cycles (type H) of metal contamination and organic pollution as well as long-term regulated contamination, e.g., some mining impacts (type F).

Xenobiotic substances and artificial radionuclides have specific patterns. The cesium-137 pattern (type I) is one example of this complexity: it peaked in all river fluxes in 1962–1963, brought on by fallout from atmospheric nuclear tests; a second peak was observed after the Chernobyl accident in 1986. The DDT flux pattern (type J) is an example of the long-term persistence of xenobiotic compounds in the environment, which continued even after its ban. Some xenobiotic fluxes are still increasing, as in the case of herbicides (type K).

HOT SPOTS OF POLLUTANT FLUXES

Careful attention is required to account for all local point sources of pollution since a single factory or mine can release enormous fluxes, equal to those found in the rest of the basin.

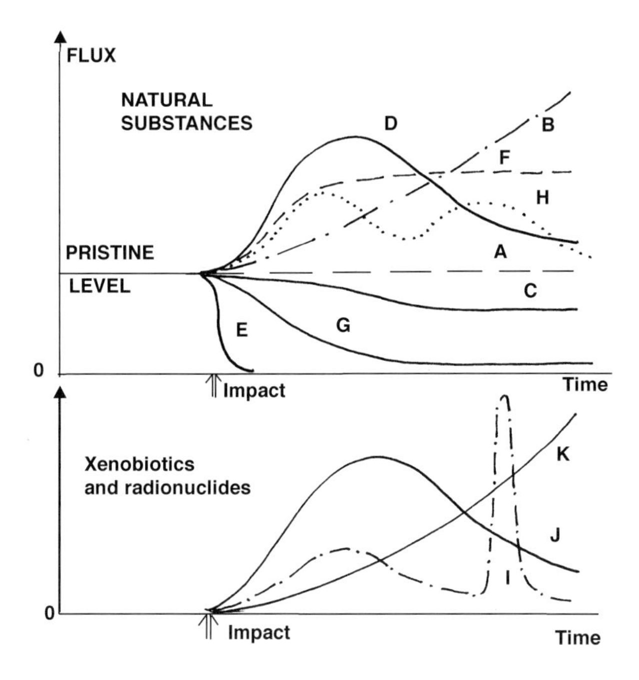

Figure 17.9 Types of river flux trends normalized to pristine fluxes (A to K, see text) since the beginning of human impacts.

Numerous examples of this can be given: the Alsace potash mines have discharged 10,000 to 15,000 metric tons of NaCl daily into the Rhine River, i.e., equivalent to the discharge of Amazon fluxes at the mouth of the rivers. In northern France, the Espierre River, a small river flowing into Belgium, received the untreated waste waters of a single fertilizer plant between 1978 and 1983, which resulted in phosphate levels between 100–400 mg P l^{-1}, i.e., 10,000 times higher than pristine levels. Although these pollution "hot spots" are the first to be reduced by basin quality managers, they should nonetheless be carefully identified, one by one, at any scale. One must remember that major mining and smelting districts can be located in very remote, unpopulated, and uncultivated areas in North America (Sudbury, Yellowknife), Europe (Severonikel), Asia (Norilsk), or in Irian Jaya. In such areas, regular monitoring may not normally be performed and environmental regulations enforced, as would be the case in densely populated regions.

Megacities can also be considered hot spots but are easier to identify. Their contributions to river fluxes of nutrients, organic carbon, and major ions can be assessed on the basis of additional per capita loadings determined from the analysis of untreated urban sewers minus distributed waters (Table 17.4). In the megacity of Paris, the waste of eight million people is collected and treated at a single plant. The average discharge of treated wastewater is ca. 25 $m^3 s^{-1}$ and is composed of 38 mg l^{-1} of BOD_5, 24.7 mg l^{-1} of $N–NH_4^+$, and 3.6 mg l^{-1} of $P–PO_4^{-3}$ (Meybeck 1998a). Thus, the amount of N and P produced by Paris, which occupies 2500 km^2, is equivalent to the natural loads of nitrate and phosphates from about 10^6 km^2 of forested western European basin prior to agriculture and urbanization. Long-term fluxes of

Table 17.4 Examples of measured per capita additional loadings in urban conditions: kg capita^{-1} y^{-1} from Ca^{2+} to TOC_B; g capita^{-1} y^{-1} for Hg to Zn.

1.	Ca^{2+}	Mg^{2+}	Na^+	K^+	Cl^-	SO_4^{2-}	HCO_3^-	SiO_2
	2.3	0.65	7.5	1.4	7.6	10.1	17.7	0.7
2.	TKN	TP						
	4.7	0.3 to 0.9						
3.	BOD_5	TSS	DOC	DOC_B	POC	POC_B	TOC	TOC_B
	19.7	28	2.6	1.9	7.3	4.7	10.0	6.7
4.	Hg	Pb	Cd	Cu	Zn			
	0.18 ± 0.05	23 ± 7	0.32 ± 0.08	14 ± 3.5	23 ± 7			

1. Average for Paris, Brussels, and Montreal (Meybeck 1998a).
2. Averages for Paris, Brussels, Frankfurt, and Vienna (Meybeck 1998a).
3. BOD_5 to TOC: Paris megacity untreated sewers. DOC, POC, TOC: dissolved, particulate, and total organic carbon, "B" biodegradable carbon in 45 days (Servais et al. 1998). These numbers normalized to the theoretical BOD_5 load of 54 g capita^{-1} d^{-1} given by the World Health Organization.
4. Average for the Seine basin (1992–1997 period); additional load of particulate metals for a total population of 15 million people and a river particulate flux of 0.7×10^6 t y^{-1} (recalculated from Meybeck 1998a).

particulate metals minus natural background fluxes can be used to determine the average per capita load, particularly in basins not affected by mining, e.g., the Seine (Table 17.4).

FLUX MODELING: A PROMISING PERSPECTIVE

To model river fluxes, one must combine hydrological and biogeochemical processes, which are particularly complex for diffuse nutrient sources and for the transfer of particulate pollutants. The most simple models relate sediment, nutrients, organic carbon, and dissolved inorganic carbon transfers, expressed in t km^{-2} y^{-1}, to river runoff (mm y^{-1} or in m^3 m^{-2} y^{-1}), climate, lithology, land use, and basin relief. Such models, however, have limited predictability; they do not take anthropogenic factors into account but rather allow only for the spatial distribution of these fluxes linked to river runoff at a 1° or 30′ grid resolution at the global scale. With such models it is possible to identify and prioritize the regions that contribute most to the fluxes, such as the highland regions for sediment transfer, tropical regions for dissolved silica, and Western Europe for nitrate and phosphate (Vörösmarty et al. 1997c).

Geographic Information Systems (GIS) are being used increasingly in flux models at the basin scale, where both diffuse and point sources of pollutants are simulated (Garnier and Mouchel 1999). Here, the source terms are generally expressed per unit of production: in kilograms per cultivated hectare per year, or kilograms per capita per year, grams per ton of manufactured products per year, or grams per square meter per year for atmospheric inputs. Pollutant routing within soil, groundwaters, and surface waters is then simulated for processes such as sedimentation, organic degradation, biological uptake, gaseous escape to the atmosphere.

In the Seine River basin, which has been studied for over ten years within the PIREN–Seine program (Meybeck 1998a), a set of nutrient and organic carbon models has been developed (Poulin et al. 1998), to simulate riverine concentrations and loads of NO_3^-, NH_4^+, PO_4^{-3} algal organic carbon on hourly to decadal scales (Billen and Garnier 1999; Garnier and Mouchel 1999). All of the above-mentioned processes were taken into account. The River Seine, characterized by some of the highest levels of nitrate and phosphate, has a very low velocity in spring and has been highly eutrophied (chlorophyll peaks may exceed 200 $\mu g\ l^{-1}$). Billen and Garnier (1999) considered seven different environmental scenarios for both phosphorus reduction and production as well as the rate of organic matter treatment. When these scenarios were applied to their general model ("Riverstrahler"), they found, for the Oise and Marne rivers (two Seine tributaries), that average spring chlorophyll — present levels around 80 $\mu g\ l^{-1}$ — was reduced by 50% when, and only if, the daily P load was reduced by 50–60%. For the Seine itself, a 70% reduction was observed — a drastic P-reduction scenario that would probably take more than ten years to achieve under normal conditions once the decision was made and funds secured (Table 17.5). Average inputs from the Seine to its estuaries have also been simulated according to these various scenarios of N and P *abatements* in sewage treatment plants.

The model outputs are generally validated by comparision to actually measured fluxes (Figure 17.10). Once validated (see B in Figure 17.10), they can be used to explore future scenarios for the next 20–50 years (see Figure 17.10, B1 and B2): (a) climate change impact on precipitation, runoff, and temperature, (b) changes in irrigation and other water use, (c) changes in land use induced by climate change or initiated by national or international regulations (e.g., European Union agricultural policy) or by the economic market. Environmental protection scenarios can also be explored.

Billen and Garnier (1997) have also modeled a virtual river (the Physon, one of Paradise's streams) for nutrients and algal growth. The Physon model has three components: river, estuary, and coastal zone. Retrospective scenarios of long-term response of basins to impacts (e.g., deforestation, urbanization) can be explored (see Figure 17.10, B3) including the ultimate response of the coastal zone after the turbid-estuary filter.

Prospective scenarios can also be envisaged but require the input from economists, policy makers, environmental engineers, and other specialists. Reconstruction of the past evolution of river basins requires related information on riverine dynamics (such as low and high water hydraulics), land use and cover from agriculture, population distribution, as well as

Table 17.5 Evolution of total nitrogen (TN) and phosphorus (TP) fluxes ($10^9 g^{-1}\ y^{-1}$) at the mouth of the Seine according to three environmental scenarios (Garnier et al. 1998).

	TN fluxes	TP fluxes
Situation in 1991	82	7.9
1. Dephosphatization in all sewage treatment plants > 2000 inhabitant-equivalent	82	2.0
2. As in Pt. 1 plus P detergent ban	85	1.2
3. Denitrification in all sewage treatment plants > 2000 inhabitant-equivalent	54	7.8

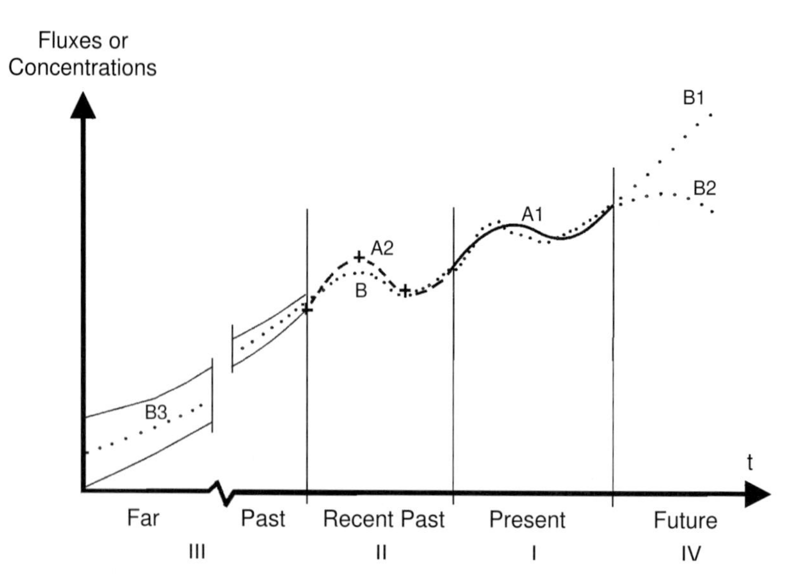

Figure 17.10 Application of validated models for the reconstruction of past riverine evolution (fluxes and/or concentrations) and for the projection of future evolution. A1: present-day full records; A2: sparse information from recent past; B: model validation for the present period; B1 and B2: future trends from models according to two contrasting scenarios; B3: past reconstruction from the model and its confidence range.

industrialization and can be verified through sedimentary records. However, this necessitates an enormous input from environmental historians, geographers, archeologists, and paleolimnologists. The simulation of these past and future trends is an exciting challenge for future environmental studies.

ACKNOWLEDGMENTS

For fruitful discussions and input, I wish to thank Max Thibault (INRA, France), Alain Thomas (Université Paris 6), Françoise Elbaz-Poulichet (Université Montpellier), G. Van Dijk (The Netherlands), B. Dröge (Bundesanstalt für Gewässerkunde, Koblenz), A. Ficht (SNS Rouen), and my students and colleagues from the PIREN–Seine Program.

REFERENCES

Billen, G., and J. Garnier. 1997. The Physon River plume: Coastal eutrophication in response to changes in land use and water management in the watershed. *Aquat. Microb. Ecol.* **13**:3–17.

Billen, G., and J. Garnier. 1999. Nitrogen transfers through the Seine drainage network: A budget based on the application of the "Riverstrahler" approach. *Hydrobiologia* **410**:139–150.

Bryan, C.F., D.A. Rutherford, and B. Walker-Bryan. 1992. Acidification of the lower Mississippi River. *Trans. Am. Fish. Soc.* **121**:369–337.

Chapman, D., ed. 1992. Assessment of the Quality of the Aquatic Environment through Water, Biota and Sediment. London: Chapman and Hall.

Claude, J., G. Francillon, and J.Y. Loyer. 1977. Les alluvions déposées par l'oued Medjerda lors de la crue exceptionnelle de mars 1973. *Cah. Orstom Sér. Hydrol.* **14**:37–109.

Colombani, J., and J.C. Olivry. 1984. Phénomène exceptionnel d'érosion et de transport solide en Afrique aride et semi aride. *Intl. Assn. Hydrol. Sci. Publ.* **144**:295–300.

Dynesius, M., and C. Nilsson. 1994. Fragmentation and flow regulation of river systems in the Northern Third of the world. *Science* **266**:753–762.

Garnier, J., et al. 1998. Developpement algal et eutrophisation dans le réseau hydrographique de la Seine. In: La Seine en son bassin. Fonctionnement écologique d'un système fluvial anthropisé, ed. M. Meybeck, pp. 593–626. Paris: Elsevier.

Garnier, J., and J.M. Mouchel, eds. 1999. Man and River Systems. Dordrecht: Kluwer.

Hock, B. 1986. Trend evaluation of water quality changes in Hungarian rivers. *Intl. Assn. Hydrol. Sci. Publ.* **157**:277–284.

Ibanez, C., N. Prat, and A. Canicio. 1996. Changes in the hydrology and sediment transport produced by large dams on the lower Ebro river and its estuary. *Reg. Rivers R & M* **12**:51–62.

ICPR. 1987. International Commission for the Protection of the Rhine against Pollution. Reports. http://www.iksr.org/icpr/13uk.htm.

Idlafkih, Z. 1998. Transport des ions majeurs, éléments nutritifs, carbone et métaux trace dans la Seine. Importance des crues. Ph.D. diss, Univ. Pierre et Marie Curie, Paris.

Justic, D., N.N. Rabalais, R.E. Turner, and Q. Dortch. 1995. Changes in nutrient structure of river-dominated coastal waters: Stoichiometric nutrient balance and its consequences. *Estuar. Coast. Shelf Sci.* **40**:339–356.

Kimstach, V., M. Meybeck, and E. Baroudy, eds. 1998. From Dniepr to Baïkal. A Water Quality Assessment of the Former Soviet Union. London: E. and F.N. Spon.

Livingstone, D.A. 1963. Chemical composition of rivers and lakes. Data of geochemistry. Prof. Paper 440G, G1-G64, Chapter G.. Reston: U.S. Geol. Survey.

Meade, R. 1988. Movement and storage of sediment in river systems. In: Physical and Chemical Weathering in Geochemical Cycles, ed. A. Lerman and M. Meybeck, pp. 165–179. Dordrecht: Kluwer.

Meade, R.H., and R.S. Parker. 1985. Sediment in rivers of the United States. In: National Water Summary 1984, Water-Supply Paper 2275, pp. 49–60. Reston: U.S. Geol. Survey.

Meybeck, M. 1996. River water quality, global ranges, time and space variabilities. *Verh. Intl. Verein. Theor. Angew. Limnol.* **26**:81–96.

Meybeck, M., ed. 1998a. La Seine en son bassin. Fonctionnement écologique d'un système fluvial anthropisé. Paris: Elsevier.

Meybeck, M. 1998b. Man and river interface: Multiple impacts on water and particulates chemistry illustrated in the Seine river basin. *Hydrobiologia* **373/374**:1–20.

Meybeck, M., R. Helmer, U. Forstner, and J. Chilton. 1989. Global waters quality assessment. In: Global Assessment of Fresh Waters Quality — A First Assessment, ed. M. Meybeck, C. Chapman, and R. Helmer, pp. 271–292. Oxford: Blackwell.

Meybeck, M., A. Ragu, and A. Pasco. 1994. Evaluation des flux polluants dans les rivières: Pourquoi, comment, à quel prix. Etude Inter Agences 28. Orléans: Agence de l'Eau Loire Bretagne.

Poulin, M., et al. 1998. Modèles: Des processus au bassin versant. In: La Seine en son bassin. Fonctionnement écologique d'un système fluvial anthropisé, ed. M. Meybeck, pp. 679–720. Paris: Elsevier.

Ramirez, A., A.W. Rose, and C. Bifano. 1988. Transport of carbon and nutrients by the Tuy river, Venezuela. *Mitt. Geol. Paleont. Inst. Univ. Hamburg* **66**:137–146.

Salomons, W.S., and A.J. De Groot. 1977. Pollution history of trace metals in sediments, as effected by the Rhine River. Publ. No. 184. Delft: Waterloppkunding Laboratorium, Delft Hydraulics Laboratory.

Servais, P., et al. 1998. Carbone organique: Origines et biodégradabilité. In: La Seine en son bassin. Fonctionnement écologique d'un système fluvial anthropisé, ed. M. Meybeck, pp. 483–529. Paris: Elsevier.

Toms, I.P. 1987. Developments in London's water supply system. *Arch. Hydrobiol.* **28**:149–167.

Trefry, J.H., S. Metz, R.P. Trocine, and T.A. Nelsen. 1985. A decline of lead transport by the Mississippi river. *Nature* **230**:439–441.

Tsirkunov, V.V., I.K. Akuz, and A.A. Zenin. 1998. The lower Don basin. In: A Water Quality Assessment of the Former Soviet Union, ed. V. Kimstach, M. Meybeck, and E. Baroudy, pp. 375–412. London: E. and F.N. Spon.

Turner, R.E., and N.N. Rabalais. 1991. Changes in Mississippi river water quality this century and implications for coastal food webs. *BioScience* **41**:140–147.

Valette-Silver, N.N.J. 1993. The use of sediment cores to reconstruct historical trends in contamination of estuarine and coastal sediments. *Estuaries* **16**:577–588.

Van Dijk, G.M., and E.CL. Marteijn. 1993. Ecological rehabilitation of the River Rhine. In: Ecological Rehabilitation of the Rivers Rhine and Meuse, ed. G.M Van Dijk and E.C.L. Marteijn. Report 50. Bilthoven: Rijksinstituut voor Volksgezondheid en Milieuhygiene (RIVM) and Lelystad: Rijksinstituut voor Zuivering van Afvalwater (RIZA).

Vollenweider, R.A. 1968. Scientific Fundamentals of the Eutrophication of Lakes and Flowing Waters. Technical Report DA5/SCI/68.27. Paris: OECD (Organisation for Economic Co-operation and Development).

Vörösmarty, C., M. Meybeck, B. Fekete, and K. Sharma. 1997a. The potential impact of neo-castorization on sediment transport by the global network of rivers. *Intl. Assn. Hydrol. Sci. Publ.* **245**:61–282.

Vörösmarty, C.J., et al. 1997b. The storage and aging of continental runoff in large reservoirs systems of the world. *Ambio* **26**:210–219.

Vörösmarty, C., R. Wasson, and J. Richey. 1997c. Modelling the Transport and Transformation of Terrestrial Materials to Freshwater and Coastal Ecosystems. Report 39. Stockholm: IGBP (International Geosphere-Biosphere Programme).

Walling, D.E. 1999. Linking land use, erosion and sediment yields in river basins. *Hydrobiologia* **410**:223–240.

Walling, D., and B. Webb. 1985. Estimating the discharge of contaminants to coastal waters by rivers: Some cautionary comments. *Mar. Poll. Bull.* **16**:488–492.

18

The Use of Models in Integrated Resource Management in the Coastal Zone

L.A. DEEGAN,[1] J. KREMER,[2] T. WEBLER,[3] and J. BRAWLEY[2]

[1]The Ecosystems Center, Marine Biological Laboratory, 7 MBL Street, Woods Hole, MA 02543, U.S.A.
[2]The University of Connecticut, Storrs, CT 06268, U.S.A.
[3]The Social and Environmental Research Institute, Leverett, MA 01054, U.S.A.

ABSTRACT

Understanding and predicting the response of ecosystems to disturbance is central to management. This goal is often facilitated by the development of ecological models that have the potential to improve the competency of policy decisions. However, the full potential of models in the policy forum has not been realized. Among the factors limiting the effective use of models in management are the lack of: (a) appreciation for models as tools for thinking about a problem (not simply generating an answer), (b) presentation of model outputs in an appropriate context, and (c) comprehension of the inherent uncertainty of model predictions.

INTRODUCTION

In recent years, ecological models have come to play a central role in many areas of environmental management. Models are used to understand environmental processes, to predict impacts of pollution, and to evaluate trends in environmental quality. A model is typically viewed as a rational, objective means of integrating complex information to predict future conditions. The expectation is that models predict — with precision, scope, and accuracy — the consequences of policy alternatives. Unfortunately, models often do not live up to these expectations. Models are often particularly poor at precise, point estimates of response variables (Korfmacher 1998; Oreskes et al. 1994). It is also unfortunate that the precision of model estimates declines with increasing trophic level (Radford and Ruardij 1987) because

Science and Integrated Coastal Management
Edited by B. von Bodungen and R.K. Turner

these are often the endpoints of interest to the public and managers. Creating more complex models does not solve these problems. Although it seems intuitive that the more a model incorporates all the processes and mechanisms known to incur in the "real" world, the closer it should come to predicting the impacts of a particular perturbation, this has not been the case. Often, more complex, data-intensive models are less stable and more difficult to test (Oreskes et al. 1994).

Despite these inherent difficulties with models, we believe that models are potentially powerful tools for assisting in policy decisions. However, we suggest that the way that we present and use models in the policy arena should reflect the strengths and limitations of models. Instead of developing more complex models, we may need to use simpler models in sophisticated, clever ways. In this paper, we use a model that we are developing to illustrate a few approaches to presenting model output that reflect the strengths of modeling and discuss the importance of uncertainty of model predictions.

INTRODUCTION TO THE PROBLEM

The coastal zone, including near-shore waters and adjacent uplands, is heavily used for a variety of purposes. Home to diverse and productive communities of plant and animals species, coastal habitats are among the most biologically productive areas in the world. They serve as nurseries for many fish and shellfish. Because of this, they are highly prized locations for residential development, for recreation, and as sources of food for human consumption. However, the multiple ways we use the coastal zone are not always compatible. Conversion of uplands from natural ecosystems to human-dominated systems, such as agricultural or urban/suburban land uses, causes profound changes in ecosystems, including alteration of water balance and changes in nutrient availability and mobility, both by direct fertilization and by changing controls on nutrient cycling processes. These changes in the uplands affect the delivery of sediment, organic matter, and nutrients to coastal ecosystems.

Nitrogen loading associated with land-use activities in the coastal zone and its concomitant impacts on estuarine ecosystems is a premier policy issue for most coastal communities. The consequence of elevated nutrient loading to estuaries, often referred to as eutrophication, follows the classic response of ecosystems to stress (e.g., altered primary producers, nutrient cycles, and loss of secondary production; Rapport and Whitford 1999). However, characteristics of particular watersheds and receiving estuaries vary widely, hampering attempts to predict effects of a particular upland development on a specific estuary. Coastal zone managers and planners need such information to guide prudent land-use decisions, yet can seldom expend the time and funds necessary to complete detailed studies of local sites. This is a common problem in management, and one solution is to use a model to provide information on which to base a decision.

We have developed an estuarine ecosystem response model to aid in understanding how several biological and physical factors influence the range of ecological responses to nutrient inputs. The model is based on our latest understanding of these complex ecosystems. Importantly, we have chosen relatively simple physical and biological formulations instead of complex site-specific hydrodynamics and ecology. This "reduced form" modeling approach allows the model to be run on small computers and requires less parameterization than more complex models. The goal is to develop a model that is readily transferred to other systems and amenable to management applications when extensive and expensive scientific studies are unavailable or not feasible.

Our approach is to link a nitrogen-loading model to a process-based, but empirically driven, ecological response model that predicts phytoplankton and eelgrass productivity, system metabolism, probability of anoxic events, and fish abundance and diversity. Empirical data to build the model derives from studies in Waquoit Bay, Massachusetts, U.S.A., sampled from 1992–1996. The model begins by estimating the amount of nitrogen moving from the land to the coastal waters based on land use in the upland watershed (Valiela et al. 1997). The model estimates how water quality, low oxygen events, and algae and submerged aquatic vegetation will change as land use in the watershed changes. This model further estimates how fish populations may change as a result of the changes in the ecosystem (Deegan et al. 1997). Eelgrass habitat quality and quantity are important components in the model relating nitrogen loading to eutrophication effects. The model estimates the amount of the bay suitable for eelgrass growth (determined from depth of the water, light penetration through the water, and light requirements of the eelgrass) under pristine conditions as well as under various scenarios of increasing turbidity by phytoplankton growth. The empirical relationships of eelgrass density and fish abundance and diversity that are incorporated in the model allow effects of nitrogen loading on these important ecosystem properties to be predicted. In addition, the model predicts the likelihood of anoxic events by estimating time-dependent total system metabolism (benthic + pelagic) and comparing it to dissolved oxygen availability.

We developed the model using data from a single subestuary of Waquoit Bay (Childs River estuary) with response and environmental variables from 1992; we call this the reference run of the model. The parameter set for the model equations was developed by tuning the model based on this run to obtain the best match between the observed and simulated distributions of response variables. Subsequently we used this same parameter set and physical characteristics (e.g., water column depth) specific for other sites in Waquoit Bay to make predictions for the Quashnet River and Sage Lot Pond estuaries for the same environmental year. We then compare the relative performance of the model output to observational data to evaluate the reasonableness of the model.

Initial comparison of the observed and model results for phytoplankton biomass and dissolved inorganic nutrients indicates that the model performs reasonably well with respect to the observational data (Figure 18.1). In comparing model and empirical results we look for general consistency of patterns produced by the model with observations in the field, rather than detailed numeric agreement. However, this comparison has identified several gaps in our knowledge that are potentially significant to understanding the complete response of these systems to nutrient inputs. Until resolved, these shortcomings limit the use of the model for specific management applications. However, they can be used to illustrate our points about the use of model outputs in a management context.

USE OF MODEL RESULTS

We believe that the most useful presentation of model output provides the results in an appropriate context and gives a sense of how certain we can be of the predictions. We suggest that

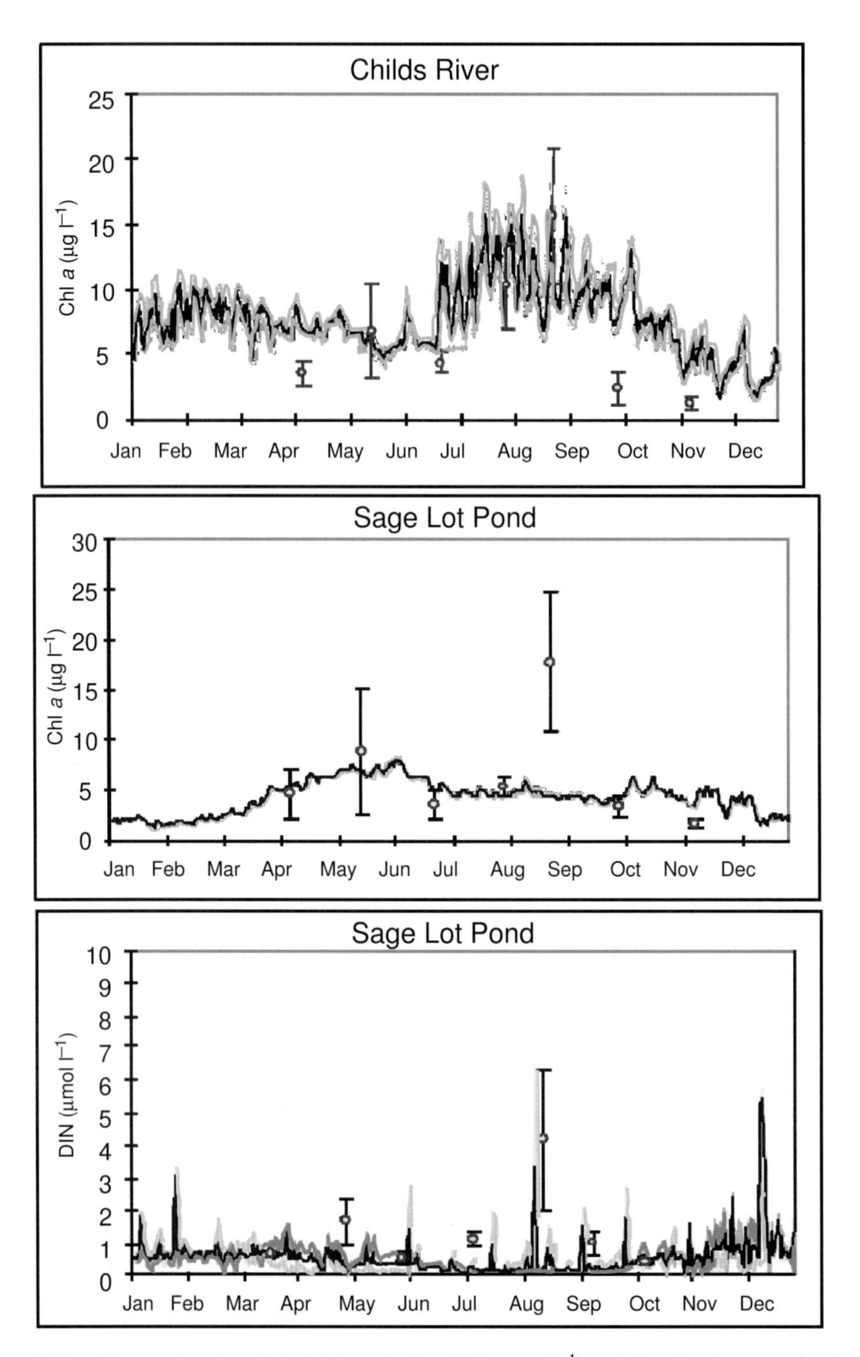

Figure 18.1 Observed and modeled chlorophyll *a* (Chl *a*, $\mu g\ l^{-1}$) and dissolved inorganic nitrogen (DIN, $\mu mol\ l^{-1}$) in Waquoit Bay subestuaries. Top: Childs River Chl *a*. Middle: Sage Lot Pond Chl *a*. Bottom: Sage Lot Pond DIN. The solid lines are modeled concentrations for the surface (light lines) and bottom layers (dark lines); the circles with bars are the mean (± standard error, SE) observations for 1992 field data averaged across several sites within a subestuary.

the best use of model outputs is not the way they are usually used: a point-by-point comparison of model output to observed data for a specific location. A more useful approach is to present modeling output and "What if?" scenarios in the context of broader comparison with other similar coastal ecosystems. We also feel it is essential to translate detailed model results into a form that can be used by decision makers who are not modelers or scientists. We suggest that by presenting our results in an appropriate context and being honest about the uncertainty involved in our predictions, we help the users of model output to make better management decisions. Some examples of our intent are discussed below.

Appropriate Context

We feel it is essential to present model results in a context that is useful to decision makers. We suggest three approaches: (a) compare data and model output to summary results for other systems; (b) develop "What if?" scenarios for relevant management options; (c) provide information in a form that is relevant to the user.

Presenting measured and modeled results for a specific site in comparison to other ecosystems provides a larger context for understanding the status of a specific site and the results of management options. Comparison of modeled outputs to other real systems with which managers are familiar may make model outputs more tangible. Although point-by-point comparison of modeled and observed data (Figure 18.1) remains the most informative way to evaluate the performance of the model and further it's development, we suggest these outputs are too detailed and lack the appropriate context for use in management. The task of confirming model output by comparison to observed data is complex (Oreskes et al. 1994). Our understanding of real systems is inherently uncertain because natural processes are variable and our attempts to characterize these processes with limited sampling are prone to measurement error. Matching the appropriate space and time scales between model output and observed data is difficult and in some ways subjective (Oreskes et al. 1994). The fine level of detail needed to confirm the reasonableness of model output is usually substantially more detail than managers need to evaluate management options. Therefore, we suggest that point-by-point comparison of model and observed data is not the most satisfactory approach for using model predictions in management.

Once the detailed output is deemed satisfactory, model results should be summarized in a general form more amenable to evaluating management decisions. In this case (Figure 18.2), we use as background the summary developed by Nixon (1992) for annual average phytoplankton chlorophyll *a* concentration versus dissolved inorganic nutrient input for a variety of coastal ecosystems and mesocosms. The simulated N-loading series for Childs River spans the range of nitrogen loading levels observed in many systems. The model predictions are illustrated with ellipses that represent the confidence level of the predicted estimate for both chlorophyll *a* and nitrogen load. Placing predicted chlorophyll *a* response in this context should assist managers in placing the consequences of a particular management action to control nitrogen loading for a specific location in a broader context.

The use of model scenarios that vary the inputs (in this case nitrogen loading, N-load) is an example of how a simulated gradient suggests a trend that has intuitive meaning and places model output in a context relevant to a potential management action (Figure 18.3). Because a series of model runs is internally consistent, such comparisons are useful for evaluating

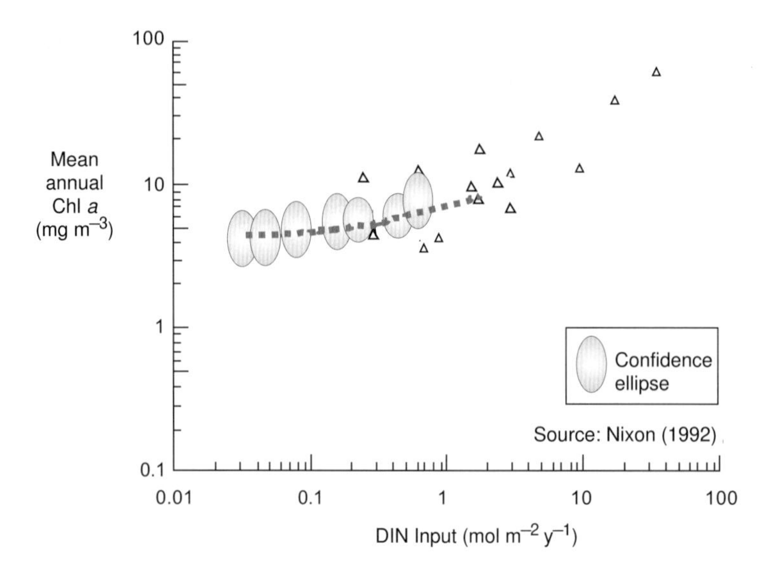

Figure 18.2 Waquoit Bay model results for chlorophyll *a* (ellipses connected by dotted line) plotted on a summary graph of the relationship for a range of ecosystems worldwide. The other data (triangles) are from Nixon (1992). The Waquoit Bay modeled output is represented as ellipses which represent the confidence limits on both input (the DIN loading, mol m^{-2} y^{-1}) and response variable (chlorophyll *a*, mg m^{-3}) for a hypothetical N-loading series for Sage Lot Pond. At the time of this publication, the ellipses are for illustrative purposes only and do not represent true confidence limits.

management options. For example, we run a series of different nitrogen-loading scenarios based on a range of nitrogen concentrations for the incoming river or groundwater resulting from land-use change (e.g., conversion of forest to housing tracts). We can hold all other inputs in the model constant (e.g., river discharge, temperature, and light regime) and compare the effects of changing only one factor (nitrogen load). This provides a clearer description of the response of the ecosystem to changing N-load than comparing actual interannual variability in ecosystems. This is because annual changes in N-load are often caused by changes in other ecosystem drivers, for example, river discharge, that not only change N-load, but also change other characteristics of the system (e.g., average salinity or degree of stratification of the water column) that may have confounding effects. The series of results is combined into a scenario that places the present (observed and modeled) case in the context of past and future levels of loading and makes a comparison. The results can be compared to summarized observed data for various sites (in this case, the three subestuaries in Waquoit Bay) for which field data are available (Figure 18.3) and also placed in a broader context of other ecosystems (e.g., Figure 18.2).

It is also important to present results in a form that is needed by users. For example, sufficient dissolved oxygen (DO) is critical to sustaining healthy biological communities and is therefore an important endpoint often used to help make management decisions. However, DO is a dynamic and highly variable attribute of aquatic ecosystems and is very difficult to monitor on appropriate space and time scales. Even short periods of low DO (hypoxia or anoxia) can have long-lasting effects on biological communities. Because DO is the result of

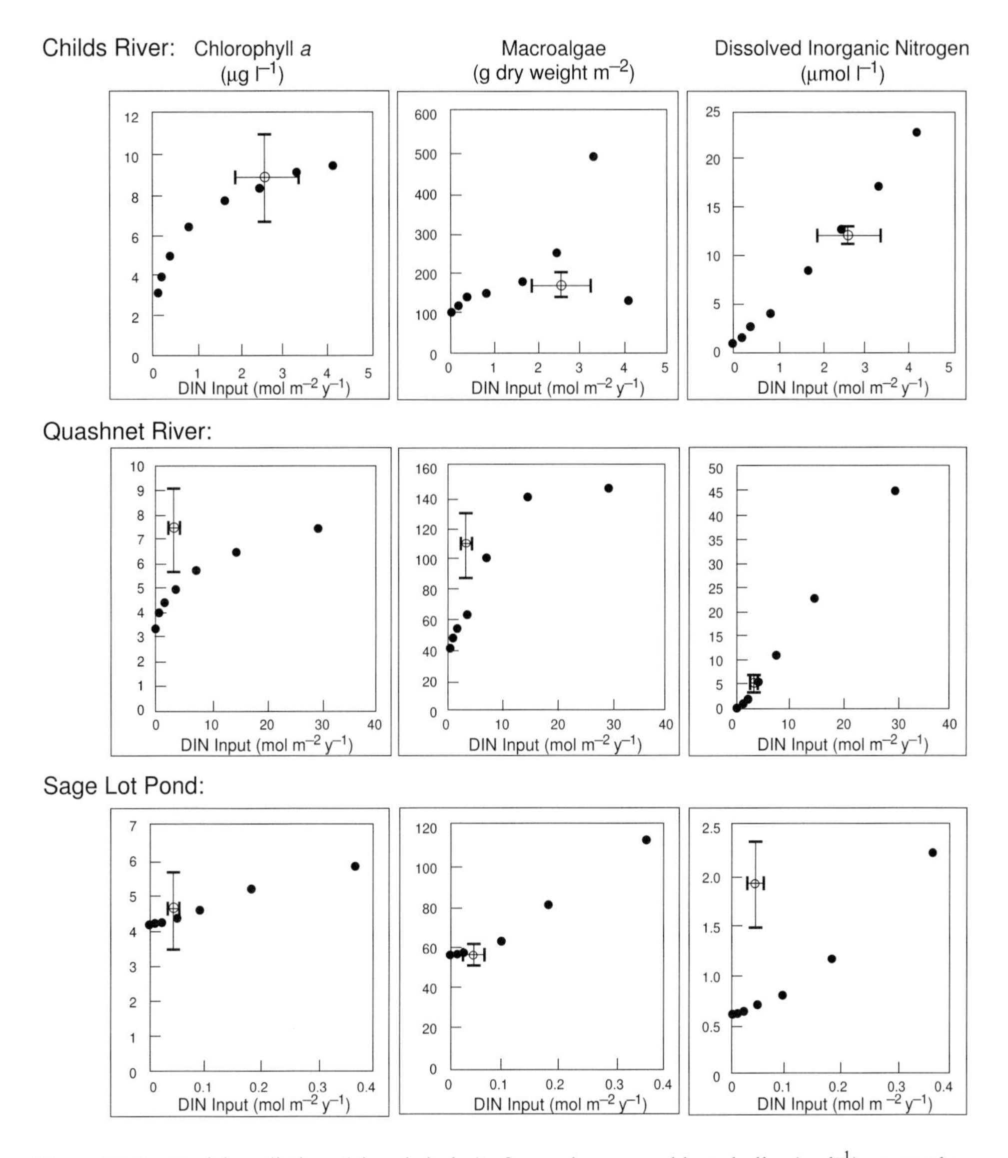

Figure 18.3 Model predictions (closed circles) of annual average chlorophyll *a* (µg l^{-1}), macroalgae abundance (g dry weight m^{-2}), and dissolved inorganic nitrogen (DIN) concentration (µmol l^{-1}) for eight nutrient loading scenarios in three subestuaries of Waquoit Bay. The mean (± SE) observed 1992 data are shown along with error bars on the estimated DIN loading for each subestuary from the land-use N-loading model for each site.

several very dynamic processes, predicting the oxygen content in the water on a given day is extraordinarily difficult and comparison to field data is problematic. In addition, plots of the daily oxygen level or rates of respiration are not very useful to a manager trying to decide if a particular management option will cause more or fewer fish kills due to hypoxic events. An alternate approach is to define an indication of potential low DO by comparing the respiration demand to the stock of DO present in the water column at atmospheric equilibrium. When total system respiration during the night exceeds the DO stored in the water column, the occurrence of low DO is likely. When the predicted low DO value drops below an ecologically important threshold, a potential low DO "event" with significant ecological consequences is possible. The model computes total system metabolism, including daily values for DO production and respiration by the community (Figure 18.4 top), thus allowing an estimate of the risk of a low DO event to be calculated for each day (Figure 18.4 middle). To make this result more understandable, the number of days per month of potential risk can be plotted, allowing direct comparison of a number of scenarios (Figure 18.4 bottom). This approach puts the focus where it belongs — on the probability of a low DO event occurring, not on matching the day-by-day observed and predicted DO. To a manager it is not important to know if the anoxic event occurred on a specific day, but whether it might occur at all, or if some management decision will increase the probability of low DO events occurring. A unique aspect of this approach is its prediction of the probability of an event. The probabilistic nature of model predictions is an important and usually underemphasized aspect of model output that planners should understand if they are to make realistic, scientifically informed decisions.

The Matter of Uncertainty

It is essential, in our view, to provide a sense of the confidence surrounding model estimates. Stating model estimates of a response variable in terms of a single, certain outcome is not only misleading but is also intellectually dishonest. Models cannot estimate response variables with certainty. There is often substantial uncertainty in model coefficients and parameters and initial conditions. Model outputs are products of mathematical equations. The parameters are often derived from observed data and as such these parameters (and inputs) are random variables with probability density functions (pdfs), therefore, the outputs of these equations are random variables with probability density functions. Procedures for uncertainty and error estimation in modeling have been described and include a variety of techniques such as first order error analysis, regional sensitivity analysis, Monte Carlo simulations, methods of moments, and maximum likelihood estimators (e.g., Beck 1987). Estimating these pdfs or confidence limits is very difficult to do, which may explain why it is rarely done. We, unfortunately, have no simple solution to offer. For the relatively simple N-loading model (Valelia et al. 1997), we undertook a formal error analysis (Collins et al. 2000). Based on the pdfs of the data used to estimate the variables used in the calculation (e.g., N-release per capita, N-retention by septic systems), we computed confidence limits on the estimated mean loading rate, and, more importantly, the pdf of the inferred population of loading rates. Most model developers use these techniques to minimize prediction uncertainty and increase model resolution; however, they rarely extend their analysis to quantifying the uncertainty of their predictions.

Applying these techniques to complex simulation models of entire ecosystems is a daunting task, but it must be done. Although we have not yet conducted an uncertainty analysis on our model, we suggest that graphs that indicate confidence bounds (e.g., the hypothetical confidence ellipses in Figure 18.2) are more useful and honest representations of model output than are simple point estimates. It is possible that model predictions in general may only be

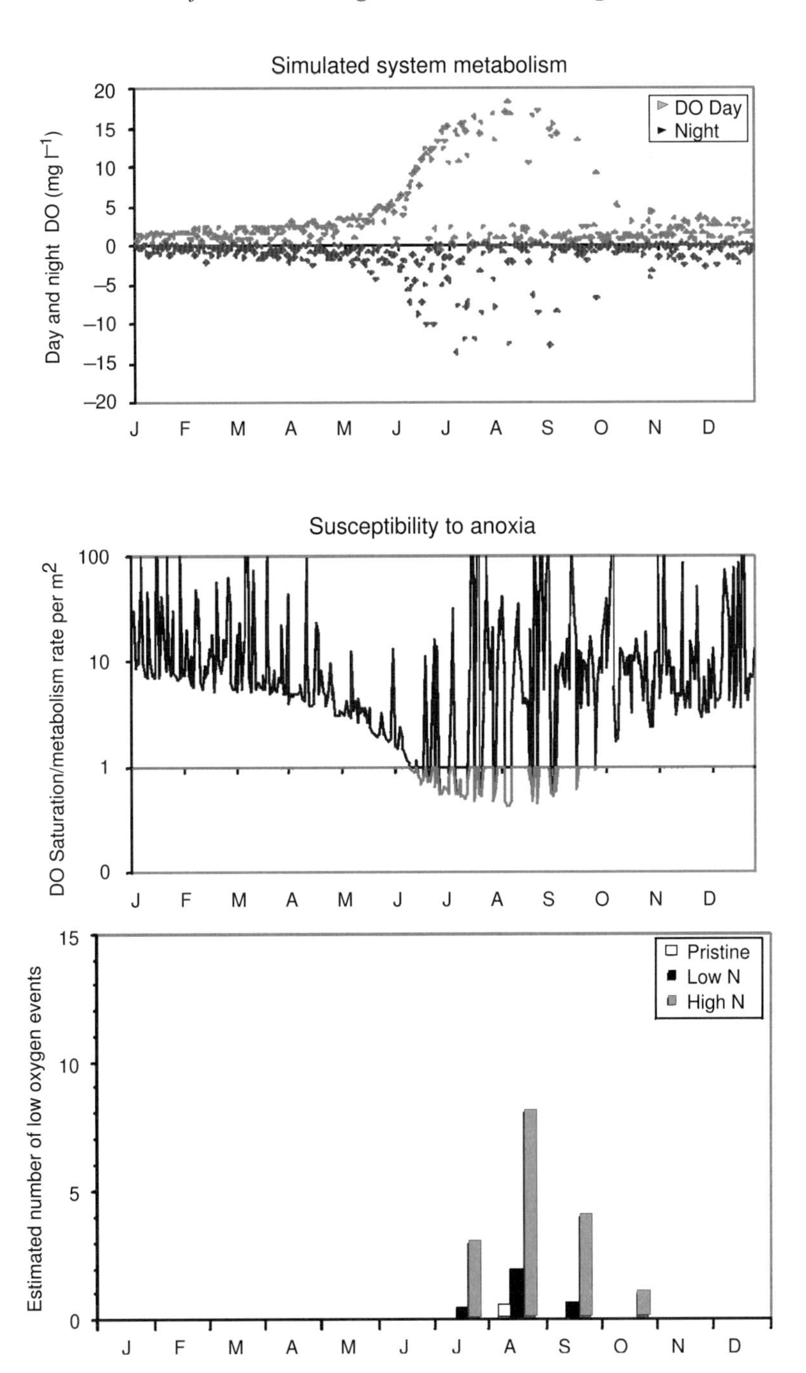

Figure 18.4 Examples of presentation of dissolved oxygen simulation results for Waquoit Bay. Top: Daily output of community production and respiration. Middle: Ratio of total respiration demand to stock of oxygen in the water column as a measure of "risk of anoxia." Bottom: Comparison of the risk of low oxygen event for three hypothetical N-loading runs (Pristine, Low N-load, High N-load).

accurate to within a factor of two or three and these predictions should be used with full consideration of these uncertainties.

Presenting the uncertainty of model predictions in an understandable and defensible way in a public forum is not a trivial task. By admitting the uncertainty in our predictions, we may avoid the pitfall of having the model thrown out of the decision-making process because the promised precise result did not occur as predicted (Haan et al. 1990, Ludwig et al. 1993). For example, the practice of estimating maximum sustainable yield (MSY) without any explanations of the uncertainty involved or the probably bounds on the estimate led to many problems. Difficulties occurred when the MSY number was exceeded and the fishery did not collapse and when the fishery collapsed before MSY was obtained (Ludwig et al. 1993). Fisheries scientists knew that the estimates of both MSY and landings had some uncertainty and were not overly surprised by these results. The results were within the "uncertainties" they knew about. Fishers thought scientists did not know what they were doing and rejected all further advice. Unfortunately, if the users see only that the prediction is not "very good," they may not be willing to use the results of the model at all. By presenting both the model results and uncertainty in a general context as described above, this difficulty can be somewhat alleviated. This approach encourages users to pay attention to the general trend, not the point estimate.

TOWARD MORE EFFECTIVE USE OF MODELS IN MANAGEMENT

Increasing reliance by managers on ecological models to address coastal ecosystem response to change is almost a certainty. Integrated ecological models have the potential to improve the competence of communities in making policy decisions. Among the factors limiting the effective use of models in management are the lack of: (a) appreciation for models as tools for thinking about a problem (not simply generating an answer), (b) presentation of model outputs in an appropriate context, and (c) comprehension of the inherent uncertainty of model predictions.

It is often stated by scientists and modelers that the most powerful use of models is heuristic: models are useful for strengthening our thinking about complex interactions, for exploring "What if?" questions, placing the response of a specific system in a larger context, and offering confirmation of our judgment. We need to bring this philosophy into the realm of public policy, instead of simply offering the point estimates of model output as specific, accurate, and objective guidelines for what must be done to protect a resource.

We also need to develop new tools and methods of describing the uncertainty of model outputs. More certainty is required in model predictions if the model is being used to manage a specific place or time, than if it is being used to assess trends or guide management principles. Educating decision makers and the public about the full range of uncertainties may result in more appropriate use of model results and better management decisions. Developing confidence estimates for outputs from complex models is a critical area that needs more attention.

ACKNOWLEDGMENTS

The ideas presented in this paper are some of the common themes that have arisen during discussions with colleagues regarding the interactions of science and management of natural systems and our own attempts to convince managers to use our models. We thank our colleagues Jason Wyda, Amos Wright, and Jeff Hughes for their comments on drafts of the manuscript. We would like to acknowledge the support of The Mellon Foundation, The Environmental Protection Agency (EPA R825757-01-0), and The National Science Foundation (LTER-OCE-9726921, DEB 9815598).

REFERENCES

Beck, M.B. 1987. Water quality modeling: A review of the analysis of uncertainty. *Water Resour. Res.* **23**:1393–1442.

Collins, G., J.N. Kremer, and I. Valiela. 2000. Assessing uncertainty in estimates of nitrogen loading to estuaries for research, planning, and risk assessment. *Env. Manag.* **25**:635–645.

Deegan, L.A., et al. 1997. Development and validation of an Estuarine Biotic Integrity Index. *Estuaries*: **20**:601–617.

Haan, C.T., J.B. Solie, and B.N. Wilson. 1990. To tell the truth — Hydrologic models in court. In: Transferring Models to Users, ed. E.J. Janes and W.R. Hotchkiss, pp. 337–348. Bethesda: Am. Water Resources Assn.

Korfmacher, K.S. 1998. Water quality modeling for environmental management: Lessons from policy science. *Policy Sci.* **31**:35–54.

Ludwig, D., R. Hilborn, and C. Walters. 1993. Uncertainty, resource exploitation, and conservation: Lessons from history. *Science* **260**:17; 36.

Nixon, S. 1992. Quantifying the relationship between nitrogen input and the productivity of marine ecosystems. *Pro. Adv. Mar. Tech. Conf.* **5**:57–83.

Oreskes, N., K. Shrader-Frechette, and K. Belitz. 1994. Verification, validation, and confirmation of numerical models in the earth sciences. *Science* **263**:641–646.

Radford, P.J., and P. Ruardij. 1987. The validation of ecosystem models of turbid estuaries. *Cont. Shelf Res.* **7**:1483–1487.

Rapport, D.J., and W.G. Whitford. 1999. How ecosystems respond to stress. *Bioscience* **49**:193–203.

Valiela, I., et al. 1997. Nitrogen loading from coastal watersheds to receiving estuaries: New methods and application. *Ecol. Appl.* **7**:358–380.

19

Governance and Sustainable Fisheries

J.M. McGLADE

Dept. of Mathematics & CoMPLEX, University College of London,
Gower St., London WC1E 6BT, United Kingdom, U.K.

ABSTRACT

Fisheries are important throughout the world, contributing to human welfare in terms of income generation, food security, and cultural needs. Because many of the world's fisheries are overexploited, their role in the future development of local communities is now of increasing concern. Overcapacity in fishing fleets and the misuse of fish resources have been contributory factors, but it is the globalization of markets which has done more in recent years to distort local prices and redistribute the net benefits of fish in terms of food security and poverty alleviation. This is not a new phenomenon; as far back as the sixteenth century, countries such as England and Holland went to war over access to their fish resources — culminating in Hugo Grotius's 1609 publication of the essay *Mare Liberum* on the freedom of the seas (Grotius 1609/1916). Now, at the end of the twentieth century, legal issues over fish trade are dealt with by the World Trade Organization or regional bodies such as the European Parliament; locally, however, acts of piracy and illegal fishing still remain the norm. Attempts by such large-scale governmental organizations to regulate or control these issues are therefore viewed with suspicion by many nongovernmental organizations, as no form of global governance has been provided to meet the challenge of fairness, equity, and sustainability of local resources. A shift in thinking is critically needed.

This chapter discusses the various elements required to achieve such a shift, foremost amongst which is the task of rethinking the problems of the past few decades to see them with new eyes. In particular, it examines the concepts behind sustainability, the role of scientific evidence and the nature of knowledge systems, uncertainty, risk, and precaution in relation to societal choice, forms of governance, participatory and regulatory regimes, and frameworks of power as they relate to fisheries. From these, the conclusion is drawn that without a fundamental change in our views on governance, and hence management, many of the world's fisheries will fail to survive or remain viable in the future.

Science and Integrated Coastal Management
Edited by B. von Bodungen and R.K. Turner

INTRODUCTION

Sustainability is an evolving body of values and ideas — a sense of preferred direction which human populations wish to take. Most of all, it concerns a process that enables people to realize their potential and improve the quality of their lives in ways that protect and enhance the planet's potential for future generations. The Brundtland Report (WCED 1987) definition of development — to "meet the needs of the present without compromising the ability of future generations to meet their own needs" — implies living on the earth's income rather than reducing its natural capital, consuming renewable resources within the limits of replenishment, and preventing the loss of the capacity for renewable resources to generate natural wealth. Yet the principles of sustainability which lie within our current conventions and frameworks, such as those generated by the World Commission on Economic Development (1987) and the UN Conference on the Environment and Development (1992), contain a fundamental dichotomy — that of growth versus development. This dichotomy arises because growth, which is generally defined as an increase in standard economic indicators such as gross national product, is not necessarily linked to development, which is about continued improvement in human well-being. Reconciling these two basic aspirations is, therefore, the first step in moving towards a new approach to sustainability in fisheries.

Fisheries can only be part of future solutions when all the elements of the system in which they are embedded are accounted for; these include the ecosystem, society and its institutions, as well as the global and local economies. We should remember that on a global scale, scientific evidence about the state of fisheries resources is still very limited. We therefore need to look far more closely at indigenous knowledge of fisheries and go beyond the arid debate as to whether or not such information is "generalizable" to a position where the outcome of different forms of local and global governance under different local conditions can be anticipated.

Which institutional structures will promote such a long-term view is not yet clear: how should communities best "think globally and act locally," solve issues of intergenerational access, and ensure that they will be here today and here tomorrow? Given that we need to provide some, if not all, of the protein needs for the two-thirds of the world population that lives in the coastal zone from the sea, a radical change in the governance of fisheries is needed. Without it, this challenge is likely to be unmet. To achieve such a change we must harness a number of large-scale socioeconomic forces, including international trade and transboundary human migration, and tackle the vagaries of the consumer preference, perverse subsidies, and fluctuations in our monetary systems. This does not mean — as many well-meaning observers have concluded — that there is a compelling need for the immediate creation of a central public authority to place limits on sovereignty or bring pressure to bear on individual states to comply with rules designed to ensure cooperation over fisheries.

Governance is a social function whose success is vital to our future viability; it centers on the management of complex interdependencies among actors (individuals, corporations, interest groups, and public agencies) who are engaged in interactive decision-making and, therefore, take actions that affect each other's welfare (Young 1994).

Governments, on the other hand, are organizations with complex material assets, often professing political ideologies, that we commonly take for granted as providing governance. In many settings, governments are poor providers of governance, undertaking instead social engineering or rent-seeking on behalf of special interest groups. When one looks to world government to provide the wherewithal to address environmental problems, we have often overlooked the very genuine advances that have been made in international governance provided by issue-specific institutions or regimes, such as the Montreal Protocol, the Antarctic Treaty System, and the Convention on Biological Diversity. However, the path to international agreement, even in these well-defined areas, is less than easy; those involved in interactive decision-making frequently fail to reach consensus on the provisions of regimes, even when the mutual gains to be had are substantial, and the cost of conflict high. More often than not, environmental challenges, such as catch limits of certain fish species, highlight the difficulties of finding appropriate ways of conceptualizing the problem and so devising ways of addressing issues of sustainability and equity. There are those who wish to spread the benefits of economic growth to the developing world, and others who see that sustainable development requires a profound change in the materialistic lifestyles of people in both developed and developing countries. All such arguments will retain their validity until issues of governance are resolved.

Fisheries are of continuing interest in and of their own right to scientists, policymakers, and local communities; however, sophisticated analyses of the governance issues are only just beginning to materialize. The overexploitation of fish resources has a tangible quality that most people find easy to grasp; an understanding of the role of governance in determining the course of local and global changes requires more subtlety and sophistication. Specifically, it requires an appreciation that changes in fisheries are as much anthropogenic as environmental in origin, a realization that social institutions are major determinants of the course of human/resource relationships, and an awareness that institutional arrangements or governance systems are at work globally, despite the absence of a central government in this social system. The focus on issue-specific regimes is one approach, because it is still far from self-evident how institutions come into existence in an anarchic social setting. In this chapter, however, I shall concentrate on the underlying precepts and structures that currently make up the governance of fisheries and ask how such regimes could be made more effective. In doing so, it will be necessary to look at the nature of fisheries problems, some of the key socioeconomic issues and ecological processes involved, and the types of regimes that have grown up around the world in response to globalization and change.

SCIENTIFIC EVIDENCE AND KNOWLEDGE SYSTEMS

Scientific Evidence

As we move from the late twentieth century into a new millennium, it is argued, by many, that we are witnessing a change from a system based upon the manufacture of material goods to one primarily concerned with information. In fisheries, as more generally, this information is highly biased towards answering specific questions such as the state of the resource and the extent of a particular market.

Evidence from around the world suggests that our current scientific methods and associated data used to provide advice for fish stock management and knowledge about stock status are failing systematically. The main reason is that the underlying models do not address some of the most important problems in fisheries, i.e., the complex dynamics of marine ecosystems, the unreliable nature of prediction in any spatially extended dynamical system, the propagation of errors and uncertainties in the models as a result of imperfect information about the system and its functioning, and the long- and short-term risks of overexploitation associated with overinvestment, cultural dependence, and technical innovation in fisheries.

Yet in most of today's fisheries institutions, there is still a belief that the effects of interventions can be predicted. This supposition occurs because the existing models allow managers to simulate, or in a crude way anticipate, the future. This, however, implies not only that all the interactions within the system evolution are adequately understood, but also those processes that will direct its forward evolution. Remarkably, most planning decisions have ignored these issues, concentrating instead on a highly restrictive view of what is actually happening. Thus we see situations where fisheries scientists and managers have been forced unremittingly into a role where they are trapped by their own knowledge and where they might think they know what the system is doing. They rarely know why or even how it is doing it (e.g., in the east coast fisheries of Canada; Allen and McGlade 1986).

The scientific basis for fisheries management is built around a series of models that are used to estimate the effects of reproduction, recruitment, and mortality on fish populations (see McGlade and Shepherd 1992). The different models have their own strengths and weaknesses, depending on the parameters they aim to estimate and the techniques used to solve them; however, all are focused on the biological aspects of commercially important fish stocks rather than their status within the marine ecosystem or the marketplace. Each model requires a range of data, including such elements as the catch of a population or stock, aggregated in some form, for example, by area or fishery; the implicit assumption is that the stock boundaries are known and that emigration and immigration are not at issue. However, sampling systems rely heavily upon the veracity of the reported origin and size of landings, so as regulatory infringements increase with increasing biological and socioeconomic pressures, the use of primary data sets at finer resolutions cannot be assumed to produce reliable results.

Generally, the data used are sparse and insufficiently resolved to provide detailed insights into any of the real dynamics of gear-species interactions, ecosystem changes, and short-term fluctuations in the abundance of different species. Also, while the genetic basis for some commercially important species have been established, estimates of the movement of individuals within metapopulations, the identification of stocks potentially at risk of extinction, and the fine-scale genetic structures of other species using molecular techniques are largely unknown.

Sensitivity analyses of the effects of systematic errors reveal that the length of the time series available and the historical level of fishing mortality limit the usefulness of many of the models. For example, recruitment and biomass estimates are sensitive to errors in the independent index of stock size used for the calibration of specific models, and errors affecting the age composition of the catch can lead to an overestimation of recruitment in years when recruitment is low, and to an underestimation in years when it is high. The statistical methods used to evaluate the output of the models are *ad hoc* and do not assume any explicit relationship or mechanism between effort and fishing mortality, thus limiting their use as predictive tools. Thus the general conclusion by scientists at the International Council for the Exploration of the Seas is that despite the increase in data over time, forecasts are not becoming any more or less precise.

Finally, it is important to remember that the majority of fisheries models are aimed at single species. Multispecies models have been developed, but the necessary data are not available in most regions and are not generally used for generating advice. New approaches, developed to look at fluxes of materials in ecosystems, are beginning to emerge (Christensen and Pauly1992). Their application to fisheries problems has led to some startling results: it is estimated that one quarter of all primary production is required to produce the fish currently landed and that fishing itself has led to a change in the trophic structure of marine ecosystems —the concept of fishing down the food chain—evidenced in a decline in the size and trophic level of fish being eaten by the top predators (Pauly and Christensen 1995).

To date, the models used in fisheries management have led to technical controls that have systematically failed to constrain or restrict capacity or exploitation. The reasons are manifold but reflect the fact that fisheries are part of a highly complex spatially extended system about which we know very little. It must also be remembered that the outputs from these models lead directly into technical instruments (e.g., quotas and gear restrictions), which are generally analyzed in a policy vacuum. This means that they not only can impede the use of other more effective measures for managing stocks but can also lead to highly uncertain outcomes (McGlade 1989) and stock collapse (Healey 1997; Hutchings et al. 1997a, b). By placing such a strong emphasis on the biological rather than human or economic aspects of fisheries (Allen and McGlade 1987) and by concentrating only on commercially important species, fisheries managers have not succeeded in generating effective governance of fisheries or policies.

Knowledge Systems

One of the major difficulties in coastal zone and fisheries management is that it is highly interdisciplinary, involving fields of varying states of maturity and with very different practices in their theoretical experiments and social dimension (McGlade 1999). Those involved in planning and fisheries policy often find themselves having to use inputs from research areas with which they are potentially unfamiliar, making it difficult to apply the same level of judgement as in their own core field. The result can be a dilution of quality control on the information gathering process and a weaker quality assurance of results (see also Emeis et al., this volume).

In the past, scientists have tended to develop a healthy prudence about passing judgement on the results of others in areas outside their own expertise, with the result that any interference in others' fields is discouraged. Unfortunately, in an interdisciplinary policy-related area such as fisheries, this approach is counterproductive, because criticism, the lifeblood of science, is likely to become weakened. It is not surprising then that most fisheries policy and planning institutions have been unable to respond to crises or change, as in many cases the organizations are suffering from a chaotic mixture of information, analysis, and interpretation and no structure in which to incorporate all the various forms of scientific, interdisciplinary, and indigenous knowledge.

The postmodern outlook, popularized by Lyotard (1985), sees a plurality of heterogeneous claims to knowledge, in which science does not have a privileged place. The British Prime Minister, Sir Winston Churchill, preempted this by coining the phrase, "science should be on tap not on top." This is not to say that sound science should be given second place, but rather that when differing bodies of evidence, alternative paradigms, or different disciplinary perspectives are in tension, it has yet to be determined which criteria should be applied in deciding on what makes up sound science and hence policy. Indeed, even adopting this process calls into question the hegemony of science and technology. In this decade alone, we have seen a crisis of confidence in science over the issues of the release of genetically modified organisms and the infection of cattle with bovine spongiform encephalitis, where the idea of science as the final arbiter of fact has been brought into question.

Giddens (1990) has argued that there is an even deeper problem: one of a need to understand that individuals and societies have an increasing sense of being caught up in a universe of events which cannot be fully comprehended and seem outside any local or individual control. In this sense, Giddens disagrees with Habermas (1987), who believes that abstract systems colonize a preexisting life-world, subordinating personal decisions to technical expertise. Instead, Giddens suggests that the embedding of modern institutions is affected by day-to-day life. Moreover, technical expertise is continuously reappropriated by lay agents as part of their routine dealings with the system. No one can become an expert in the sense of the possession either of full expert knowledge or of the appropriate formal credentials in more than a few small sectors of the immensely complicated knowledge systems that now exist.

In the premodern environment, local knowledge which individuals possessed was rich, varied, and adapted to the requirements of living in the local milieu (Geertz 1983). Individuals in modern contexts are just as knowledgeable about their milieus, but receive enormous feedback from a number of sources, some of which are technical and some cultural. Thus, we can see a type of *second-order science* emerging, one in which individuals must rely on other peer groups and experts to be able to evaluate the information within their own domains of expertise properly.

This type of interaction between expertise, experience, and reappropriation by lay persons is especially important in activities such as fisheries, where direct scientific evidence is generally missing, but where there is often a wealth of local know-how and historical information. For the individual fisherman, this does not necessarily add up to a feeling of security or control over day-to-day exigencies; it does help, however, to identify levels of trust in particular systems and known experts in the community. In this sense, the concept of an expert as part of the system of governance has to be broadened to include those who have particular knowledge about a system, but who do not necessarily walk the corridors of power. As will be discussed below, this is an important concept in determining the level of user participation in any governance model.

Dealing With Uncertainty and Fuzziness

In policy-related issues, the question of what constitutes sound science is exacerbated by uncertainty as well as the societal dimensions described above. Science is judged by the public, including bureaucrats, on its performance in sensitive areas such as the economic returns on foreign aid, returns from the exploitation of natural resources, the dumping of hazardous wastes, the dangers of oil spills, and environmental pollution. All involve much uncertainty, as well as inescapable social and ethical aspects, so simplicity and precision in predictions or even setting safe limits are not always feasible.

Yet policymakers and the public tend to expect straightforward information to use as input into their own decision-making process. In such circumstances, the maintenance of confidence among policymakers, planners, and the community becomes increasingly strained, with the "pure" scientist often caught in the middle. The problems become manifest at several levels, the simplest one being the representation of uncertainty in only qualitative estimates. Any fisheries scientist knows that a prediction such as a "one in a million" chance of the collapse of a stock should be hedged with statements of many sorts of uncertainty so as to caution any user as to the reliability of the numerical assertions. Yet if these were all expressed, policies would become tedious and incomprehensible; if omitted, then the same policies could convey a certainty unwarranted by the facts.

In addition to low-frequency hazards, there are also problems relating to higher probability events, such as the failure of an investment/development program, diffused hazards (e.g., the contamination of certain species of shellfish), or possible large-scale environmental perturbations (e.g., climate change). The dilemma is that any definite advice is liable to go wrong: a prediction of danger will appear alarmist if nothing happens in the short term, while reassurance can be condemned if it retrospectively turns out to be wrong. Thus the credibility of a purely science-based approach is endangered by giving any scientific advice on inherently uncertain issues. If a scientist prudently refuses to accept vague or even qualitative expert opinions as a basis for quantitative assessments and declines to provide definitive advice when asked, then science itself is regarded as obstructionist, not performing its public function, and its legitimacy is called into question.

In areas outside policy and planning, scientists have learned to cope with such problems. One way has been simply to use error bars when estimating variables. However, errors can derive from uncertainties in a wide range of processes and objects, e.g., in the instruments themselves, calibration, design, lack of skill, and general confusion about the theoretical foundations of particular measurements. When a problem becomes more and more complex, simple inexactness cannot fully describe the situation, and uncertainty must be dealt with explicitly. Uncertainty is not merely the spread of data around some arbitrary mean known with confidence but rather a systemic form of error that can swamp an otherwise easily calculated random counterpart. Because uncertainty cannot be removed, it has to be clarified.

The errors associated with data points represent the spread, i.e., the tolerance or random error in a calculated measurement, whereas confidence limits refer more to risk. For example, in a risk analysis of future scenarios resulting from different policies, confidence limits are reflected in estimates when they are qualified as optimistic, neutral, or pessimistic. An assessment based on historical estimates of some fish stock thus acts as a qualifier on the numbers used and on the spread of data points. An assessment represents unreliability and relates to our knowledge about the processes involved, whereas the spread represents inexactness and relates to our knowledge of the behavior of the data.

An important concept is *ignorance*; this is a measure of the gaps in our knowledge. These gaps may simply be anomalous results that are exposed when a new advance in understanding occurs or simply reflect the maturity of the subject. The boundaries of ignorance are very difficult to map. One approach has been to assess the pedigree of a particular field from which the quantity derives. For example, in the case of the theory of relativity, there was a progression from an embryonic field in 1905 through to the 1950s, when experimental results had corroborated the theory and all but cranks had accepted it. Fisheries management on the other hand relies on data that are highly qualitative and heterogeneous. Well-structured theories, common in many branches of science, are conspicuous by their absence. Thinking that we can make exact predictions under highly complex circumstances is thus likely to be premature, leading those involved in decision-making towards a misdirected sense of concreteness in overall policy judgement.

To help make more explicit some of the uncertainties in the management of fisheries, we also need to incorporate linguistic phrases, such as catches are getting worse, or fish are smaller. This can be approached through the use of fuzzy logic, an extension of logic developed by Zadeh (1965) and now applied to a number of situations, including coastal zone management and fisheries (McGlade and Hogarth 1997; McGlade 1999; Mackinson et al. 1999). Fuzzy logic and expert systems based on fuzzy rules provide a means of coping with decision processes involving imprecise data. Their use is particularly relevant where local communities are providing indigenous knowledge to support fisheries management regimes, and in dealing with complex coastal zone conflicts.

Societies have many different ways of dealing with uncertainties, especially when they are entailed in risky endeavors; religion and magic are often used to translate uncertainty into relative security. But where risk is known as risk, this mode of generating confidence in hazardous arenas is by definition unavailable. Widespread lay knowledge of risk leads to an awareness of the limits of expertise and forms one of the public relations problems for fisheries managers who seek to sustain lay trust. To support the trust in experts requires a blocking off of ignorance of the lay person; however, realization of the areas of ignorance that confront experts themselves may weaken the trust that others have in them. Fisheries managers rarely take the true nature of the risks explicitly into account while conducting their analyses and often conceal this fact from the clients. More damaging than this, however, is the situation where the full extent of a particular set of dangers and the risks associated with them is not realized by the experts. In this case, what is in question is not only the limits of or gaps in expert knowledge, but an inadequacy which compromises the very idea of expertise. Just such a situation arose in the case of the collapse of cod stocks off Newfoundland (Hutchings et al. 1997a, b; Healey 1997).

RISK, PRECAUTION, AND SOCIAL CHOICE

Risk and Precaution

Fishing companies and communities are faced with an alarming array of risks arising from a variety of sources. These include environmental uncertainty, changes in resource distribution, regulatory changes, supply and demand fluctuations, and geopolitical instability. For many in the world, the only possible adaptive reaction is one of pragmatic acceptance (Lasch 1985) and a concentration on survival. This represents not so much a withdrawal from the issues of the world, but rather a type of participation that maintains a focus on day-to-day problems and tasks. A fishing family, which relies totally on the small amount of fish from the coral reef adjacent to their house to survive, is not going to worry unduly about the price of cod or tuna in Europe. Only if alternative livelihoods were found for such a family could preferences and issues about precaution and sustainability of the fish resources be explored. Even among governments, pragmatic acceptance is common because so much goes on in fisheries that is outside their control: temporary gains are all that are planned or hoped for, and the enormous investment of human resources, financial capital, and institutional reputation, which can render certain exploitation trajectories effectively irreversible, are allowed to run on until either the resource or the price collapses.

There are a number of approaches that could be used to embed the analysis of risk properly into fisheries management and thereby open up the possibility for the precautionary principle to be adopted; these include decision trees, value trees, multicriteria analysis, sensitivity analysis, and scenario building. However, to be effective, they need to be accompanied by procedures to involve interested groups, such as consensus conferences, citizen's juries, scenario workshops, focus groups, and deliberative polls. Few of these have been applied in the case of fisheries, for the very good reason that many fishing communities lie at the periphery of society and are disenfranchised.

Still, embedding the concept of precaution into fisheries management is only likely to succeed where there are real — not just perceived by the experts — risks of serious irreversible damage. Lack of scientific certainty is unlikely to be the reason for postponing cost-effective measures to prevent environmental degradation, rather a lack of detailed analyses that show otherwise. This is of course contrary to the general notions of precaution, first enunciated in the *Vorsorgeprinzip* of German environmental policy, and taken up by the Food and Agriculture Organisation of the United Nations (FAO) in its Code of Conduct for Responsible Fisheries (FAO 1995) and Technical Guidelines (FAO 1997) as an outcome of the 1992 Rio Declaration of the UN Conference on Environment and Development (Principle 15). The implications of the precautionary approach to capture fisheries have been looked at and involve the application of prudent foresight, requiring *inter alia*:

- consideration of the needs of future generations and avoidance of changes that are not potentially reversible,
- prior identification of undesirable outcomes and of measures that will avoid them or correct them promptly,
- the initiation of any necessary corrective measures without delay, such that they should achieve their purpose promptly, on a time scale not exceeding two or three decades,
- prioritization in conserving the productive capacity of the resource where the likely impact of resource use is uncertain,
- harvestable and processing capacity to be commensurate with estimated sustainable levels of a resource and that increase in capacity be further contained when resource productivity is highly uncertain,
- all fishing activities must have prior management authorization and be subject to periodic review,
- an established legal and institutional framework for fishery management within which management plans that implement the above points are instituted for each fishery, and
- appropriate burden of proof by adhering to the requirements above.

Responsible fisheries require that critical socioeconomic factors be understood, but it is not at all certain that these are known for the vast majority of fisheries. Thus the risks associated with adopting a precautionary approach as a motivating principle in fisheries management are not only likely to arise from failing to control exploitation or to place burdens on parties responsible for damaging activities but also failing to adopt a number of associated concepts, i.e., to acknowledge the limitations of science; have humility about knowledge and anticipate surprise; recognize the vulnerability of the natural environment; uphold the rights of those who are adversely affected by overexploitation; take into account availability of alternative livelihoods; consider the complexity of behavior in different organizations; pay attention to the variability of local and other contextual factors; assign equal legitimacy to different value judgements and adopt long-term, encompassing, and inclusive perspectives in assessment (Renn et al. 1999).

Social Choice

To turn the concept of precaution into a reality for fisheries will involve social choice about a set of incremental measures: clear appraisal techniques (e.g., peer review of science, validation of framing assumptions of consensus workshops, freedom of information about the full range of options); capacity building; development of strategies for markets, monitoring, conservation, and surveillance; introduction of appropriate financial instruments (e.g., incentive schemes, removal of perverse subsidies, introduction of take-back schemes); and legal provisions (e.g., property rights, safe minimum standards, personal legal responsibility on individual decision-makers, forcing targets).

There is a need to distinguish the type of resource and the property rights associated with it. A well-worn debate in fisheries is the tension between defining fisheries as common and noncommon pool resources (Berkes 1989), the former being a class which is difficult to bound or divide either through extraction or use, the latter being those which are bounded and to some extent private. Economists generally use three broad categories of factors of production: natural (i.e., free for all to use), human, and capital resources. However, natural resources are not necessarily free: use may impose costs on the use of other resources in the ecosystem. These externalities are the social costs of production and are not accounted for in the marketplace.

The two traditions of environmental economic thought that frequently emerge to address this issue are those of Pigou (1946) and Coase (1960). Pigou's work makes extensive use of the metaphor of external cost, wherein the price mechanism cannot assure the efficient use of natural resources due to the distorted price signals perceived by individuals and can in fact exacerbate the degradation of the ecosystem—a situation referred to as market failure. On a political level, this approach leads to state intervention, with the efficient use of resources to be ensured by the imposition of appropriate regulations. The approach of Coase refutes the idea of market failure and emphasizes instead the significance of property rights. Inefficient use of resources is interpreted as a consequence of unspecified property rights. The most important aspect of this for fisheries lies at the institutional level, by which many ecological problems are solved through moderate state intervention, i.e., via specification of rights and liabilities.

There is, however, the additional problem that investment returns for fisheries are likely to be much less than could be obtained from a standard form of financial investment. The individual could thus be justified in market terms in depleting the resource completely. Discounting the future simply adds to the problem because most ecological resources would have almost no monetary value for a rational decision-maker assuming a high discount rate. If one assumes a very low discount rate, then the resource is essentially removed from the market. A better solution would be to set a high value in the present so that a nontrivial future rate is obtained and the correct message about fisheries expressed. However, these approaches, which are gaining favor within many of today's institutions, assume objective efficiency as an external perspective. It implies that there is sufficient understanding of the markets so as to be able to determine it externally through fixed goals.

Even if external costs and property rights have been accurately interpreted, there still remains the problem of determining costs and liabilities. A different perspective would be to presume that market failure cannot be determined independently of the subjective opinions that individuals have about the magnitude of external costs. This phenomenon is observed in many fishing communities, where there is a distinct tenacity to hold onto the resource base as a means of existence despite economic inefficiencies. Such an internal perspective, generated by both public and political processes, implies that straightforward economic models, involving the calculation of maximum economic yield, are likely to be inappropriate, as there will be a difference between that defined for the industry and that defined for the community.

Most actions taken within a fishery are influenced by social choice (Arrow 1963), that is, decisions of investment, profit, and the participation of other individuals and social groupings, such as the fishing brotherhoods, or cofradias, of northwestern Spain or the fish producers' organizations found throughout the European Union. Rational decisions, as implied by the models above, are unlikely to be possible, implying that much of fisheries management is premised on an erroneous assumption of human behavior. Instead, we can introduce a model which does not simply assume that individuals have preferences which can be ordered, thereby determining a type of utility function, i.e., an instrumental sense of rationality, but one in which actions can be concerned with deciding on, creating, or exploring the ends pursued. People in this model are less certain about their objectives and the environment in which they operate, less autonomous but more active and enquiring. This more elusive idea of the individual is of course much less mathematically tractable than the rational choice version of standard resource economics, so consequently there are fewer theorems and elegant proofs. Nevertheless it is possible to use these ideas to distinguish two types of action: the procedural role (or rule-bound) and the expressive (or existential or autonomous). In the procedural role, people use rules of thumb to avoid the costs of acquiring information that would better enable them to take a course of action; the use of rules constitutes a significant shift away from the instrumental model and when shared, as in norms, they become a source of reason in their own right. In a wider sense, rule following marks the irreducible social and historical location of individual action because, when shared, they form the building blocks of a society's culture.

The expressive form complicates the relation between action and objectives further by making people self-reflexive; they become capable of deliberating and choosing ends they wish to pursue. Developed by Kant, Marx, Weber, Habermas, and Elster (1983), it is this form of rationality that is important to welfare and resource economics and which is highly relevant to the embedding of the precautionary principle into fisheries. Here, rationality is taken up with establishing the value of the ends pursued; action is thus as much an expression of beliefs as it is the execution of a plan to implement them.

Anthropologists are quick to remind us that shared beliefs are to some degree arbitrary when they act as a system for communication; this is because individual choices cannot always be understood solely in terms of individuals acting to satisfy their own ends. This has a direct bearing on any analysis of a fishery because it implies that we need to understand not only prices and cost/profit structures of local markets but also the behaviors involving such things as imputed shadow prices, repeated or infrequent decisions, emotional and mechanical ends, individual levels of wealth and poverty, status, and age. This implies that understanding how communities will react to differing levels of scarce resources is unlikely to be open to precise economic study. Instead, such approaches as *social network analysis*, which can help to disclose the reciprocity arrangements that exist, need to be undertaken.

Reciprocity arrangements can be negative (i.e., exchange practices which resemble self-interest), balanced or equivalent (i.e., practices and rates relying on tradition), or generalized (i.e., in the realm of mutual help and solidarity). The specific relationship between reciprocity arrangements and modern economic transactions is especially important in establishing fisheries management regimes, because in many fishing communities the significance of balanced and generalized reciprocities appears to resist erosion in an otherwise "economic society" promoting self-interest.

Another approach to the problems of dealing with *Homo economicus* in a world based on fluctuating and vulnerable natural resources is to take a property rights view and introduce usufructuary rights. This corresponds to a quite different understanding of social practice and may have a high potential for eliminating the problem of private rates of discounting that from a sustainable resource management perspective are too high.

Even the introduction of such rights is not enough to address all problems. For example, solutions based on property rights, permits, and other similar instruments are often difficult to reconcile with community-based management. To create a conformity between ecological parameters, precautionary principles, and economic sustainability, it is necessary to have institutions that are organizationally flexible. Such flexibility relies on responsiveness and innovation. On the industrial side, it mainly requires firms with intensive links to the local and regional environment, and on the management side it relies on good lines of communication and implementation. This is especially true for fisheries, where conservation is not an attribute of single units of production but rather the outcome of a symbiosis and cooperation between producers, suppliers and producer services, and local authorities in a region.

GOVERNANCE, POWER, AND THE REGULATORY LADDER

To help our understanding of the role of institutions and individuals in creating effective governance of fisheries progress, it is important to examine the different forms of governance and the ways in which representation, participation, feedback, transparency, and accountability exist within them. Such criteria are important in determining the viability of institutions and should be used as a foundation upon which to establish policies and regulations. Just as more standard economic and environmental indicators are derived using a variety of methods, institutions need their own normative methods by which to be assessed. Rigorous indicators of performance, such as the degree of cooperation versus competition, or the response to different levels of risk and uncertainty, are required not only for internal monitoring but also for inter-system comparison. To achieve this, an analytical framework is set out below.

Structures and Approaches to Governance

Institutional infrastructures critically affect our ability to deal with the complexities associated with use of resources and any consequential or externally generated social or environmental pressures. Often it is a matter of governance rather than insufficient environmental information impeding progress.

The term governance has a number of meanings: it can be the activity or process of governing, a condition of ordered rule, those people charged with the duty of governing, or the manner/method/system by which a particular society is governed. The current use of governance does not treat it as a synonym of government, but rather signifies a change in the meaning of government. At least six separate structures of governance exist: the minimal state; corporate governance; new public management; good governance; sociocybernetic system (Kooiman 1993); and self-organizing networks. The last type is particularly interesting in relation to the governance of resources.

Self-organizing networks involve the transformation of a system of local government into a system of local governance, involving complex sets of organizations drawn from the public and private sector. Governance is thus about managing networks. The key to understanding the importance of this type of governance comes from the observation that integrated networks resist government steering; they develop their own policies and mold their environment. This leads to interdependence between organizations (governance is broader than government), continuing interactions between network members caused by the need to exchange resources and negotiate shared resources, game-like interactions, and a significant degree of autonomy from the state. The *hollowing out of the state* generated by this form of governance is characterized by privatization and limits on the scope of public intervention, loss of functions by central and local government, loss of function at national and regional levels, and limits being set on the discretion of public servants.

One manifestation of this type of governance is the fact that over the past decade, nongovernmental organizations (NGOs) and their memberships have grown nearly sixfold, from five thousand to nearly thirty thousand. Although these have existed for generations (e.g., in the 1800s, the British and Foreign Anti-Slavery Society played a key role in abolishing the slavery laws), changes in this decade have given them renewed vigor. The end of communism, the spread of democracy, technological change, and economic integration have all contributed to this development. Along with this growth, a new phenomenon, known as "NGO swarm" has also emerged. This is a process of amorphous groups linking together to tackle a single issue; examples of this already exist in fisheries, such as the push for dolphin-friendly tuna fishing and the attacks on sealing and whaling leading on from the wider backlash against wearing furskins.

In managing fisheries resources, it is therefore necessary to choose between different forms of governing structures. None is intrinsically good or bad for allocating resources authoritatively or for exercising control and coordination, and the choice is not inevitably a matter of ideological conviction, rather practicality. The interplay between local conditions and the global system and the approach to governance will determine the most appropriate form. Given a world where governance is increasingly operative without government, where lines of authority are increasingly more informal than formal, where legitimacy is increasingly marked by ambiguity, people are increasingly capable of holding their own by exploring when, where, and how to engage in collective action.

Forms of Action and User Participation

The social subsystem is comprised of social and formal rules. Social rules are simply the ways of doing things in everyday life, but they are not generally reducible to formal prescriptions or legal rules or even physical entities; they are expressed and interpreted in natural languages (Giddens 1984; Burns and Flam 1987). They can be considered as emergent properties of human interactions (Bhaskar 1978). For example, there are important differences in the way that fisheries management is defined and enforced and the way certain tasks are carried out in the day-to-day operations of personnel.

It is evident that the expectations of different users, beneficiaries, and managers of fisheries and coastal resources need to be broadly convergent and, at the same time, be consonant with the prevailing environmental setting. Otherwise, the ability to optimize resource use and avoid or resolve conflicts is likely to be strongly impaired. The ability to meet all these demands depends on several factors, in particular the approach to governance that is adopted. Two approaches in fisheries management are commonly followed: *communicative and consensual action* and *instrumental action* (McGlade and Price 1993). The two approaches treat communication in radically different ways, reflecting the difficulties often encountered by those in power to reach consensus on policy objectives.

Within *communicative and consensual action*, moral rules or norms are established and form the set of consensual actions of the participants. Local people largely determine and implement the actions, although these are usually symbolically mediated under some other guise. In a number of fishing communities, moral rules, norms, and guiding principles are provided by a religion, deity, or faith (Berkes 1989).

Instrumental action is where a technical rule is developed, which then leads to an action. A good example from fisheries is the perceived desirability of establishing an exploitation level that will not harm the parent stock (FAO 1993, 1997); instrumental actions include the adoption of appropriate mesh sizes or the enforcement of quotas. Unlike the first approach, both the instrumental action and underlying rule may become enshrined and go largely unchallenged by resource users and beneficiaries. This aspect has been highlighted over a number of years, where some of the more widely accepted technical rules have been seen to fail (McGlade and Shepherd 1992; Healey 1997; Hutchings et al. 1997a, b). One of the reasons for their failure is that while such actions can help to achieve stable markets and availability of supplies, they exacerbate the problems of conservation by first potentially improving the profitability of a fishery and then attracting more participants, thereby reducing its profitability. As efficiency improves, fishers will often try to evade regulations, thereby nullifying the positive effects of an instrumental action.

It is clear that the effectiveness of any form of governance depends on good communication, coordination, and integration between the various institutions, users, and beneficiaries. Time and again the importance of this has been underestimated in fisheries, leading to widespread dissatisfaction and scepticism about the ways and forms of intervention in management. The criticisms, which come mainly from those interested in the practical forms of interventions, stress the low level of bureaucratic effectiveness and a diminishing acceptance of much of what government agencies are doing. Ideas of decentralization, networks, and community-based management are often promoted as an alternative, the assumption being that at lower levels of governing, the close contact of those governing and those being covered in terms of problems and potential solutions will give the best results. This is the antithesis to governance via communicative action, where a real dialogue and communication is established throughout the governing process and structure.

User Participation

The type of governance most likely to succeed is thus one in which the complexity, dynamics, and variety of ecological and social processes present in the system are taken into account. Just as aspects of problems hang together in quite specific ways, so do solutions. Governance should therefore be formulated in terms of interaction levels. Virtually no significant sociotechnical problem can be handled by a single participant, either private (market) or public (bureaucratic/political). Solutions to these major problems, especially in the field of environmental resources, depend on some form of interaction between semipublic, public, and private actors, organizations, and authorities. Successful governance requires continuous interactions between participants on different levels with different responsibilities and different but interdependent tasks.

The key issue is the question of autonomy of the state and the way in which users interact with the state in developing and changing existing systems of governance. The extremes of the typology are complete state intervention and, at the other end, self-determined fisher power. Models produced by Hance et al. (1988) and others provide a useful guide to intermediary areas. The first category is *state intervening*, where the government acts without communicating with fishing communities. The next category is *informing*, in which participants listen and the government agency talks via hearings and meetings. There is little effort to bring the participants into any policy planning until late in the process. The next category is *first-level consulting*, in which fishing communities have demanded a right to be heard and the government in response asks for limited input but prefers not to listen. Public hearings are a typical example of this procedure. The extension of this leads to *second-level consulting*, where members of the fishing community are asked for detailed input to which government indicates its willingness to listen and discuss as it develops policy. Confusion often arises between these two levels because of expectations by participants that they are involved in the latter when government is operating under the former. An intermediate area, which is equally frustrating for participants, is *representing*, where individuals or representatives are asked to serve as advisors on committees simply to legitimize programs.

The next category, *comanaging*, involves power sharing and represents a partnership between government agencies and citizens with a stake in the common pool resource. Some of the recent success attributed to comanagement is only a reflection of individuals talking to one another, building up on existing frameworks of power rather than a real shift in autonomy. However, successfully comanaged fisheries that are well documented include the inshore fisheries of Japan (see Ruddle 1989) and the Lofoten winter cod fisheries in Norway (Jentoft and Kristoffersen 1989) and Nova Scotia (Apostle et al. 1993). An extension of this category is community-based management, which relies on self-involvement in the management of resources, where the community takes responsibility for the monitoring, conservation, and enforcement, and where the whole community is involved, including those not directly participating in the fishery. An example of this can be found in the Caribbean island of Saba, where the establishment of a Marine Park Area with designated fishing areas is supported by the whole island.

The final category is *fisher power,* where resource exploitation is undertaken in the absence of reference to government or state. This supposedly unregulated situation may, however, be the only one possible in many developing country fisheries, given the lack of resources or low priority to fund a state agency. It may or may not lead to overexploitation (*sensu* Hardin's tragedy of the commons), depending on the need for local food resources and the presence of local rites and customs.

Local communities in modern, highly developed, and well-diversified societies are dependent on and integrated into a wider socioeconomic and political system. In a limited world, however, there is need to restrain exploitation of resources to a locally sustainable level. The central question is which resource management responsibilities and powers should be placed at which levels of organization within the public and or private domain.

Frameworks of Power

Two figures who played a key role in mapping out our understanding of power and hence the rights of governments were Hobbes and Machiavelli. Where Hobbes (in *Leviathan*) (Hobbes 1651) and his successors endlessly legislated on what power was in terms of sovereignty and community, Machiavelli (in *The Prince*) (Machiavelli ca. 1514) and his successors interpreted what power could do, thus initiating a concern for strategy and organization. In fisheries we can derive a number of regimes for fisheries (generalizing from McCay 1993):

- *Laissez-faire:* there is virtually no governance nor is there any effective market-based regulation. This is typical of a situation that would be caricatured as a tragedy of the commons if exploitation pressures were high and where the power lies very much with the individual.
- *Market regulation*: market mechanisms work more effectively; governance is required to uphold private property claims and other conditions. This regime may lead to a misplaced faith in market remedies where externalities and long-term, indirect ecological effects are concerned. Here, power is achieved through the collective behaviors of those with direct vested interests.
- *Communal governance:* the existence and potential of user-governance and local-level systems of common pool resource management is highlighted, irrespective of whether rights are common or private or a mixture. Collective action challenges may be met and is therefore of particular relevance where the interests of people and the sustainability of their resource use is not well served by governments and where privatization is not feasible or politically acceptable.
- *State governance:* the state of government has a central role with respect to common pool resources, whether state owned or not. State property can be property owned outright and used exclusively by agents of the state, or property deemed public over which the state exercises governance. The power lies very much with the sovereign state.
- *International governance:* there are features and challenges to common pool management under state governance, such as the lack of centralized enforcement, as defined by Young (1994). The free-rider syndrome is a very difficult issue for international institutions as is their reliance on persuasion and indirect penalties for breaking the rules. The International Whaling Commission and Tuna Commissions are examples of this.

- *Virtual governance:* this form is characterized by a group of individuals who meet to determine their own priorities and reactions in response to policy. They have no legal power and may not have a direct vested interest, but no state governance is undertaken without reference to their collective opinions. As discussed above, associated with this is the emergence of swarms among NGOs.

Governing and the Regulatory Ladder

If we look once again at governing as the active process whereby effective governance emerges, we can identify three stages: *first-order governing*, wherein problems and opportunities are identified; *second-order governing*, in which there is a balancing of needs and capacities and chances for cooperation are sought; and *meta-governance*, which takes forward governing via three modes — hierarchical, market-based, and participation. The regulatory ladder needs to be constructed alongside these processes so as to better ensure their success. Thus international conventions, such as the Montreal and Kyoto protocols, are often about the need to identify what the problem actually is and what opportunities exist for solving it. Legal structures take such protocols one step further so that regulations can be enacted and enforced. Societal pressure can then be brought to bear via "naming and shaming" offenders, although ultimately hypothecation of taxes and mitigation are likely to be the best approaches to many issues of sustainability in fisheries, with communication and participation at the heart of any successful regulatory process.

CONCLUSIONS

We are faced with a global paradox: as the world integrates economically, it is becoming ever more obvious that its component parts are becoming more numerous and smaller and yet more important. The global economy is growing while the size of its critical parts is shrinking. Each nation must adopt a strategy consistent with its resources and capacities to achieve its individual goals and, at the same time, cooperate regionally and internationally to organize collective solutions to global issues of sustainability. In a world of increasingly interlinked institutions, societies, and economies, coordinated efforts and shared responsibilities are essential.

Even though world food production has grown faster than population, every day one in five people in the developing world cannot get enough food to meet their daily needs; in 17 African countries the figure is two or three people out of five. Most of those affected are in low-income food-deficit countries that lack the hard currency to buy enough food on the international market to make up the shortfall, but there are also food shortages in more affluent, developed countries, in cities as well as in rural areas. Given the proximity of two-thirds of the world's population to the coastal zone, fisheries will inevitably continue to play a role in providing the only source of protein for many people. However, to achieve this, there will have to be a radical shift in our understanding of how these resources should be managed and exploited in order that the greatest benefit for all can be achieved.

In this chapter, I have laid out the various elements needed for such a shift. The task is more one of rethinking the problems of the past few decades to see them with new eyes, rather than placing more and more layers of control onto the system. We also need to accept that there is a second-order science, which is more relevant to solving key problems in fisheries. This second-order science requires the integration of socioeconomics into the base-line of scientific evidence, so that risk and precaution can be properly dealt with. This means that the values, ends, and sometimes expression of beliefs about fish resources need to be properly embedded in the analysis of management, and a distinction made between technical rules and the processes that lead to participation and consensual action.

As with other renewable resource-based industries, fisheries need to be examined in a disaggregated way (i.e., at international, national, regional, local, household levels), for the real constraints to growth and development lie at the boundaries between levels. At the local level, people who depend solely on natural resources, such as local inshore fisheries, generally have no adapted institutions or technology to enable them to bypass adverse economic or environmental conditions. At the international level fish are traded as a commodity for foreign exchange, via institutional structures that have little to do with conserving the resource itself.

In setting out a framework for sustainability of fisheries, it has not been possible to stress the critical role of women, the need for revitalization of rural and coastal areas, elimination of unsustainable patterns of consumption and production, and stabilization of world trade. However, these are also elements of the solution. Conflict, terrorism, corruption, and environmental degradation also contribute significantly to the problems of fisheries. There is clearly a need for urgent action at the level of individual governments, to develop enabling environments and institutions with policies that ensure peace, social, political and economic stability, equity and gender equality, as well as at the level of the international community itself to ensure active cooperation.

The future of fisheries will be determined by the actions that national governments undertake, often in collaboration with NGOs and members of civil society and with external aid where necessary, to improve the well-being of their citizens. Governments are thus responsible for creating enabling environments for private and group initiatives to devote their skills, efforts, resources, and investment towards the common goal of security for all. Experience has shown that this needs to be undertaken with the cooperation and participation of all parts of society in mutually agreed forms of governance. Without these changes, fisheries are likely to fall far short of their real potential as a common resource.

REFERENCES

Allen, P.M., and J.M. McGlade. 1986. Dynamics of discovery and exploitation: The case for the Scotian Shelf groundfish fisheries. *Canad. J. Fish. Aqua. Sci.* **43**:1187–1200.

Allen, P., and J.M. McGlade. 1987. Modelling human behaviour. *Euro. J. Op. Res.* **30**:147–167.

Apostle, R., B. McCay, and K. Mikalsen. 1993. Overcapacity and privatisation: The case of ITQs in the Scotia-Fundy groundfish fisheries. Paper presented at the 12th Ann. Conf. of the Intl. Soc. for the Study of Marginal Regions.

Arrow, K.J. 1963. Social Choice and Individual Values,2nd ed. New Haven: Yale Univ. Press.

Berkes, F., ed. 1989. Common Property Resources, Ecology and Community-based Sustainable Development. London: Belhaven.

Bhaskar, R. 1978. On the possibility of social scientific knowledge and the limits of naturalism. *J. Theor. Soc. Behav.* **8**:1–28.

Burns, T.R., and H. Flam. 1987. The Shaping of Social Organisation. Social Rule System Theory with Applications. London: Sage.

Christensen, V., and D. Pauly. 1992. ECOPATH II — A software for balancing steady-state models and calculating network characteristics. *Ecol. Model.* **61**:169–185.

Coase, R.H. 1960. The problem of social cost. *J. Law Econ.* **3**:1–44.

Elster, J. 1983. Sour Grapes: Studies in the Subversion of Rationality. Cambridge: Cambridge Univ. Press.

FAO (Food and Agriculture Organisation of the United Nations). 1993. Reference points for fishery management: Their potential application to straddling and highly migratory resources. FAO Fisheries Circular No. 864 FIRM/C864. Rome: FAO.

FAO. 1995. Code of Conduct for Responsible Fisheries. Rome: FAO.

FAO. 1997. Technical Guidelines for Responsible Fisheries. 4. Fisheries Management. Rome: FAO.

Geertz, C. 1983. Local Knowledge. New York: Basic Books.

Giddens, A. 1984. The Constitution of Society. Cambridge: Polity Press.

Giddens, A. 1990. The Consequences of Modernity. Cambridge: Polity Press.

Grotius, H. 1609. Mare Liberum. Of the Freedom of the Sea. Latin and English edition, trans. Ralph van Deman Magoffin. Oxford: Oxford Univ. Press, 1916.

Habermas, J. 1987. The Philosophical Discourse of Modernity. Cambridge: Polity Press.

Hance, B.J., C. Chess, and P.M. Sandman. 1988. Improving dialogue with communities: A risk communication manual for government. New Brunswick: Environmental Communication Program, New Jersey Agric. Exp. Station, Rutgers Univ.

Healey, M.C. 1997. Comment: The interplay of policy, politics, and science. *Canad. J. Fish. Aqua. Sci.* **54**:1427–1429.

Hobbes, T. 1651. Leviathan, or the Matter, Forme and Power of a Common-wealth Ecclesiasticall and Civill. Printed for Andrew Crooke, at the Green Dragon in St. Paul's Church-yard, London. London: Penguin Books, 1985.

Hutchings, J.A., C. Walters, and R.L. Haedrich. 1997a. Is scientific inquiry incompatible with government information control? *Canad. J. Fish. Aqua. Sci.* **54**:1198–1210.

Hutchings, J.A., C. Walters, and R.L. Haedrich. 1997b. Reply: Scientific inquiry and fish stock assessment in the Canadian Department of Fisheries and Oceans, and: Reply: The interplay of policy, politics, and science. *Canad. J. Fish. Aqua. Sci.* **54**:1430–1431.

Jentoft, S., and T. Kristoffersen. 1989. Fisherman's co-management: The case in the Lofoten fishery. *Human Org.* **48**:355–365.

Kooiman, J., ed. 1993. Modern Governance. London: Sage.

Lasch, C. 1985. The Minimal Self. London: Picador.

Lyotard, J.-F. 1985. The Post-modern Condition. Minneapolis: Univ. of Minnesota.

Machiavelli, N. 1514. 1961. The Prince. (Italian original title *Il Principe* published by Feltrinelli, Milan, in the volume of the *Opere* containing *Il Principe e Discorsi*). London: Penguin Books, 1981.

Mackinson, S., M. Vasconcellos, and N. Newlands. 1999. A new approach to the analysis of stock-recruitment relationships: "Model-free estimation" using fuzzy logic. *Canad. J. Fish. Aqua. Sci.* **56**:686–699.

McCay, B. 1993. Beijer Discussion Paper Series No. 38. Stockholm: Beijer Intl. Institutes of Ecological Economics.

McGlade, J.M. 1989. Integrated Fisheries Management models: Understanding the limits to marine resource exploitation. *Am. Fisheries Soc. Symp.* **6**:139–165.

McGlade, J.M., ed. 1999. Advanced Ecological Theory: Principles and Applications. Oxford: Blackwell Science.

McGlade, J.M., and A. Hogarth. 1997. Sustainable management of coastal resources. *Hydro* **1**:6–9.

McGlade, J.M., and A.R.G. Price. 1993. Multidisciplinary modelling: An overview and practical implications for the governance of the Gulf region. *Mar. Poll. Bull.* **27**:361–377.

McGlade, J.M., and J. Shepherd. 1992. Techniques for biological assessment in fisheries management. Berichte aus der Ökologischen Forschung, vol. 9. Jülich: Forschungszentrum Jülich.
Pauly, D., and V. Christensen. 1995. Primary production required to sustain global fisheries. *Nature* **374**:255–257.
Pigou, A.C. 1946. The Economics of Welfare. 4th ed. London: Macmillan.
Renn, O., et al. 1999. On science and precaution in the management of technological risk. Ispra: European Community Forward Studies Unit, ETSO Network, JRC (Joint Research Centre) of the European Community.
Ruddle, K. 1989. Solving the common-property dilemma: Village fisheries rights in Japanese coastal waters. In: Common Property Resources, Ecology and Community-based Sustainable Development, ed. F. Berkes, pp. 168–184. London: Belhaven.
WCED (World Commission on Environment and Development). 1987. Our Common Future. Oxford: Oxford Univ. Press.
Young, O.R. 1994. International Governance: Protecting the Environment in a Stateless Society. Ithaca: Cornell Univ. Press.
Zadeh, L.A. 1965. Fuzzy Sets. *Info. Control* **8**:338–353.

20

Inventing Governance Systems That Respond to Coastal Ecosystem Change

S.B. OLSEN

Coastal Resources Center, University of Rhode Island, Graduate School of Oceanography, Narragansett, RI 02882–1197, U.S.A.

ABSTRACT

Coastal governance comprises the policies, laws, and institutions that together respond to the transformations of coastal ecosystems that are being brought about by anthropogenic forces. Coastal governance is adaptive and dynamic and must be rooted in sustained learning. This chaper examines the tensions between a governance process that integrates participatory democracy with the generation of knowledge on the processes of ecosystem change that it attempts to address. The term ecosystem, as used here, includes ecological, economic, and institutional components. Recommendations are offered for how the linkages between governance process and knowledge on ecosystems function and change might be strengthened. Particular attention is given to developing nations in the tropics, where the pace of coastal ecosystem change is most rapid and governance institutions are particularly fragile.

INTRODUCTION

The GESAMP report, "The Contributions of Science to Integrated Coastal Management" (GESAMP 1996), offers a simple conceptual framework for tracing the evolution of integrated coastal management (ICM) initiatives and analyzing the contributions of the natural and social sciences in each step of the process (see Figure 20.1). GESAMP reinforced that knowledge from both the natural and social sciences is required if coastal governance issues are to be analyzed and acted upon effectively.

There are many descriptions of the phases or steps by which coastal governance initiatives evolve (Chua and Scura 1992; GESAMP 1996; Cicin-Sain and Knecht 1998; UNEP 1995). GESAMP selected the most essential and stripped-down description, which emphasizes that

Science and Integrated Coastal Management
Edited by B. von Bodungen and R.K. Turner

ICM Policy Cycle

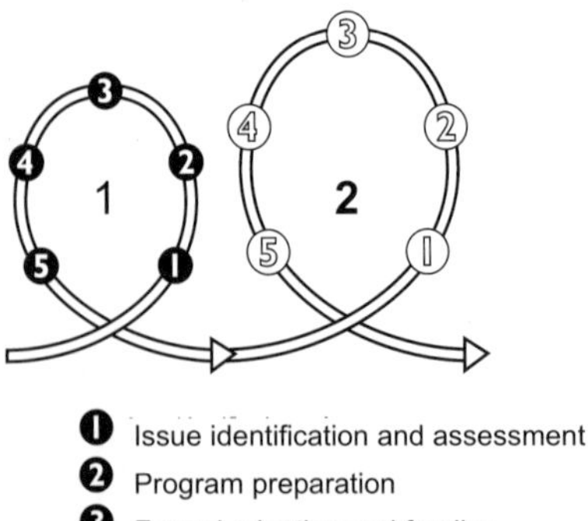

Figure 20.1 The steps in the ICM cycle. From GESAMP 1996, as adapted in Olsen et al. (1998).

the process is a cycle of learning that proceeds from awareness of a set of problems and opportunities (Step 1) to their analysis, to the formulation of a course of action (Step 2). Next comes the politically charged time (Step 3) when a society, be it a village or a nation, commits itself to new behavior and allocates the resources by which the necessary actions will be implemented. In most settings, Step 3 involves formalization of a policy and plan and the allocation of funds. Implementation of the actions is Step 4. Evaluation of successes, failures, learning, and the reexamination of how the issues themselves have changed rounds out an ICM cycle as Step 5. These five steps may be completed in other sequences, as for example, when an initiative begins with enactment of a law (Step 3) that provides the mandate for analyzing issues and developing a detailed plan of action (Steps 1 and 2). Altering the sequence, however, often comes at the cost of efficiency, as when it becomes apparent that the authorities provided by law are inadequate for implementing the actions that are required. Progress and learning are greatest when there are many feedbacks between the steps that make up a "generation" of coastal management (Olsen et al. 1996, 1999). This deceptively simple policy cycle is useful because it draws attention to the interdependencies between the steps within a generation and between successive generations of management. Diagramming coastal governance at different geographic scales helps diagnose issues and priority needs in a specific place. This is important because much of the planning and analysis (Steps 1 and 2) currently being conducted on coastal change and its implications fails to result in the meaningful implementation of new policies or plans of action. In many regions, particularly in the tropics where coastal change is most rapid and governance processes and structures are weak, what we see are the fragments of many unconnected cycles.

ICM is a form of adaptive management, a concept that first appeared in the natural resource management literature in the mid-1970s (see Holling 1979). ICM requires understanding the interplay between social processes and ecosystem change. To be successful in the face of complexity and uncertainty, ICM initiatives need to be flexible, adaptive, and have the capacity to learn. Adaptive management can overcome some of the obstacles from which traditional management suffers (Gunderson et al. 1995; Holling and Sanderson 1996; Imperial et al. 1993). It is designed to cope with the uncertainty and complexity of natural and social systems by creating spaces in which reflection and learning can occur, allowing management processes to take corrective action and modify behavior in light of new information. As Berman (1980) observes, "The ideal of adaptive management is the establishment of a process that allows policy to be modified, specified, and revised — in a word adapted — according to the unfolding interaction of the management process with its institutional setting."

The practice of adaptive management is based on the ideas that (a) projects and policies are inevitably experiments and should be designed and administered as such; (b) information has value not only as a basis for action but as a product of action; (c) actions can and should be taken in the face of uncertainty and complexity; and (d) management of ecosystems essential to humankind will continue for as long as humankind exists and there is no "final" solution to the management problem (Healey and Hennessey 1994). Some of the strategic changes in attitude that advocates of adaptive management promote are shown in Table 20.1.

THE TWO PILLARS OF INTEGRATED COASTAL MANAGEMENT

Lee (1993) has probed the interplay between knowledge of how ecosystems function and respond to anthropogenic forces and the processes of governance in democratic societies. In this paper, Lee's insights are adapted to conjure up the image of knowledge and a governance process as constituting the two pillars of ICM. The first pillar is a *governance process* that examines the interests of the many stakeholder groups, negotiates plans, policies, and decision making, and then applies enforcement mechanisms that are transparent and accountable to those affected by its actions. The second pillar is the generation and incorporation of the *reliable knowledge* that allows the manager to understand, and sometimes to forecast, the consequences of different courses of action. Lee emphasizes that such knowledge does not flow

Table 20.1 Conventional versus adaptive attitudes about management and policy analysis. Adapted from Walters (1986).

Conventional	Adaptive
1. Seek precise predictions and promote programmed management	1a. Embrace alternative approaches to resolving problems and addressing pertinent issues
2. Presume certainty in seeking best action	2a. Evaluate feedback and learn from failure as well as success and apply those lessons to future program decisions
3. Emphasize short-term objectives	3a. Promote long-term objectives
4. Minimize conflict among stakeholders	4a. Highlight difficult trade-offs and conflicts and build space for multiple viewpoints, consensus building, and negotiation
5. Seek equilibrium	5a. Expect and profit from change

only from "the sciences" but rather is the product of the scientific method. Thus, a learning-based approach requires the objective and careful analysis of management policies and management actions to specific issues addressed by an ICM initiative (e.g., specific measures to address a problem of overfishing or eutrophication) and to generations of the ICM cycle. From this perspective, management policies and actions need to be viewed as experiments based upon clearly stated hypotheses and evaluated by suitable indicators selected to probe the purposes and expected outcomes of the policies that are implemented. As stated by Lee (1993), "Without experimentation, reliable knowledge accumulates slowly, and without reliable knowledge, there can be neither social learning nor sustainable development."

Healey and Hennessey (1994) have examined the utilization of scientific information in the management of estuaries in the U.S. They concluded that while an adaptive approach that integrates "science" with all the steps in the management cycle is the best option, it is seldom practiced. Often, science and management proceed as "two solitudes" with little meaningful interaction. In some cases, there is episodic contact in which collaboration is typically restricted to the initial steps of issue analysis and planning. The sustained interaction and cross-fertilization recommended by GESAMP (1996) and detailed by Lee (1993) is unfortunately rare in current ICM practice.

An organizing framework for applying the adaptive, learning-based approach to the analysis of ICM initiatives that addresses an analysis of each step in the ICM process and the outcomes of ICM initiatives at different stages of maturity is described in Olsen et al. (1996). A "Manual for Assessing Progress in ICM" (Olsen et al. 1999) and an initial application of this framework to selected case studies is available (Olsen et al. 1998). These documents focus upon the first pillar — the governance process itself. This chapter is focused more on the second pillar — "reliable knowledge" — and particularly on the difficulties inherent in bridging the two pillars of ICM with a resilient lintel. To where does this gateway lead? The hope is that it signals the pathway to sustainable forms of human development. Herman Daly (1996) has suggested that the defining feature of sustainable development is "the replacement of quantitative growth expansion (growth) with qualitative improvement (development) as the path of future progress." Of the many definitions in circulation, this one appears to best capture the essence of the sustainable development concept.

THE DIMENSIONS OF COASTAL CHANGE

According to recent estimates (Cohen et al. 1997), almost half of the world's population lives within 150 km of a coastline on less than 20 percent of the planet's nonpolar land. Demographic trends suggest that the proportion of the world's human population that will live in coastal regions will increase within the next half century at a time when the numbers of people is likely to grow from the current 6 billion to nearly 10 billion. By the mid-1990s, twelve of the world's fifteen megacities were coastal. Coastal regions attract human populations because they are focal points for economic growth. They account for more than half of humanity's infrastructure for manufacturing, transportation, energy processing, tourism, communications, and other services and probably, therefore, a similarly disproportionate share of global consumption and waste production. High population growth, combined with endemic poverty in some regions and increasing consumption in others, is producing losses in important qualities of coastal ecosystems — often the very qualities that attract people to them. The symptoms include declining water quality, degradation or destruction of critical habitats, decline and collapse of fisheries, and losses in biodiversity. These losses combine to generate user conflicts and pose unprecedented challenges for institutions with coastal governance responsibilities (Olsen et al. 1998).

Within this context of sustained change, the challenge is not to define and achieve a static and optimal "mix of products and services." It is rather to engage in a dynamic, iterative process that works to modify societal behavior so that it can adapt more efficiently to rapidly changing circumstances. A definition of ICM that emphasizes these features (GESAMP 1996) is:

> "A continuous and dynamic process that unites government and the community, science and management, sectoral and public interests in preparing and implementing an integrated plan for the protection and development of coastal ecosystems and resources."

THE FIRST PILLAR OF ICM: THE GOVERNANCE PROCESS

Considering the enormity of the challenges that confront those attempting to promote responsible and effective responses to coastal change, it is useful to consider the differences between management and governance. Management is the process by which human and natural resources are harnessed to achieve a known goal within a given institutional structure. Governance, on the other hand, generates the fundamental goals and the institutional process and structures that are the basis for planning and decision making. Governance, therefore, sets the stage within which management occurs (Olsen and Christie 2000). For most coastal ecosystems, neither the goals nor the institutional structures for progressing towards sustainable forms of coastal development have yet been invented. Those engaged in negotiating the goals and inventing the institutions to achieve them are therefore engaged in governance, not mere management. However, since the term "management" is embedded in the terms *integrated coastal management* and *adaptive management*, it is difficult to consistently use the two appropriately. Yet the differences are important.

In this chapter, I use the term *ecosystem* to include both nature and its associated human society (see Ngoile et al., this volume). This inclusive definition is a feature of the models of ecosystems offered by Costanza et al. (1997) and the analyses of Lubchenco (1998). As defined by Slocombe (1993), the "ecosystem" implies an overt, systems approach in which human societies are viewed as one element of the planet's living systems. The focus is therefore upon coherent, self-defined, and self-organizing units comprising interdependent ecological, economic, and social components.

Both coastal governance and coastal ecosystems must be conceived as nested systems that range across spatial scales. This requires the successful application of the Subsidiarity Principle which, when applied to coastal ecosystems, calls for placing power and responsibility for planning and decision making at the lowest practical level in the governance hierarchy. Thus, coastal governance can and should be formulated, implemented, and adopted at the scale of individual communities or municipalities as well as at the state (provincial), national, and regional scales. Within individual nations there are major challenges at present that lie in effectively linking the goals, institutional structures, and decision-making processes at the community level with national policy, national plans, and national institutions. In developing nations, one approach to instigating coastal governance termed comanagement, or the two-track approach, adopts a strategy of formulating actions simultaneously at the community and national scales. This strategy typically begins a national coastal management initiative with pilot projects at selected sites that define and analyze the issues that must be addressed and formulate new approaches to resolving them at that small scale. The large amounts of money often invested in these initial pilots is justified on the assumption that success will be replicated and eventually produce a coherent and effective coastal management plan and decision-making procedures that encompasses the nation as a whole. National actions must in turn be integrated to address regional and global problems that require collective action at a multinational scale. This is where international conventions and the initiatives sponsored by the Global Environmental Facility (GEF) come into play. Yet the current reality is that ecosystem governance of all kinds, and coastal governance in particular, are not nested across scales and are full of contradictions and gaps. What does one do? There are arguments among those who advocate top-down approaches and those who argue for bottom-up. I suggest that a combination of the two is required.

The coastal ecosystems (containing both social and ecological components) that are the subject of coastal governance are also nested across spatial scales. Ideally, national and regional ICM should encompass areas that extend from the headwaters of catchments to the outer limits of Exclusive Economic Zones (GESAMP 1996). In practice, much of the best ICM is currently at far smaller scales and addresses only fragments of ecosystems (as in shorefront management that is the focus of many coastal zone management programs in the United States) or individual coral reef/lagoon/seagrass/mangrove systems or estuaries. Efforts to manage large systems are fraught with difficulties as illustrated by efforts in the Baltic and Black Sea (see Mee as well as Elmgren and Larsson, both this volume). Yet some successes are seen at larger scales, such as in the Chesapeake Bay Program and the binational program that has successfully addressed the drastic losses in the qualities of the North American Great Lakes that became apparent in the 1960s. From a global perspective, governance initiatives which attempt to define the changes in human activity needed to conserve the qualities of enclosed seas and gulfs, show much activity in problem analysis and negotiation but little in reversing downward trends. A multitude of small, but linked, local level actions will be needed to produce the desired results. Here again, the long-term goal is to recognize that ecosystems, like governance, are nested together and thrive when there are abundant positive feedback loops across spatial scales.

THE SECOND PILLAR OF ICM: RELIABLE KNOWLEDGE

As suggested by Lee (1993), the second pillar is knowledge on how coastal ecosystems function and change. It is essential to recognize that such "reliable knowledge" must be drawn from both natural and the social sciences and incorporate knowledge derived from the experience of the governance itself. Thus, the knowledge that informs ICM comes from a diversity of sources and may specifically include "traditional knowledge" when this is shown to be based on sound interpretations of observed phenomena and when it offers insights into the beliefs and values of a society.

Science contributes to the coastal management process by providing objective information that makes the debate over contentious issues an informed debate. Science can inform on the status and trends of coastal ecosystems and the causes and consequences of change. Science gives legitimacy to particular policy options or lines of argument. Science does not, however, provide all the answers. However intensively and extensively data are collected, however much we know of how the system functions, the domain of our knowledge of specific ecological and social systems is often small when compared to that of our ignorance. An important challenge of ICM is to cope with the unknown, the uncertain, and the unexpected and to design decision-making systems that can function in the presence of gaps in information and understanding. The reality is that managers must make decisions, whether or not unequivocal scientific information is available. The challenge is to identify, locate, and organize information in ways that will make it accessible and usable in the ICM decision-making process.

THE DIFFICULTIES OF LINKING THE TWO PILLARS OF ADAPTIVE MANAGEMENT

Sustained success in adaptive management requires linking the two pillars, and this involves many difficulties. First, there are major discontinuities in the scales of time and space within which governance cycles and ecosystem change occur. Second, there are the differences in the attitudes, the reward structures, and interests of those who identify with the policy process and those who pursue the rewards of a career as a scientist.

Linking Across Temporal Scales

Holling (1995) describes the process by which ecosystems evolve as a cycle comprising phases of exploitation, conservation, release, and reorganization. This sequence of phases plays out over many scales of time and of space. Thus, the cycle of change in an abandoned field or a pond may be observed over a few decades or less while the completion of a cycle in a watershed may span centuries or millennia.

Perhaps the greatest tension between the two pillars of adaptive management lies in the mismatches between cycles of learning in governance and the cycles by which the coastal ecosystems that are the subject of governance change. One mismatch lies in time. In coastal regions, the expressions of anthropogenic ecosystem change are similar or identical in both tropical and temperate regions, and the impacts on society have been repeatedly documented. The ubiquitous pattern includes reductions in the permanent vegetative cover in a watershed, changes in the volumes, quality, and timing of flows of fresh water to estuaries, reductions in the area of wetlands, destruction or degradation of estuarine habitats of crucial importance to fish and wildlife, overfishing, and severalfold increases in nutrient loading. In temperate regions, the sequence of anthropogenic change has often progressed over many human lifetimes, and "the way it was" (i.e., within the consciousness of the society and those engaged in resource management) spans only a small segment of a cycle of ecosystem change. In the tropics, the sequence is frequently being telescoped into a few decades, and the drama of what is occurring and its implications for society are more obvious. In both settings, the temporal mismatches between the cycles of coastal ecosystem change and cycles of coastal governance are usually large. Yet it is crucially important to locate coastal governance initiatives within the longer-term cycles of ecosystem change. The usual practice is to examine trends in the ecosystem for time periods only somewhat longer than the likely time span of the governance initiative. In politically stable settings, it is common for trends of ecosystem change to be examined over two to four decades, and the governance response is designed to modulate that fraction of the full cycle as if it was isolated from the larger process. In the tropics, where governmental systems tend to be less stable, management initiatives are typically conceived and executed as four- to ten-year "projects." Rather than contributing to a coherent and sustained program of governance, these projects are usually conducted in isolation from one another and are justified and evaluated on impacts discernible in a handful of years. Estimating trends over longer periods may be difficult but is not impossible. Where governance of ecosystems at an intermediate scale is vigorous and has assembled strong constituencies (the Great Lakes, Chesapeake Bay, The Netherlands), placing governance as a response to change over longer periods has been crucially important.

Linking Across Spatial Scales

Equally important are mismatches in spatial scale. Protecting an individual wetland and regulating the outflows of polluting industries along an estuary has a minor impact when nonpoint pollution flows and change to the hydraulics of a watershed by the construction of dams, logging operations, and the like neutralize the impacts of localized efforts. Thus, better integration of both reliable knowledge and governance across scales of space and time is essential. Modeling can fill this need especially when it helps integrate our understanding of the implications of coastal ecosystem change and governance options across scales of space and time.

Involving Scientists in the Governance Process

The second cluster of difficulties lies in the realm of differences in the values, knowledge, and skills of managers, planners, and politicians in the first pillar and scientists in the second. These differences and their consequences have recently been examined in several reports and articles (e.g., GESAMP 1996; NRC 1995, 1997). The consequences of the differences can be examined as they relate to the different actions associated with each step in the ICM policy cycle and how these, in turn, affect the success of completing a full cycle of coastal governance and linking it to a subsequent generation of effort. One of the many hurdles lies in understanding the importance of political salience. Herein lie the reasons why so many analyses of coastal issues and technically sound coastal management plans fail to gain the official endorsements and win the institutional commitments and funds required for implementing a plan of action. These crucial events play out in Step 3. Success lies first in gaining a place on the political agenda, defined by Kingdon (1995) as "the list of subjects or problems to which government officials, and people outside of government closely associated with those officials, are paying some serious attention at any given time." To win such attention, a coastal governance initiative must be perceived by a sufficiently large and powerful constituency as important. Furthermore, the proposed course of actions must be seen to be feasible and its likely consequences sufficiently attractive to reward those responsible for making it happen. Judgments are made as much on the basis of values and beliefs than "the facts of the matter." Furthermore, formal adoption of a new coastal governance program typically affects the distribution of authority and power among institutions, interest groups, and politicians. This triggers defensive behavior and bureaucratic maneuvering. This process is highly distasteful and mysterious to many natural scientists. It does not help that scientists are not professionally rewarded for becoming involved in these critical elements of the governance process (Lubchenco 1998).

CONDITIONS THAT FOSTER SUCCESSFUL LINKING

Boesch (1996, 1999) has identified factors that contribute most directly to a successful interpretation of science into a sustained coastal governance process. These, with minor modifications, can be stated as follows:

1. *Evidence of significant change.* Coastal management initiatives are triggered by change in a coastal ecosystem. Such change can be in the form of both threats and opportunities and must be perceived to be significant enough to warrant the attention of society.
2. *Reliable and valid indicators have been selected* and information on those indicators has been assembled to analyze and communicate the processes of change that the initiative will attempt to address.
3. *A measure of consensus exists within the scientific community.* Action will seldom be justified in the face of major, well-substantiated differences on the significance of the change, its likely causes, and its potential implications for society.
4. *Forecasting capabilities exist*, often in the form of models that serve to motivate and guide management actions.
5. *Effective and feasible responses* to change are known, operationally feasible, and the constituencies that will work to implement them are present.

It is instructive to place this set of preconditions for the success of the integration of science within the six preconditions critical to the successful implementation of a program identified by Mazmanian and Sabatier (1979, 1981). These preconditions suggest that only the first two of the six success factors concern the relationship between the first and second pillar:

1. Clear and consistent policy objectives
2. Convincing science in support of those objectives
3. Sufficient jurisdiction and authority
4. Good implementation structures
5. Competent and committed staff
6. A priority position on the policy agenda

A highly respected social scientist, Elenor Ostrom (1999), who analyzed the governance of common property resources, reaffirms the crucial importance of indicators that provide reliable and valid information to characterize the condition of a resource or ecosystem. She also emphasizes the importance of predictability — even if it is limited to predicting that the resources or ecosystem qualities of concern will be present long enough to warrant a management initiative.

These insights on preconditions to success in both science and governance help us understand why progress is so elusive in the vast majority of the world's coastal regions. Particularly in developing nations, reliable documentation of change early on in the transformation process is seldom available, even when such change is obvious to local residents. Reliable data that traces changes in, for example, vegetative cover, water quality, the abundance of estuarine fish and shellfish, frequently does not exist. Similarly, Mazmanian and Sabatier's success factors for implementation can rarely be met along tropical coastlines and are also difficult to achieve in many politically stable, wealthy nations. Worse yet, "reliable and valid indicators" have been identified in only a few instances. What are indicators for the condition of an estuary, a wetland, or watershed? To be useful, the protocols for collecting and analyzing the necessary data must be agreed upon and achievable at a reasonable cost. We are far from agreeing on what indicators should be tracked and what frequencies of sampling are necessary to estimate change across a sequence of scales. In the tropics, where agreement on such basics is most urgently needed, models of ecosystems, or of components of ecosystems that can be used to forecast "what if" scenarios, are rarely available to coastal managers. Little wonder that scientists tend to feel isolated from the ongoing governance process and decision making that determines the trajectory of coastal development.

FIRST STEPS IN CONSTRUCTING THE LINTEL THAT LINKS THE TWO PILLARS

The lintel that must join coastal science with coastal governance is absent, or weakest, in developing nations, where anthropogenic change to coastal systems is proceeding most rapidly. Reflecting on the enabling conditions that are conducive to progress towards effective ecosystem governance, three categories of actions emerge as potentially fruitful if we are to overcome the current difficulties:

1. *Selection of Indicators for the Documentation of Trends in Ecosystem Quality.* There are three interrelated needs:
   a. To broker a measure of consensus with the community of natural scientists concerned with coastal ecosystem change on a set of valid and reliable indicators that can be used responsibly and at a reasonable cost to document change in the qualities of coastal ecosystems.
   b. To develop standardized visual methods for conveying trends of coastal change to the public and to decision-makers that document and compare both the societal and natural components of coastal ecosystems.
   c. To promote the formulation and testing of indices of coastal ecosystem condition that combine sets of social and environmental variables into composite measures.

There are techniques for conveying change in natural components that display complex data in easy-to-grasp pictorial forms. These may be adaptable to developing country contexts. For example, in the U.S., the Environmental Protection Agency (EPA) has developed a simplified "report card" format that color-codes variables indicative of ecosystem conditions in estuaries and rivers. Red is used to highlight variables that are of greatest concern or are trending in an undesirable direction. Orange warns of variables that are in transition from one condition to another while green denotes acceptable and stable. The Dutch Ministry of Transport, Public Works, and Water Management (1989) has used the "AMOEBA concept" (Colijn and Reise, this volume) to portray graphically shifts in the abundance of categories of life forms since a baseline year in the 1930s.

In the social sciences, a consensus has emerged on reliable and valid indicators for assessing trends in the condition of a human population (child mortality, life expectancy, literacy). There is, however, no consensus among natural scientists on a comparable set of indicators for gauging for the condition of the "natural" components of coastal ecosystems. These indicators are urgently needed. Creating the conditions for sustained data gathering and data analysis on such indicators is another priority. This requires designing systems for tracking change in ecosystem qualities that allows for nesting analysis across spatial scales. In many developing nations, a combination of citizen monitoring and remote sensing technology could prove to be a powerful combination; however, to my knowledge, this has not yet been attempted as a strategy for documenting change in coastal ecosystems.

If such indicators had been selected, and trend data for them had been assembled, we could make the critically important comparisons between trends in the "natural" elements of ecosystems with trends in the associated human population. At present, convincing natural scientists that both data sets are important and such comparative analysis is useful is surprisingly difficult. For example, an effort has been underway for several years to negotiate a consensus among the community of coral reef scientists so that trends in reef condition worldwide can be documented in a similar enough manner to permit the aggregation of data on condition and trends at regional and global scales (McManus and Vergara 1998). It is proving difficult, however, to persuade the natural scientists involved in this effort to integrate social and governance variables into this system.

Another step is to integrate sets of variables for both "natural" and societal components into indices of ecosystem condition. Here again, social scientists and public health professionals have generated various indices, such as the United Nations' Human Development Index, that are useful when attempting to draw conclusions on progress or its absence. However, indices appropriate for tracking change in the condition of coastal environments, or that combine the natural and societal components of coastal ecosystems, are not being used in coastal governance practice.

2. *Developing Forecasting Capabilities.* In developing nations, where economic growth at almost any cost predominates, it is critically important to increase the scope of the messages that natural scientists are delivering to politicians and society. Too often these messages are a confirmation and documentation of past mistakes and "bad news" that comes from analyzing change that has already occurred. Greater efforts are needed in responsible forecasting—especially where such forecasting integrates across the natural and societal components of ecosystems. When natural scientists do get involved in governance decision making, they are often placed in the position of opposing development options because they see negative consequences for biodiversity, water quality, or habitats important to species other than our own. Often the linkages and interdependencies of their concerns with the societal variables that are usually of greatest concern to politicians and bureaucrats are weak or missing. Using simple trend projections and models to forecast the impacts of typical options in the development path could increase the salience of the messages being delivered by the scientific community into the governance process.

Experience in the governance of estuaries and their watersheds in the United States has demonstrated the power of such analysis. The EPA (1992) has developed ecological risk assessment frameworks and these are examples of simulation models for such variables as water quality and the impacts of hurricanes. In developing countries, where the magnitude and pace of change is greater, such tools are almost unknown. Where they do exist, they usually have been developed to assess the impacts of one-of-a-kind engineering proposals, e.g., a major dam project proposed by an international institution such as a development bank.

3. *A Typology of Contexts for Coastal Governance*. As the density of people in coastal regions and the intensity of human activities both increase, coastal ecosystems appear to progress through a predictable sequence of governance contexts along a natural to engineered ecosystem continuum. If this hypothesis is correct, a typology of coastal governance contexts could be developed to guide the science and the governance that are most likely to be appropriate and feasible for different contexts. An important task is to examine the winners and losers—in terms of both the natural ecosystem and the human society—that are associated with each step along the continuum.

In conclusion, the construction of a stronger connection, or lintel, that unites the governance process (ICM) with reliable knowledge (the sciences) will require agreement on the indicators by which the changing condition of ecosystems can be assessed. Such indicators will be most salient to all those involved in coastal governance when they integrate societal with environmental parameters, either as indices or by visual techniques that reinforce the interdependencies between the societal and the environmental realms. It is the acceptance and use of such objectively verifiable and reliable indicators that provides a shared language and sense of common purpose among professionals collaborating in more mature endeavors such as public health and economic development.

REFERENCES

Berman, P. 1980. Thinking about programmed and adaptive implementation: Matching strategies to situations. In: Why Policies Succeed or Fail, ed. H. Ingram and D. Man. Beverly Hills: Sage.

Boesch, D.F. 1996. Science and management in four U.S. coastal ecosystems dominated by land–ocean interactions. *J. Coast. Conserv.* **2**:103–114.

Boesch, D.F. 1999. The role of science in ocean governance. *J. Ecol. Econ.* **31**:189–198.

Chua, T.E., and L.F. Scura, eds. 1992. Integrative Framework and Methods for Coastal Area Management. ICLARM (International Center for Living Aquatic Resources Management) Conf. Proc. 37. Manila: ICLARM.

Cicin-Sain, B., and R.W. Knecht. 1998. Integrated Coastal and Ocean Management Concepts and Practices. Washington, D.C.: Island Press.

Cohen, J.E., et al. 1997. Letter: Estimates of coastal populations. *Science* **278**:1209c–1213c.

Costanza, R., et al. 1997. An Introduction to Ecological Economics. Boca Raton: St. Lucie Press.

Daly, H.E. 1996. Beyond Growth: The Economics of Sustainable Development. Boston: Beacon.

EPA (Environmental Protection Agency). 1992. Framework for Ecological Risk Assessment. EPA/630/R-92/001. Washington, D.C.: EPA.

GESAMP (Joint Group of Experts on the Scientific Aspects of Marine Environmental Protection). 1996. The Contributions of Science to Integrated Coastal Management. Reports and Studies No. 61. Rome: FAO (Food and Agriculture Organisation of the United Nations).

Gunderson, L.H., C.S. Holling, and S.S. Light. 1995. Barriers and Bridges to the Renewal of Ecosystems and Institutions. New York: Columbia Univ. Press.

Healey, M., and T. Hennessey. 1994. The utilization of scientific information in the management of estuarine ecosystems: The case of Chesapeake Bay. *Ocean Coast. Manag.* **23**:167–191.

Holling, C. 1979. Adaptive Environmental Assessment and Management. New York: Wiley.

Holling, C. 1995. What barriers? What bridges? In: Barriers and Bridges to the Renewal of Ecosystems and Institutions, ed. L.H. Gunderson, C.S. Holling, and S.S. Light. New York: Columbia Univ. Press.

Holling, C., and S. Sanderson. 1996. Dynamics of (dis)harmony in ecological and social systems. In: Rights to Nature: Ecological, Economic, Cultural, and Political Principles of Institutions for the Environment, ed. S. Hanna, C. Folke, and K. Maler. Washington, D.C.: Island Press.

Imperial, M., T. Hennessey, and D. Robadue. 1993. The Evolution of Adaptive Management for Estuarine Ecosystems: The National Estuary Program and its Precursors. *Ocean Coast. Manag.* **2**:147–180.

Kingdon, J.W. 1995. Agendas, Alternatives, and Public Policies, 2d ed., New York: Collins College Publishers.

Lee, K.N. 1993. Compass and Gyroscope: Integrating Science and Politics for the Environment. Washington, D.C.: Island Press.

Lubchenco, J. 1998. Entering the century of the environment: A new social contract for science. *Science* **279**:491–497.

Mazmanian, D., and P. Sabatier. 1979. The conditions of effective implementation: A guide to accomplishing policy objectives. *Policy Anal.* **5**:481–504.

Mazmanian, D., and P. Sabatier. 1981. The implementation of public policy: A framework for analysis. In: Effective Policy Implementation, ed. D. Mazmanian and P. Sabatier, Lexington: Lexington Books.

McManus, J.W., and S.G. Vergara, eds. 1998. ReefBase: A Global Database on Coral Reefs and Their Resources. Version 3.0. Manila: ICLARM.

Ministry of Transport, Public Works, and Water Management. 1989. Water in the Netherlands: A Time for Action. National Policy Document on Water Management. The Hague: Ministry of Transport, Public Works, and Water Management.

NRC (National Research Council). 1995. Science, Policy, and the Coast: Improving Decisionmaking. Washington, D.C.: National Academy Press.

NRC. 1997. Striking a Balance: Improving Stewardship of Marine Areas. Washington, D.C.: National Academy Press.

Olsen, S., and P. Christie. 2000. What are we learning from tropical coastal management experiences? *Coast. Zone Manag. J.* **28**:5–18.

Olsen, S., K. Lowry, and J. Tobey. 1999. A Manual for Assessing Progress in Coastal Management. Coastal Resources Center Report No. 2211. Narragansett: Univ. of Rhode Island.

Olsen, S.B., J. Tobey, and L. Hale. 1998. A learning-based approach to coastal management. *Ambio* **27**:611–619.

Olsen, S.B., J. Tobey, and M. Kerr. 1996. A common framework for learning from ICM experience. *Ocean Coast. Manag.* **37**:155–174.

Ostrom, E. 1999. Self-governance and Forest Resources. Occasional Paper No. 20. Jakarta: Center for Intl. Forestry Research.

Slocombe, D.S. 1993. Implementing ecosystem-based management: Development of theory, practice, and research for planning and managing a region. *BioScience* **43**:612–622.

UNEP (United Nations Environment Programme). 1995. Guidelines for Integrated Management of Coastal and Marine Areas with Special Reference to the Mediterranean Basin. Regional Seas Reports and Studies No. 161. Geneva: UNEP.

Walters, C. 1986. Adaptive Management of Renewable Resources. New York: Macmillan.

Standing, left to right: Jeff Benoit, Michel Meybeck, Bodo von Bodungen, Jacqueline McGlade, Stephen Olsen
Seated, left to right: Kay Emeis, Linda Deegan, Virginia Lee, Alison Gilbert

21

Group Report: Unifying Concepts for Integrated Coastal Management

K.-C. EMEIS, Rapporteur

J.R. BENOIT, L.A. DEEGAN, A.J. GILBERT, V. LEE, J.M. MCGLADE, M. MEYBECK, S.B. OLSEN, and B. VON BODUNGEN

INTRODUCTION

Experience and expertise gained in recent years call for an attempt to identify and document the principles that advance the effective integration of the sciences into Integrated Coastal Management (ICM). Three overarching principles emerged from our discussions as we worked to address the assigned topic of "Unifying Concepts for Integrated Coastal Management":

1. ICM needs to strengthen science policy and systems of governance so that they can respond to expressions of ecosystem change at all social and natural scales.
2. The ICM process requires full participation of scientists, ecosystem managers, and user interests in all steps of the policy cycle to foster improved understanding, communication, and decision making.
3. The subsidiarity and precautionary principles are equally important to the effective evolution of ICM. Subsidiarity is the democratic principle that decisions should be made at the lowest level of society that is practical and consistent with the overall public good. (No decision affecting the lives of others should be undertaken by a government without mandate or by a corporation without authority by government granted by charter or legislation.) The precautionary principle can be defined as follows (Dovers and Handmer 1995): "Where there are threats of serious or irreversible environmental damage, lack of full scientific certainty should not be used as a reason for postponing measures to prevent environmental degradation. In the application of the

Science and Integrated Coastal Management
Edited by B. von Bodungen and R.K. Turner

precautionary principle, public and private decisions should be guided by (a) careful evaluation to avoid, wherever practicable, serious or irreversible damage to the environment; and (b) an assessment of the risk-weighted consequences of various options."

These overarching principles are the distillate of discussions that focused on the following five questions.

QUESTION 1: If we accept that ICM should be an expression of adaptive management, by what means should its practice be conducted as public policy experiments designed for the accumulation of reliable knowledge?

This question centers on measures that can correct the current tendency to decouple natural science and social science in the latter steps of the ICM cycle described by GESAMP (1996) and outlined by Olsen (cf. Figure 20.1, this volume). We discussed how the scientific method (empiricism) can be applied to all steps of the ICM process to maximize learning. Specifically, how can (natural and social) sciences contribute more effectively to Steps 3 (adoption of plan) and 4 (implementation of plan) of the ICM policy cycle? We also discussed how completed cycles of coastal governance can be evaluated (Step 5) as "policy experiments" to assess objectively the strengths and weaknesses of the policies and actions that have been implemented and the adjustments that should be made to an ICM program as it proceeds into a new cycle.

Ideas Presented in the Background Papers

Healey and Hennessey (1994) characterize the utilization of scientific information in ICM as occurring in three modes:

1. The "two solitudes," where science and ICM proceed as unrelated endeavors in which specific management issues are addressed by managers and science provides only a "general enlightenment function." Policy development clearly benefits from science but does not benefit from the direct involvement of scientists. Here, the scientific basis of ICM is relatively weak.
2. The "episodic" mode, where the impact of science varies with the specific stages of the governance process. Both science and scientists are directly engaged in some stages but are excluded from others. Scientific influence is typically strong in the early phases (Steps 1 and 2) and weak in the latter phases (Steps 3, 4, and 5) of the policy cycle. This appears to be currently the dominant mode of scientific interaction with coastal governance.
3. The third mode, termed the "Active Adaptive Management Model," applies the power of the scientific method to enhance learning in all steps of the ICM policy cycle. Lee (1993) sees adaptive management as the most efficient means for generating reliable knowledge drawn from the objective and verifiable interpretation of completion of the policy cycle when these are viewed as "experiments." Lee argues that "without experimentation reliable knowledge accumulates slowly, and without reliable knowledge

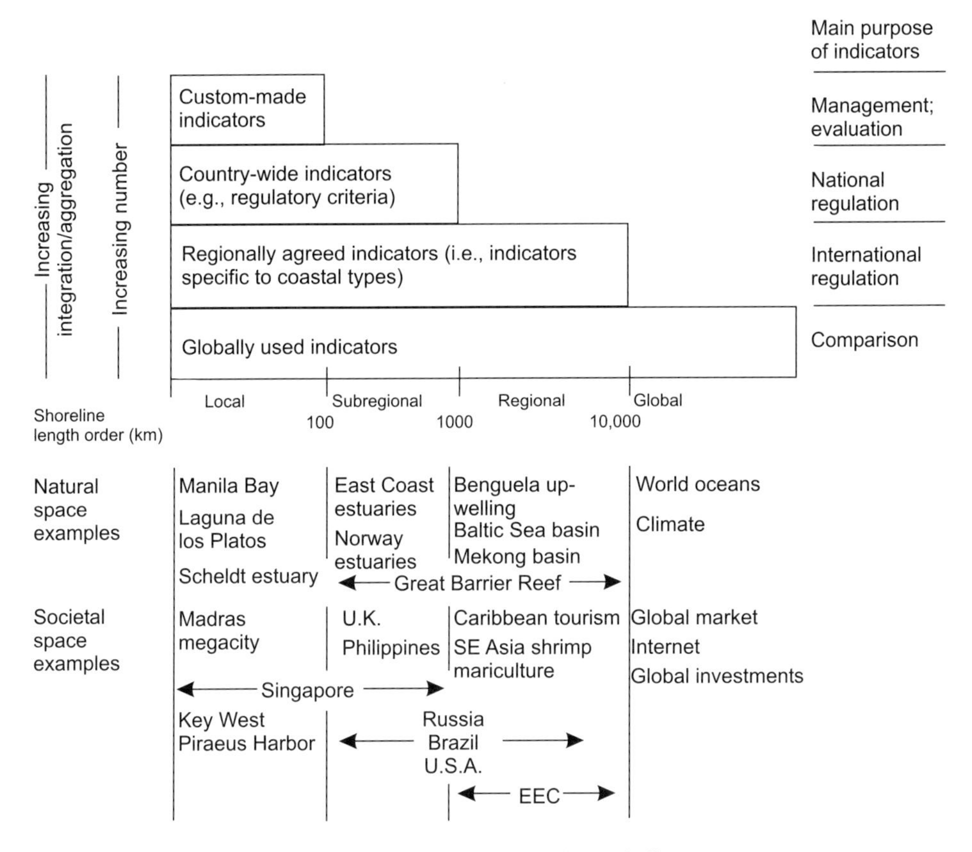

Figure 21.1 Spatial dimensions and the nature of coastal zone indicators.

> there can be neither social learning nor sustainable development." Olsen (this volume) builds on these ideas and suggests that adaptive management rests on two pillars. The first is a process of governance rooted in the traditions of participatory democracy while the second is the generation of reliable knowledge. The governance process provides for the negotiation of conflicting values and conflicting needs and can yield decision making at a societal scale that is resilient and sustainable, because it is perceived to be just and holds out hope for a desirable future.

Several papers prepared for this conference (including Olsen and Ngoile et al., both this volume) argue that the adaptive management of coastal ecosystems must define the "ecosystem" as including both people and their associated environment. This inclusive definition underscores the importance of contributions from both the social and natural sciences to ICM.

Topics of Discussion

We recognized that resistance to the adaptive management model is often present among both scientists and managers. Some scientists believe that it is professionally inappropriate to become actively involved in the decision-making process (Steps 3 and 4) since this invariably involves making judgments based as much on values as objective "facts." Within the management community there are those who believe that policy implementation (Step 4) involves the application of "blueprints" that require only slight adjustments to meet the conditions of a particular place. From this perspective, ICM is the application of technical fixes to technical problems rather than a sustained process of adaptation and learning. Furthermore, a learning-based approach requires examining and highlighting failures as well as successes. This can be distasteful to those responsible for the design and implementation of an ICM project or program.

This led to a discussion of likely resistance among the public and within government agencies to ICM activities presented as public "policy experiments." The difficulty is overcome when ICM actions are designed and presented to encourage "learning." The public will usually welcome an approach to problem solving that permits periodic reviews of progress and adjustments to the management strategy. The public accepts this approach in medicine. New advances in cancer research, for example, are eagerly anticipated, and the testing of new drugs on sample populations is accepted as necessary and desirable even though the consequences for those involved may be severe. Several workshop participants were concerned that scientific uncertainty on the outcomes of an ICM policy or course of action is unacceptable to both the public and to decision makers when decisions on ICM policy and courses of action are made. Here again the "art" of successful ICM practice lies in presenting such uncertainties in a manner that is both responsible and inspires confidence in the course of action being proposed.

Several instances of ecosystem management experiments were offered that have been accepted by the public. For example, the Himmerfjärden sewage treatment plant on the Baltic Sea (Elmgren and Larsson, this volume) conducted a large-scale, year-long experiment to simulate a change to the N:P ratio likely to result from a 50% removal of nitrogen from the plant effluent. This was done after discussion with all relevant management bodies, from the Swedish Environmental Protection Agency to the Governing Board of the sewage treatment plant and after discussion with the main local environmental nongovernmental organizations (NGOs), as well as provision of detailed information to the local press and other media. When the feared blooms of phytoplankton failed to occur and biomass was decreased by 31%, the decision was made, with wide public support, to invest in nitrogen neutralization technologies in the sewage treatment plant.

Research Needs and Opportunities

The many ideas for research offered on this topic can be grouped into three categories.

Documentation and Analysis of ICM Initiatives

If we accept that ICM is a sustained process of learning, then the outcomes of ICM projects and programs need to be more fully and critically documented, analyzed, and disseminated. It is particularly important to do this in a manner that encourages comparison across ICM projects and programs rather than focusing on individual initiatives, as is the usual practice. It was noted repeatedly, that the funding required for such evaluative research is difficult to obtain. Those involved in the European Union's funding process expressed particular frustration, which re-enforced the perception echoed in other groups that the incentive/disincentive structures for advancing interdisciplinary, whole-system research in support of effective ICM needs to be critically examined and changed.

A central feature of ICM is the identification and analysis of options for addressing the opportunities and problems created by coastal change. It was recommended that well-documented case studies of the options considered as a response to specific expressions of coastal change (e.g., degradation of coral reefs, land reclamation/dike building, overexploitation of an estuary-dependent fishery) would be very useful. Elucidating the societal, economic, and environmental consequences of different courses of action is particularly important. This led to discussion on the need to develop typologies of contexts for coastal governance interventions. It is important to compare the like with the like. Such typologies need to address both the ecosystem types, the scale at which processes and ICM interventions are implemented, the degree to which anthropogenic forces have altered the system, and the capacity of the ICM institutions involved to act effectively.

Indicators and Indices for the Documentation and Analysis of Coastal Ecosystem Change

Many participants addressed the urgent need for negotiating a consensus on sets of valid and reliable indicators that can be used to document trends in coastal change. Such indicators must address change in both the coastal environment and the associated human society. It will be particularly useful when such indicators are combined into indices of coastal ecosystem condition. Experience in other fields is instructive. It has been very useful to formulate indices in public health and economics, even though the process of negotiating what to measure can be arduous. For example, the Human Development Index (HDI) emerged from a series of expert consultations sponsored by several United Nations agencies. It has subsequently proved useful when making comparisons in the direction of societal change in different places. Some participants felt that major difficulties will be encountered, particularly among natural scientists unwilling to accept the validity of making broadly applicable generalizations on the likely consequence of expressions of coastal change at large geographic scales.

The "Entrainment" of Scientists into the Adaptive Management Process

It would be useful to examine the strengths and weaknesses of different organizational structures (e.g., advisory committees) that have been used to integrate scientists better with managers and user groups throughout all stages of the policy cycle. At the global scale, the International Panel on Climate Change (IPCC) offers an instructive body of experience. In the United States, the Estuary Management Program has utilized, with varying success, a standard structure of committees both to analyze issues and to shape the management strategies that respond to them. This led to some discussion of the need to adjust the manner in which scientists are educated. The ideal has been expressed as the "T-shaped" scientist who has the depth in a specific field of natural or social sciences, but also a sufficient understanding of other disciplines to engage effectively in interdisciplinary research and the assessment of management options. The topic of educating for better coastal management has been addressed in several international workshops.

Periodic Summary Statements on Knowledge about Important ICM Topics

Another opportunity lies in periodically synthesizing the "state of the art" in the specific scientific disciplines as they apply to the analysis of common ICM issues (e.g., changes to the trophic structure of fish populations and the effects of changing the volume, timing, and quality of freshwater inflows to estuaries).

Communicating the Implications of Coastal Ecosystem Change to the Public

There is a need to develop and utilize "social marketing" techniques to inform and inspire the changes in human behavior at the societal scale that may be required to address important coastal management issues. Social marketing is a sophisticated field of critical importance to the implementation of public health strategies, for example, to control AIDS and promote effective family planning. In coastal management the challenge lies in developing simple messages that speak to the heart as well as to the mind. Such messages must be grounded on an interpretation of the best available scientific information. The "AMOEBA method," developed by the Dutch ministry of Transport and Public Works (Colijn and Reise, this volume), and the United States EPA's "report cards" on the condition of coastal ecosystems are examples of visual techniques used to convey the implications of complex data sets on trends in the condition of coastal ecosystems.

QUESTION 2:
How can we define/prescribe the most appropriate governance model?

This question addresses the problems inherent in mechanically transferring proven concepts to other cases where boundary conditions are different.

Ideas Presented in the Background Papers

The background papers present examples for all forms of governance in the continuum of possible modes. Several papers touch on *international* governance, including the use of conventions, and provide evidence that a new form of governance is gaining ground where adaptive management is emerging as the accepted mode.

One example is from the Baltic Sea (see also Elmgren and Larsson, this volume). A "Periodic Assessment of the State of the Environment of the Baltic Sea Areas" is compiled every five years by the Helsinki Commission (HELCOM) as part of its management process. This serves to focus attention on the degree to which environmental goals are being achieved, as opposed to the implementation of future management action. The assessments, the fourth of which is expected to be published in Spring, 2001, have gradually evolved in breadth and depth but still suffer from some problems of data comparability and availability. They are an important basis for virtually all decisions on Baltic environmental problems within HELCOM as well as outside it. This example indicates that international governance is moving towards an evaluation of mature subcycles of ICM policy. In these new generations of ICM policies, efforts are made to identify and involve stakeholders at the earliest possible stage of the process.

In some *regional* cases, such as the Black Sea (Mee, this volume) or the Benguela Current Initiative (O'Toole et al., this volume), the emergence of adaptive management as the preferred governance method is obvious and structures are forming to support this governance model. The Great Barrier Reef (Crossland and Kenchington, this volume) is a case where adaptive management is inherent in the chosen path, but it is focused on a specific treaty that is geographically linked and involves one nation.

Several papers discussed *national* programs and the impression gained from them is that in many cases the remnants of unfinished or aborted previous ICM cycles contribute to an evolving agenda; frequently, only parts of cycles are present. Several examples are presented that show a diversity of governance models operating on *state* or *subnational* levels, with a high degree of variance in perceived success.

Topics of Discussion

Below, we delineate several crucial issues that determine the optimal or most appropriate choice of a governance model.

Definition of Stakeholders

Definition of stakeholders is by no means straightforward, and weighting of opinions is a problem. Is a democratic approach the optimal way to arrive at decisions, in what way should vested interests be recognized, and what is the role of NGOs and scientists? There is a tendency in issue-specific questions to use a prescribed set of rules; however, it is obvious that this is not the only path to follow. The problem is not alleviated by a lack of agreement on the definition of the characteristics of a stakeholder. One of the major barriers to effective governance is the disempowerment of stakeholders afflicting many societies, especially those in transition from centrally planned governments. The lack of an effective driving force for change (e.g., NGOs, scientists, community-based organizations) has resulted in minimal implementation of existing policies and laws. An example of consequences of this problem is the demotion of the Ministry of the Environment in Russia to a state committee with limited powers and budget. The environment is perceived as irrelevant.

Instead of focusing effort on propping up this failed structure or its components, greater effort needs to be placed on giving objective and understandable information to the stakeholders, especially at the community level, and allowing them to work out their own solutions. Economic aid needs to embrace this reality by placing emphasis on such instruments as environmental impact assessments accompanied by adequate public information campaigns.

Tested and Accepted Ways to Allow Optimal Evaluation of Options

How can the stakeholders be provided with the best possible scientific background to enable them to see the consequences of possible options? The challenge is to find procedures for stakeholders' participation in the decision-making process without derailing the management processes through controversies. The evaluation requires that critical scientific information on the consequences of decisions is provided to stakeholders. In an optimal situation, stakeholders should be able to make sound decisions and find appropriate mechanisms of governance themselves, thus reducing the need for prescribed solutions. Scientists are often themselves stakeholders as well as the suppliers of information for managers. Lack of consensus among this group, while it is the "default situation" as far as many scientists are concerned, needs to be taken into account without derailing the decision-making process. Resources are needed to facilitate the communication of scientific knowledge and dissent, as well as to assist scientists in reaching consensus.

A positive example for the raising of public awareness and interest in scientific issues is the Solent Forum, a coastal group operating in the Solent area (south central English coast) that brings together a wide range of coastal stakeholders. The forum organized the Solent Science Conference to heighten awareness of the most recent research related to the Solent. It brought together a wide range of new material, allowed discussion and debate on both funding and future research, and will disseminate the results in book form. It was considered to be highly valuable by managers and scientists for information dissemination, awareness-raising, and supporting the ongoing dialogue between them.

Use of Evaluation/Valuation Techniques to Permit an Optimized Decision

What is the role of monetary valuation within the evaluation of management options? On what basis are different management options compared? The role of the scientist is to support the evaluation and not to make the decision. In the process, each stakeholder is confronted with the implications of achieving his/her own objectives, as well as the objectives of other stakeholders.

If there were an objective means to evaluate the consequences of different options of governance in terms of their gains and losses, the stakeholder decision would be relatively objective and easy. However, the method is fraught with problems, from the different objectives which each stakeholder would wish realized, to the means by which the achievement of these objectives could be measured, to unequal access by stakeholders to the evaluation process and/or its results. The central problem is the definition of the value system that is adopted. In a study investigating conflict between aquaculture and mangrove preservation (Gilbert and Janssen 1998; Janssen and Padilla 1999), the evaluation of alternatives for mangrove management considered three different evaluation objectives: conventional economic efficiency, equity, and environment. "Equity" addressed who would derive financial benefit from management alternatives (owners of fishponds tended to be wealthy individuals living elsewhere) while "environment" attempted, qualitatively, to capture the potential for irreversible effects.

Three conclusions drawn from these studies are relevant here. First, valuation studies on their own do not provide enough information to support decision making. Issues of uncertainty and irreversibility are difficult to capture adequately in monetary terms. Second, economic efficiency on its own was insufficient as the evaluation objective, because it does not address who derives benefits or suffers costs, nor does it encompass issues of uncertainty and irreversibility. Third, the decision depends heavily on the goals of the individual making that decision. For example, the local fishpond manager would consider only economic efficiency (profit making). The local government manager might consider the interest of the local population directly (equity) or look towards obtaining additional financing for its policies by license agreements, etc., with local developers (economic efficiency). Social planners at the national level would probably emphasize economic efficiency or equity issues, while the sustainable social planner would include, even give a relatively greater weight to, environmental issues and risks of irreversibility.

Coastal development and management requires implementation of better social science, and not just economics, to protect populations from ill-conceived governmental/economic perspectives. Stakeholder-based ICM needs to find a way to open up the evaluation of management options and the management process itself to more groups than is currently the case. This point is further illustrated by the issue of protecting Venice from flooding. While about $250 million have been spent on studies since 1973, the entire thrust has been the evaluation of a system of mobile gates and subsidiary works. The organization undertaking the studies (the CONSORZIO) has been committed since then to building this engineering project (at a cost of about $6 billion). The opponents of this scheme have had virtually no resources to fund research or access to the models developed by the CONSORZIO, which could be used to test alternative solutions (Bondesan et al. 1995; Zanda 1991).

Measurement of Success/Outcomes

Can scientific conflict be used to benefit ICM? If the perception were that an experiment in ICM can be conducted to test predictions of conflicting views, the antagonists will have the impression that science is advancing through the experiment and will help the process in the long run. In this mode of science in ICM, management must find the right place. Of great importance is the creation of transparency of procedure in the decision-making process. The criteria upon which a decision is based must be set in advance, and all possible outcomes must be considered and communicated. In such an experimental approach to ICM, proper milestones and goals must be established and the attainment of, or progress towards, the milestones and goals must be evaluated independently.

Learning and Adaptation Strategies

What will we do in ten years when the ICM projects currently underway are finished? What is the strategy or course of action to get new stakeholders involved? These are questions that grow in importance when it is realized that the initial generations of the ICM cycle are nearing completion in many nations.

The challenge of concluding or discontinuing specific management activities and providing room for new issues is a difficult one. The inertia in ongoing programs (seldom fully completed), the emergence of new issues, and the limits of resources often clash, posing difficulties for the adaptive evolution of ICM and the science to support it. One approach to overcome these difficulties is to establish specific target dates for policy implementation and goals, and then to conduct evaluation or assessments at these milestones. Unmet goals can be reset and new goals established. For example, the Chesapeake Bay Program has developed milestones at the arrival of its key year 2000, a broad agreement which reconnects to nutrient reductions goals but also establishes some new goals for issues now seen as of critical importance, e.g., oyster habitat restoration (D. Boesch, pers. comm.).

For science, the evaluation and reorientation phase is both a challenge and a chance to evaluate progress during the preceding cycle. However, the perception in other sectors may be that enough knowledge exists and that science is only acting to perpetuate itself. This perception will be weakened only if the scientists are honest about the importance of their work, and if the entire governance process is designed to assimilate, continually or periodically, new knowledge.

Research Needs and Opportunities

Many projects are moving from the first to the second cycle of ICM, and the accumulated experience on the appropriate ways to aid definition and choice of governance needs to be consolidated. This requires that ICM experience around the world be collected and investigated with the aim of understanding why some programs achieved their goals while others did not. The analysis should include an assessment of governance structures and sociological analysis to create typologies and explanations for successful and nonsuccessful governance systems. A limited effort in this direction has been made by the European Union (EC 1999).

In the frame of this assessment or in independent studies the optimal ways to educate stakeholders on relevant scientific issues must be explored: What are reasons and motivations of stakeholders, and how can science help to keep people engaged? This requires the design of interview and communication techniques to help elicit what stakeholders want, to communicate the consequences of decisions, to communicate the relative weighting of the different stakeholder objectives, and to help identify the limits to what is acceptable.

In a more specific area, it was felt that techniques used in the evaluation of management alternatives need to be expanded and formalized. The role of monetary information, and thereby of valuation studies, within these techniques is clear, but possibly not sufficient given the multiplicity of factors that management needs to consider. Stakeholder analysis is needed to specify the objectives of different stakeholders, as well as a range of techniques for measuring their achievement.

QUESTION 3: How should one deal with uncertainty within disciplines, between disciplines, as well as across the science–policy boundaries?

In every aspect of life, decisions must be made using uncertain information. It is, however, important to understand the origins of uncertainty and where this uncertainty is important in the decision-making process. Both the existence and degree of risk can be affected by uncertainty about the information and about the motivations of other people involved.

Ideas Presented in the Background Papers

Healey and Hennessey (1994) suggest that "science will drive the policy process only when scientific evidence is clear and uncontested." Most participants in our discussion disagreed and felt that this rarely happens. Scientific knowledge is constantly being questioned and refined such that the degree of uncertainty must be understood. Three background papers define three kinds of uncertainty: measurable uncertainty, ignorance, and complex predictions. McGlade (this volume) suggests that there are two fundamentally different kinds of uncertainty. First, there is a measurable uncertainty in a known quantity (e.g., error estimate, imprecision, averaging over space and time). Examples of this include estimating the annual average chlorophyll level. Second is ignorance — gaps in our knowledge or understanding. These are things we simply do not know anything about. An example might be that we need to know the fecundity of a new species being considered for exploitation in a fishery, but no one has ever seen an egg-bearing female. The most problematic issue is when we do not know that we are ignorant. This can be minimized by keeping the process open to differing opinions and considering nontraditional sources of knowledge. A third kind of uncertainty can be identified in the use of complex models to make predictions (Boesch; Deegan et al., both this volume). In this case, uncertainty becomes not the simple propagation of known error but is a blend of both the error of measurement and the error of understanding. The mathematical formulation of a model will only contain the relationships we know about or define as important. Thus, lack of knowledge or decisions about the interconnectedness of model components are implicit formulations of our understanding. When a problem becomes complex, simple inexactness (i.e., standard error bars) no longer adequately describes the uncertainty.

Topics of Discussion

Should Uncertainty Be Explicitly Acknowledged?

Although there was universal agreement that understanding uncertainty was a critical factor in making responsible decisions, there was some concern that discussions of uncertainty could lead to confusion or delays in making decisions. First, not everyone agreed that uncertainty should be explicitly discussed with all levels involved in making management decisions. Some felt that introducing the idea of uncertainty at the interested lay individual level would confuse already difficult concepts. The majority, however, felt that all stakeholders were fully capable of understanding uncertainty because they confront it every day in other aspects of their lives. For example, it is easy to understand that the price of purchasing insurance for their car or house is proportional to the expectation of a problem. The second major concern was that identifying uncertainty can lead to either postponing management action, even when the consequences of no action are severe, or to the exploitation of the uncertainty to promote a management strategy that would not otherwise be considered. This was considered the more serious issue of the two.

Is There a Fundamental Difference in the Degree of Uncertainty between Social and Natural Sciences?

There was a general perception that the degree of uncertainty in social sciences was as large as, or perhaps larger than, that in the natural sciences, but that uncertainty was not as generally acknowledged in the social sciences. For example, price–demand regression lines are never given with confidence bounds, while it is unacceptable to provide a similar regression relationship (e.g., nutrient loading–chlorophyll response) in the natural sciences without error estimates. In addition, it is generally recognized that some aspects of social sciences cannot be known. For example, there are no indicators that can predict a stock market crash or the state of the economy in thirty years. Yet, this information is needed to use other kinds of social science information (e.g., employment indices) in evaluating management options.

In addition, we felt that expressing natural science estimates using social science terms gives the natural science estimates a feeling of more certainty. For example, if the value of ecosystem services was placed in monetary terms by making several assumptions to convert a qualitative ranking to a quantitative scale, the estimate was more "certain" than if the same concept were expressed in words.

What Are the Implications of Uncertainty in Numerical Models?

Complex ecological models play an important role in environmental management. The expectation is that models will make predictions about the state of the environment in response to a specific management action. Models are a synthesis of currently existing knowledge tooled for a specific purpose, and they can be no stronger than the conceptual model behind them and the data used to construct them. The strongest use of models is in exploration (thinking and learning) and scenario description; however, they are most often used for prognostication. Unfortunately, the usual output of models is a single point estimate for a response variable with no estimate of the uncertainty of the value. This tends to raise the expectations of the reliability of the model output to a level often unwarranted by the level of uncertainty that went into building the model. Unfortunately, there is nothing intrinsic in the way that a model fails (i.e., the estimate provided by the model does not coincide with the actual response of the ecosystem) that provides insight into the cause of the failure. Thus, we do not know if a model fails because we have not estimated a parameter well, or if we are ignorant about a particular part of the ecosystem. Models can be very useful for understanding the potential results of a management action, but they will be most useful if their use is focused on the big picture context, not on individual details (e.g., the exact oxygen content of the water at a particular place on a particular day).

Do We Need to Communicate Distinct Kinds of Uncertainty Differently?

Understanding the different kinds and the level of uncertainty and the consequences to the decisions at hand is important in determining the risks that people and managers are willing to take. Uncertainty is a problem when a specific problem arises and where certainty is crucial; the level of certainty on noncritical issues is not important. Imprecision/variability versus lack of knowledge are fundamentally different types of uncertainty and must be considered relative to the consequences of the action (Dovers and Handmer 1995). There is a definite need to use different language to convey these different reasons for uncertainty.

Another aspect to consider in the communication of uncertainty is the degree of consensus on a subject. Lack of knowledge or uncertainty has been frequently used as a smokescreen to hide unpalatable predictions. Furthermore, even when the facts are clear and uncontested, not everyone will agree on the interpretation. The uncertainty associated with interpretation is qualitatively different than when the uncertainty of the facts is at issue. Interpretation then involves the personality (i.e., is that person or board known as a risk-taker?) and the reputation (i.e., have they been right before when others were wrong?) of the person or board offering the interpretation. This kind of uncertainty must be conveyed in a different way than uncertainty of knowledge. The format chosen by the IPCC, which explicitly states the level of consensus in the panel on a publicized statement, was considered to be a good model for conveying this kind of uncertainty.

Is There a Difference between Effective versus Complete Communication of Uncertainty?

The need to communicate uncertainty or confidence must be done in an appropriate manner for the context of both the users and the problem. We felt that extreme change should probably be communicated without much detail on uncertainty, because the detail of how the uncertainty was determined might undermine the understanding or acceptance of the real problem. When communicating with nonscientists, it may be better to provide an appropriate sense of the uncertainty, rather than the technical details of the estimation of uncertainty. An illustrative lesson was the explanation of why the beaches in some areas of the Black Sea needed to be closed to users who prior to this had never had beach standards. The analogy used likened the precautionary measure of closing the beaches to the precautions one would take around a person who has the flu. Although it is possible that you might not get sick if you behaved normally, if you changed your behavior to avoid contact with the sick person (or beach), the possibility of getting sick is lessened.

What Is the Relationship between the Precautionary Principle and Uncertainty?

Increasingly, environmental decision making is scrutinized with respect to a *precautionary principle*. The precautionary principle is a legal mechanism for managing the environmental risk arising from incomplete scientific knowledge of a proposal's impacts. The precautionary principle is applied to actions that carry with them the potential for serious or irreversible environmental change. This principle asserts that where uncertainty and doubt make it impossible to be sure about a correct decision, any errors should favor the long-term sustainability of the environment (Underwood 1997). Thus, the precautionary principle is a framework of decision making in which the concept of uncertainty is central. Under the precautionary principle, contrary to past practices, the absence of adequate scientific information can no longer be a reason for postponing or failing to take conservation measures.

A major difficulty is recognizing and admitting when lack of knowledge is the limiting step. Scientists have a vested interest in uncertainty or not knowing, because it provides an opportunity to continue their research. However, it can be difficult to admit the answer is still unknown after working hard for several years on a problem.

It is also important to consider whether the uncertainty of an estimate overlaps with an irreversible threshold. For example, dissolved oxygen (DO) in water is highly variable, but most animals have a relatively wide tolerance above a threshold. If the DO drops below that threshold, the organism usually dies. Thus, it is important to know not only the variability of the DO level, but its relationship to a threshold value.

Research Needs and Opportunities

Scientists need to develop an appropriate language to describe and effectively communicate uncertainty. Measurement error, often expressed as "±1 standard deviation," and "I don't know" are distinct kinds of uncertainty and need to be communicated differently. Defining ways to communicate the various types of uncertainty needs to be established by systematic study of communication. In addition, we need to estimate the trade-off between the effectiveness (i.e., the sense of uncertainty) and the completeness (i.e., the details) of communication uncertainty through empirical studies. It would also be useful to develop better explanations of the "scenario" use of models to explain the implications of policy. This may be as simple as calling these "What if" projections instead of "scenarios."

We also need to develop better ways of incorporating differing opinions and acknowledging scientific controversy. Currently, science tends to present information as "black or white" and this tends to encourage a "right or wrong" interpretation. This attitude permeates the "culture of science," even though the acknowledged state of the science is several shades of gray because of incomplete information or understanding. We need to develop ways of bringing antagonists together to develop a consensus on the state-of-the-art in science.

New ways to estimate the uncertainty in complex model predictions are needed. In model outputs, simple error propagation does not fully capture the kinds of uncertainty involved. Formal error propagation techniques tend to overestimate the imprecision of the predictions, while in other cases, such as ignorance, standard error propagation simply does not apply. We need to develop novel ways to link observations and monitoring with modeling to incorporate more stochastic predictions. Ecological modelers need to learn from other disciplines, such as atmospheric science where data assimilation is used to estimate model parameters (Vallino 2001). We also must explore novel ways of identifying complex interactions that may be emergent properties that arise from numerical models (Shipworth and Kenley 1999).

QUESTION 4: What do we need to learn about human behavior to provide guidance on the communication of knowledge to facilitate the acceptance of science?

This question arose from a discussion on the fact that scientific knowledge is frequently available for transfer and application in similar situations (i.e., from one watershed to adjacent or nearby watersheds). However, decision makers often will not accept and apply the knowledge until it has been recreated locally. Discussion ensued as to the motivation of individuals to accept or reject knowledge and how a better understanding of behavioral response may influence the effective use of science and communication between scientists and policy makers.

Ideas Presented in the Background Papers

The question of human behavior is most directly addressed, although briefly, by McGlade (this volume) who states that "individual choices cannot always be understood solely in terms of individuals acting to satisfy their own needs." The implication is that many external, or nonpersonal, factors influence how an individual may incorporate information leading to a policy decision.

More indirectly, von Bodungen and Turner (this volume) discuss socioeconomic drivers that may represent such external factors. Boesch as well as Crossland and Kenchington (both this volume) offer examples of integrating science with policy and user groups, and although they do not relate this directly to contributing to a better understanding of "human behavior," they do highlight "a benefit" resulting from the involvement of scientists with the process. Crossland and Kenchington particularly note the need for all interests (management, science, users) to "engage in design, coordination, interpretation, and application of outcomes of research."

Topics of Discussion

Why has available scientific information not been used effectively to inform the coastal planning, policy, and management process? The problem involves many aspects that are intrinsic to human behavior and are exceedingly difficult to generalize.

Taking Risks

One is an aspect that is a basic principle of how human society works: any given society has a majority that is adverse to *taking risk*, whereas some individuals take risk with varying amounts of success. An example given concerned sociological studies on a population of fishermen and how they act on scientific knowledge ("cartesians" in a traditional and not adventurous way, as opposed to a minority of "stochasts" who were prepared to take high risks to improve catches). The "stochasts" had intrinsic knowledge supplemented by information from fisheries scientists and were prepared to act on it, being vastly more successful at fishing than the conservative majority. In the same manner, some scientists may prefer to not expose "unfinished" or uncertain results to the public or to the decision makers, whereas other scientists are prepared to give answers based on best estimates or intuition. These scientific risk takers require a governance context that is able to profit from information and at the same time is able to evaluate the relevance of that information independently.

The Issue of Trust

The issue of *trust* — a basic requirement of successful communication — was discussed at length because it is elemental to the relationship and communication between scientists and managers. Trust is created through an interplay of culture and personal relationships. As applied to ICM, trust must occur between scientists, users, and managers. In most cases, this requires a mutual understanding of the values, feelings, curiosity, and reality of the individuals. Management frequently perceives news from scientists as bad news, meaning new work, additional problems, and a shortening of resources that can be applied to other projects; in many cases the information provided by science does not suggest a feasible course of action. Some managers may presume that enough science is available or that an issue is not salient: it takes trust to not assume that the scientists overstate a problem. And trust is a two-way street. Scientists can see that there is or may be a problem but cannot get it onto the managerial agenda because the issue is not salient, because there are overpowering counter issues, or because the interaction between science and managers is impeded by activities to slow the communication between managers and scientists (lobbying). To be prepared to act on information and to adopt pressing issues, managers must trust the motives and relevance of the information.

Communication

Problems of communication of motives and scientific issues to managers is of vital importance but is hampered by a variety of problems: scientific controversy, conflicting signals, specialization, and fragmentation clad in jargon are powerful deterrents. In addition, some issues surpass the capacity of managers to act upon scientific information (e.g., the spatial or temporal disassociation of cause and effect, e.g., in chemical time bombs or in transboundary issues), and in many developing countries there is a lack of institutional capacity to use or absorb the information. In many cases, scientists are not using the power of stimulating curiosity, feeling, and motivation effectively, even if few scientists would not agree that their choice to get involved in ICM was originally motivated by early basic experiences of feeling and curiosity.

The aspect of cultural background and different perceptions is often greatly underestimated and is frequently the reason that solutions to a given problem in ICM may work well in one context and fail miserably in another. Research at Middlesex University (Penning-Rowsell, pers. comm. 1999) investigated attitudes to beaches at a score of sites in the United Kingdom and conducted interviews with more than 4000 people. A comparison was made between beach-nourished "wilderness"-type sites and the traditional Victorian seawall and beach assemblage. In general, people nostalgically preferred the traditional seawall and small beach rather than the large, nourished beach, not least because this coincided with their understanding of what the coast "should be like" and because the large beach was seen as alien and dangerous (especially for those with children). To communicate effectively with coastal managers, scientists must understand the value systems that are held by the public that both groups are trying to serve.

In the science arena, the *reward systems* for synthesizing available information and preparing briefs on relevant issues is frequently inadequate, although such compilations or syntheses are the prime source of information for managerial decisions. There are excellent examples of well-communicated, concise, and salient information, such as the annual "State of the Baltic Sea" series prepared by the Ecosystems Group of Stockholm University, or the effort of the Ecological Society of America to publish issue documents on nitrogen sources targeted to policy makers, to the public, and to other scientists. In most cases, such successful synthesis for public information is aided by the knowledge and skill of professional journalists or science writers. Other forms of communication with the public are off-editorial columns to publish scientific knowledge.

Research Needs and Opportunities

Research is needed to evaluate the characteristics of human behavior that influence interactions and decision making between scientists, users, and managers. An immediate need is to document and assess social science case studies of successful and unsuccessful integration of science, policy, and user interests throughout an ICM process.

The ICM process needs equal participation of science, policy, and user interests throughout the entire time and cycle of the process to foster understanding and communication. Science needs to be better represented in the ICM process from the very beginning. In addition, scientists also need to be involved for targeted research while maintaining objectivity.

Maintaining equal involvement will also ensure the participation of managers and users in the preparation of relevant research proposals. To facilitate exchange and building of trust, scientists need training in communication techniques with particular attention to the use of various media. Both managers and scientists involved with ICM should attend one another's professional conferences.

QUESTION 5:
How should ICM incorporate the implications of a range of time and space scales at each step of the process?

Nature operates on a large range of scales, and scales of human activity in the coastal zone and reactions to changes in the coastal zone are greatly variable. Any ICM policy decision is thus faced with the need to reconcile a variety of process scales. How can the ICM process take into account and acknowledge the disparate scales?

Ideas Presented in the Background Papers

Two approaches of analysis are applied in coastal management (von Bodungen and Turner, this volume; GESAMP 1996): The Drivers-Pressures-State-Impact-Response (D-P-S-I-R) model of interaction analyzes the complex interactions between cause and effect of coastal zone changes. The response component is usually policy related, and time and space domains are not explicitly treated in the approach. However, the cause and effect relations of the D-P-S-I-R model have an obvious temporal and spatial component (von Bodungen and Turner, this volume). The second model is ICM, which is a policy-oriented perspective (Olsen, this volume). Both approaches share the analytical step (problem identification), and inherent in ICM is a temporal succession of processes and activities. Both approaches could thus be linked in a sequence that has the analytical steps (issue definition) from either the socioeconomic D-P-S-I-R concept or from Step 1 of the ICM cycle (see Figure 20.1 in Olsen, this volume).

Topics of Discussion

Indicators

Indicators are needed throughout the entire process of ICM. They facilitate scaling (e.g., issues), comparison (e.g., site), regulation (e.g., pressures), communication, prediction, assessment of trends, and mapping of change. Indicators can be fully qualitative in nature (e.g., a damage scale), or their role may be regulatory (e.g., emission permits, water quality criteria) with arbitrarily fixed threshold values, or they may be fully quantitative (e.g., nitrate concentrations). The establishment of commonly accepted indicators is an important step in the ICM process. This is particularly important with respect to future evaluations of success. Indicators used on global scales are usually simple ones and qualitative in nature, and are mostly used for comparison and communication. However, in other fields, global quantitative and regulatory indicators have been defined and accepted by most countries; examples include the World Health Organization's drinking water quality standards, or the HDI set forth by the

United Nations (see Olsen, this volume). The HDI is a composite index of life expectancy, literacy, and per capita gross domestic product that measures the socioeconomic development and quality of life in a region.

On a regional scale, indicators have been set up by international conventions for descriptive and regulatory purposes. At the country level, many indicators of environmental health, economic state, or for regulatory purposes have been proposed and established. On the local scale, many indicators of little scientific basis but with tremendous potential to convey the message have been established. One example presented was the annual test on how far a well-known political figure can wade out into the bay and still see his feet as an index of eutrophication.

Differing Perception of Space in Natural and Social Sciences

There is not a single set of indicators that can equally and effectively be established and used to describe and measure issues that range from local to global in scale. To complicate matters, the common understanding of the scales involved differs between natural and social sciences (see Table 21.1).

The scales relevant in natural sciences are usually limited by physical geography and functional boundaries. In social sciences, they are constrained by administrative, political, and economic boundaries. The structural discrepancy of these two spatial perceptions of the coastal zone is a major issue in coastal management. This issue can only be resolved when and if the governance structure adapts to the natural boundaries (forming, for example, river basin authorities, regional seas basin authorities).

Topology and Typology of the Coastal Zone

The definition of the coastal zone (CZ) and its limits were identified as an unresolved matter in the discussion, because the definitions vary widely depending on criteria used to define the limits (e.g., natural vs. socioeconomic). The natural variability of the CZ and the wide range of use make it distinct from the relatively homogeneous open ocean. Coastal features such as bays, fjords, lagoons, mangroves, salt marshes, estuaries, deltas, reefs, and upwelling cells have very different types of land–ocean interactions and differ in the relative importance of influence by land and sea. The entire land mass of small islands (100 km in diameter) can be

Table 21.1 Differences between natural and social sciences in common understanding of scales involved in ICM.

Scale	Natural Science	Social Science
Local (1–100 km)	Small bay, estuary, lagoon	Community, municipal to district level
Subregional (100–1000 km)	U.S. East Coast estuaries NW Europe estuaries	State, small country
Regional (100–10,000 km)	Regional seas and their catchment basins	ASEAN, EU, U.S.A., Russia

considered as part of the coastal zone in both natural and social science definitions. In the case of land-dominated features, their size is critical in ICM. Large river basins (>10,000 km^2) have linkages between the coastal zone (50 km inland from the sea shore) and very remote upstream regions that cannot be regarded as part of the CZ.

The perception and/or definition of the CZ also varies according to the view taken by natural and social sciences. A natural science approach would be to delimit the CZ by tidal range, light limitation, biomass, or morphological gradient and other measurable criteria. The social science definitions usually encompass a large range of natural environments because they are based on arbitrary choices of spatial limits, e.g., the Exclusive Economic Zone, the distance covered by a boat trip of one day, the distance from shore that can be covered by swimmers, etc. One of the main challenges of ICM in the early phase is to reconcile these different perceptions and create common scales.

From Local Drivers to Global Drivers

In the case of land-dominated coastal features, the spatial distance between a cause and an effect in the CZ has been illustrated in a number of cases (e.g., Mississippi Delta and catchment area nutrient loading; Boesch, this volume). Such connections over spatial distances are termed teleconnections. Important teleconnections are through the atmosphere (moisture, nutrient, or particle transport), through rivers (pollution, particles, nutrients), or through the ocean (suspended matter, nutrients, pollutants). Other teleconnections are related to human activities, which can take the form of toxic spills or the dissemination of harmful organisms in bilge waters, population dynamics, mass tourism, and socioeconomic developments in river basins. These last types of teleconnections are recent additions to the influences on the coastal zone and may not be perceived by many stakeholders as direct influences on their issues. The global market greatly affects mariculture and fishing, and investments made by corporations and bankers from very remote places can be key drivers of industrial and fisheries development in the coastal zone. Unless fully informed, such teleconnections and stakeholders are difficult to identify. Information exchange by a variety of media is a form of immaterial teleconnection that has strong potential to influence local decisions.

ICM is generally organized at the levels of impacts and/or use, and is not prepared to handle teleconnections — this is another major challenge of enhancing ICM in the future. Even in large countries, where river basins teleconnections are common (e.g., Russia, Brazil, Canada, U.S.A., China, Australia, India) teleconnections are often not established, ignored, or even denied when recognized. The hydrosolidarity principle, which is now invoked more and more often, is not yet accepted even within a given country. Notable exceptions of successful international agreement on the management of entire river basins are the Rhine water quality restoration between 1970 and 1990 and the Great Lakes Joint Commission between Canada and the U.S.A.

A predicted (but not yet observed) global teleconnection is the sea-level rise as a result of climate change. The causes and drivers have been identified; the state is forecast although the timing is not yet known; and the location — everywhere — is predictable with certainty. The issue is therefore well defined and is being widely debated years ahead of any impact. The course of action by ICM measures to face climate change is being established in few countries. This issue has alerted a considerable part of the population and all ICM managers, politicians, and decision makers about global-scale issues and teleconnections. If the predicted effects come to pass, they will probably change forever the way ICM is presently considered and handled.

Time and Space Scales Relationships in CZ Issues and Governance

As was the case with the typology of coastal zones, we are faced with an enormous range of time and space characteristics of processes. The spatial dimension has already been presented: from the local (typically 10–100 km shoreline) to the global. The time dimensions can be illustrated by many descriptors such as (a) the duration of typical pressures, (b) the duration of related environmental impact, (c) the recovery period of the natural and/or social environments, and (d) the duration of policies and/or regulation (Figure 21.1).

There is not necessarily always a relationship between time and space scales: very short-term events may affect an entire region (as is the case with hurricane damage in the North Atlantic) and local impacts may be caused by secular land-use change (forest clearing). After the construction period of a major dam (few years), there is an immediate subregional or even regional abatement of sediment transfer.

The longest time scales (>100 years) are found for:

- Drivers and pressures: urban development, dams, land-use change, sea-level rise.
- Impacts: coastal environment, biological invasion, complete destruction of ecosystems with very slow rates of formation, such as mangroves and coral reefs.

In the case of chemical pollution and eutrophication, the recovery period may be directly linked to the size of the coastal object. This is the case for semi-enclosed bodies such as bays, some estuaries, and fjords with sills characterized by higher water residence time and/or complex hydrodynamics. For lakes, the recovery period is generally two to three times the water residence times as a result of imperfect vertical mixing. When there is an internal source of nutrient contaminant, e.g., from interstitial waters, the recovery process is even slower. Such rules may apply to these coastal zone types as well for entire regional seas, such as the Baltic Sea and the Mediterranean Sea.

The environmental recovery after a direct alteration of biological communities through habitat destruction or overfishing is usually species dependent and ranges from a few months, or less, to several decades for the longest-lived animals (e.g., fish; Hutchings 2000). This recovery is much less size dependent although it is highly dependent on the spatial distribution of the original damage: a mosaic-type damage will result in faster recovery than large-scale isolated damage.

Societal and social issues in the CZ can also be addressed by similar temporal descriptors, including impact duration, recovery duration, and application period for any given law or regulation (see Figure 21.2).

In economies, the time perspective is essential (e.g., investment returns). In social and educational fields, we may raise the question of the number of years needed for the appropriate awareness of coastal citizens to environmental issues to emerge, for convincing farmers to shift agricultural practices, or simply to educate a graduate student, or to convert a pure research scientist into a coastal zone scientist. In some countries, ICM can actually be limited by lack of trained managers. In some former Soviet countries, where ICM used to be

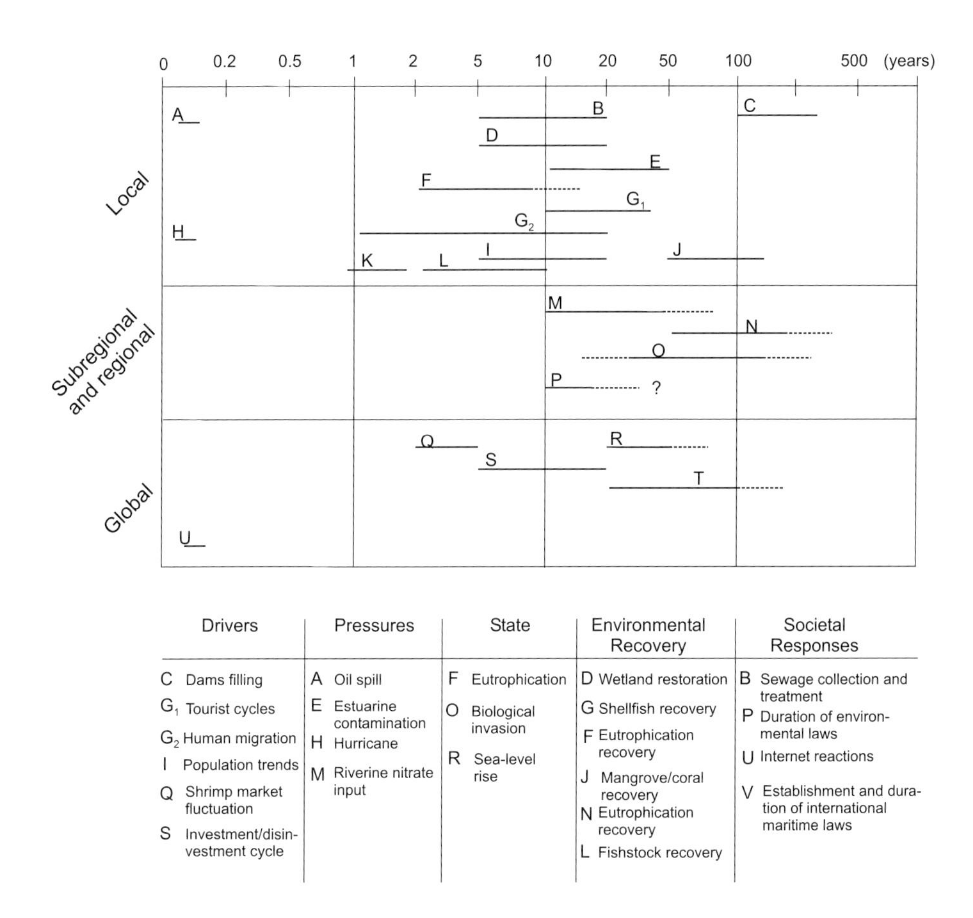

Figure 21.2 Some schematic examples of time and space scales in problems confronting ICM.

practiced by trained scientists and technicians, the present economic collapse may soon lead to a generation gap in coastal management. In fast-developing countries, the education of local citizens may be the limiting factor even if managers are available.

Matching the environmental and social issues scales with governance scales is as critical in time as it is in space. Typical scientific programs are funded for three or four years; elections occur every four or five years. The time needed to pass a law can exceed ten years, and the related regulations may take another few years to be established. The duration of environmental laws may be up to several decades, the same period needed for the establishment of some environmental criteria. ICM scientists often believe that a given problem cannot be solved quickly, whereas politicians request that it should be solved within their mandate. The CZ policy agenda also has its inherent time scale that scientists should know and consider. In The Netherlands, the political agenda has focused on different environmental issues each decade. After the catastrophic floods of the 1950s, the CZ management of the 1960s was essentially about water level control. In the 1970s, when the Rhine and Meuse were at their maximum levels of pollution, the focus was on water quality; in the 1980s the focus has been on ecology and habitat restoration.

Most issues concerning the CZ have well-defined time and space scales. If the governance domain does not at least partly overlap the time and space issue domain, then it is very unlikely that the ICM project will be successful. For example, it should take into account the full spatial dimension of all the drivers, and not just that of the coastal users. Concerning time scales, the CZ manager often faces impacts related to former pressures, occurring at a time when no D-P-S-I-R relation was made by scientists (scientist ignorance) or accepted by managers (manager blindness). It is essential to put in place mechanisms of ICM for forecasting the impacts as soon as the state change is documented by scientific survey (set up specific surveys for alarm bell ringing) or, even more efficient, as soon as the pressures are known to occur (improve environmental impact assessment), particularly for those issues that have any long-term dimensions.

There are two extreme examples of ICM: When a major accident occurs, e.g., an oil spill, the ICM loop from driver identification to program implementation must be performed within 24 hours, effectively shortcutting all stakeholders (there is a mutual consensus for action, and the issue and potential impacts are well defined), the information from scientists of all disciplines is made available, collected, and synthesized — even in a very crude manner — in real time, and the question of funding is not an issue during the evaluation stage. For the issue of sea-level rise, we already know the drivers, the pressures, most of the future state — except exact timing and amplitude — and the impacts. ICM has years ahead to build up a remediation program, and there will probably be a willingness to pay on the part of all local shareholders as well as certain others. Both examples can provide precious information on the effectiveness of existing ICM structures and procedures at local to national scales.

Shortening the Structural Lags in Coastal Management

When a completely new issue emerges (as was the case with lake eutrophication in the 1960s), a succession of lags can be observed:

1. Time for scientists to make studies and reach a consensus (knowledge lag).
2. Time for citizens to be aware of issues and be alarmed (awareness lag).
3. Time for politicians to react and make decisions (political lag).
4. Time to build up and fund the program of action (management lag).
5. Time to implement the program (technical lag).
6. Time for reestablishment of ecosystem structure and function after the management program is implemented (recovery lag).

When the knowledge and experience from other similar issues is collected, synthesized, made available, and accepted by stakeholders, many of these lags can be shortened or even eliminated through diffusion of available scientific information, citizen awareness programs, NGOs' request for action, use of existing regulations elsewhere, use of scenario models ("what if models"), or application of the precautionary principle.

An important aspect of ICM is the forecast of issues, except in coastal defense for which this was established a long time ago. We do not generally have early change detection surveys to check and question the trends of both natural and social environments. Such specific surveys should be established (e.g., micropollutant screening at selected sites and periods, surveillance of species biodiversity to check invasive species, and interpretation of river and groundwater surveys in terms of coastal zone response).

Single Issue versus Multiple Issue

ICM generally faces multiple issues that should be identified, ranked (based on appropriate indicators), and scaled in time and space. This is generally made during the D-P-S-I-R stages (see above). The response(s), i.e., program design, funding, implementation, is then established by stakeholders who make their choices. At this stage, two ICM options may be found: (a) focus on a single issue (e.g., eutrophication) or (b) consider all identified issues. The first option allows for the allocation of funds and human resources to one target, but preventive measures to control other issues may be blocked. The second option may just sprinkle efforts on many issues without any effective result.

Research Needs and Opportunities

It is essential to involve science on an equal basis with users and managers throughout the ICM process. Time and space scales of the proposed ICM process must be developed and shared with all participants in the process.

Time lags in the ICM process can be shortened by enhancing communication of what is already known at each stage.

Effective interaction between small-scale (spatial, temporal) and large-scale analysis hinges on our ability to handle these different scales so that we can estimate and predict system changes.

The most effective way of sorting for the appropriate scale of time and space in ICM is to segregate by issue. Examples are: shoreline processes (erosion/accretion), public access to the shore, estuary-dependent fisheries, and oceanic fisheries. A typology that arrays increasing issue complexity against increasing governance capacity is useful.

CONCLUSIONS

The rich variety of environments and the multitude of socioeconomic structures therein present a challenge to identify some general principles for the ICM process. Although some such principles have been formulated, our group emphasized that management should take the form of an adaptive approach to respond to the changing coastal environments, of which human societies should be regarded as an integral part.

Our general recommendations are to:

- Collect, document, and analyze outcomes of ICM actions in relation to coastal change in order to identify the best possible practice for science in the ICM process.
- Initiate research on the entrainment of scientists into the ICM process, analyze the optimal organizational structure for scientific participation, and investigate how education might increase the rate of entrainment.

- Periodically prepare syntheses of the state of science relevant to particular ICM issues; endeavor to publicize them widely, responsibly, and effectively.

REFERENCES

Bondesan, M., et al. 1995. Coastal areas at risk from storm surges and sea-level rise in Northeastern Italy. *J. Coast. Res.* **11(4)**:1354–1379.

Dovers, S.R., and J.W. Handmer. 1995. Ignorance, the precautionary principle, and sustainability. *Ambio* **24(2)**:92–97.

Dovers, S.R., T.W. Norton, and J.W. Handmer. 1996. Uncertainty, ecology, sustainability, and policy. *Biodiv. Conserv.* **5(10)**:1143–1167.

EC (European Commission). 1999. Lessons from the European Commission´s Demonstration Programme on Integrated Coastal Zone Management (ICZM), Directorates-General Environment, Nuclear Safety and Civil Protection, Fisheries, Regional Policies and Cohesion. Luxembourg: Office for Official Publ. of the EC.

GESAMP (Joint Group of Experts on the Scientific Aspects of Marine Environmental Protection). 1996. The Contributions of Science to Integrated Coastal Management. Reports and Studies No. 61. Rome: FAO (Food and Agriculture Organisation of the United Nations).

Gilbert, A.J., and R. Janssen. 1998. Use of environmental functions to communicate the values of a mangrove ecosystem under different management regimes. *Ecol. Econ.* **25(3)**:323–346.

Healey, M.C., and T.M. Hennessey. 1994. The utilization of scientific information in the management of estuarine exosystems. *Ocean Coast. Manag.* **23**:167–191.

Hutchings, J.A. 2000. Collapse and recovery of marine fishes. *Nature* **406**:882–885.

Janssen, R., and J.E. Padilla. 1999. Preservation or conversion? Valuation and evaluation of a mangrove forest in the Philippines. *Env. Resource Econ.* **14**:297–331.

Lee, K.N. 1993. Compass and Gyroscope: Integrating Science and Politics for the Environment. Washington, D.C.: Island Press.

Shipworth, D., and R. Kenley. 1999. Fitness landscapes and the precautionary principle: The geometry of environmental risk. *Env. Manag.* **24**:121–131.

Underwood, A.J. 1997. Environmental decision-making and the precautionary principle: What does this principle mean in environmental sampling practice? *Landsc. Urban Plan.* **37(3-4)**:137–146.

Vallino, J.J. 2001. Improving marine ecosystem models: Use of data assimilation and mesocosm experiments. *J. Mar. Res.* **58**:117–164.

Zanda, L. 1991. The case of Venice. In: Proc. of the First Intl. Meeting "Cities on Water," ed. R. Frassetto, pp. 51–59. Venice: Marsilo Editori.

Author Index

Barcellos, C. 203–212
Benoit, J.R. 341–364
Boesch, D.F. 37–50, 93–112
Brawley, J. 295–305
Burbridge, P.R. 253–273

Colijn, F. 51–70, 93–112
Crossland, C.J. 135–147, 165–189

Daffa, J. 191–201
de Barros Neto, V. 229–251
de Luca Rebello Wagener, A. 165–189
de Vries, I. 113–120, 165–189
Deegan, L.A. 295–305, 341–364
Dronkers, J. 165–189

Elmgren, R. 15–35, 93–112
Emeis, K.-C. 341–364
Frykblom, P. 93–112

Gätje, C. 253–273
Gilbert, A.J. 341–364

Jickells, T.D. 93–112

Kenchington, R.A. 135–147
Knoppers, B.A. 203–212, 253–273
Kremer, J. 295–305
Kulekana, J. 191–201

Lacerda, L.D. 203–212
Larsson, U. 15–35
Lee, V. 341–364

Malan, D.E. 229–251
Marins, R.V. 203–212
Martins, O. 253–273
McGlade, J.M. 307–326, 341–364
Mee, L.D. 71–91, 93–112
Meybeck, M. 275–294, 341–364

Ngoile, M.A.K. 191–201, 253–273

Olsen, S.B. 327–339, 341–364
O'Toole, M.J. 229–251, 253–273

Pacyna, J.M. 93–112
Penning-Rowsell, E. 165–189
Pethick, J. 121–133

Ramachandran, S. 253–273
Reise, K. 51–70, 165–189
Richter, C. 253–273

Salomons, W. 253–273
Sardá, R. 149–163, 165–189
Schwarzer, K. 165–189
Shannon, L.V. 229–251

Talaue-McManus, L. 213–227, 253–273
Taussik, J. 165–189
Turner, R.K. 1–14

von Bodungen, B. 1–14, 341–364
Voss, M. 93–112

Wasson, M. 165–189
Webler, T. 295–305
Wulff, F.V. 93–112

Subject Index

aboriginal hunting 137
abrasion platforms 123
adaptation 37, 38, 43, 47, 48, 175
adaptive management 23, 38, 39, 47, 102, 106, 108, 118, 132, 259, 328–333
  compared to conventional 329
  of Himmerfjärden Bay 26
  model 342, 343
  role of science in 107
Adriatic Sea 98
Agenda 21 265, 266
agriculture 2, 21, 40, 43, 78, 79, 94, 95, 100, 102, 139, 150, 282
  atmospheric inputs from 96
  through reclamation 124, 126
  reform 88
  in Sepetiba Bay 205
  as source of nutrients 71, 95
agroforestry 213, 216, 217, 220, 221
Agulhas Current 230, 240, 241
algae 297
  benthic 63, 75, 77
  Phyllophora sp. 73, 74, 87, 89
algal blooms 63, 239, 244, 245
AMOEBA method 55, 56, 336, 346
Anglian coast 121, 132, 170, 171, 184
  defense 125, 126
  future development 127, 128
  inappropriate development 131, 133
  location 122
  past management 123, 124
  sustainability 130
  tidal deltas 124, 125
Angola 229–235, 239–242, 249, 250
animal husbandry 71, 81
  intensive pig farming 81, 86
anoxia 73, 74, 259, 300, 303
  reduction in Black Sea 76
Antarctic Treaty System 309
Anthropocene 275, 276
Apo Island 216, 217
aquaculture 6, 21, 61, 88, 203, 217, 222–224
  Laguna de Bay 214, 215
Aral Sea 175
Araruana Lagoon 97
archipelago 30
  Baltic 18, 20, 22, 24
  Philippines 213
Arctic Monitoring and Assessment Programme 110
Arusha Resolution 198
assessment 181, 196, 219, 244
  of environmental impact 267
  of fish resources 214, 223
  monitoring 47, 100, 107, 237
    for Great Barrier Reef 145
    Research Assessment Exercise 181
  rapid assessment technique 77, 268
  risk 143, 197
assimilative capacity 175, 177
Atchafalaya River 41, 42
atmospheric deposition 20, 93, 106
awareness 89, 140, 141, 145, 150, 204, 219, 236, 254, 321
  promotion of 182, 261

Balearic Islands 150, 156
Baltic Sea 16, 57, 96–102, 110
  eutrophication 18
  phosphate load 19, 24
barrier islands 52, 53
beach erosion 152, 158, 193
beach management 123, 129, 149, 255
  recharge program 125
  in Spain 153–155

belief system 317–320, 324
BENEFIT Programme 229, 233, 234, 239, 249, 262
  science and technology foci 234, 235
Benguela Current Large Marine Ecosystem (BCLME) 229–233, 250, 347
  BCLME Programme 229, 241, 249
    goals of 240
    public participation in 245
    structure of 245, 248
  boundaries of 241, 242
  characteristics of 230
  ecosystem 233, 259
  Interim Benguela Current Commission (IBCC) 245, 248
  management history of 232, 233
  map of 231
  regional approach to management 237, 239
Benguela–Environment–Fisheries–Interaction–Training Programme (*see* BENEFIT Programme)
benthic algae 63, 75, 77
benthic animals 73–76
biodiversity
  Convention on Biological Diversity 309
  in Great Barrier Reef 139, 140
  loss in 331, 337
  of Spanish Coast 149
bird protection 43, 101, 127
  in Wadden Sea 60
Black Sea 71, 77, 80, 97–101, 104, 347
  Comprehensive Black Sea Assessment 100
  Environmental Programme 88
  eutrophication in 89
  high primary productivity in 75
  Istanbul Commission 72, 85, 86, 100
  map of 72
  N emissions 77, 79, 86
  Novorossiisk Bay 75
  phosphate loads 77–82
  prior to 1960s 73
  Strategic Action Plan 85, 100, 101
Blackwater estuary 128
Blakeney 128
Blyth estuary 125
Brazil, management policy 203, 204
Brundtland Report 308
Bucharest Convention 85, 86, 100
Bulgaria 74, 76
cadmium concentrations 58, 64
Canary Islands 150, 156
Cantabrian coast 149
capacity building 217–221, 240, 243, 256, 261, 262, 271, 316
carrying capacity 166, 167, 172, 175, 176
  definition of 174
Catalan coast 158
CENR Integrated Assessment of Hypoxia 100
chemical time bombs 104
Chesapeake 2000 Agreement 44
Chesapeake Bay 37, 42–46, 97–102, 107
  Chesapeake Bay Program 40, 43–47, 332, 349
  ecosystem 41
  population in area 39
Chichester Harbour Conservancy 179, 180
Childs River 297, 298
  N-loading series 299
chloride 277, 280, 283, 290
chlorophyll *a* 26, 63, 299, 300
  models of 298, 301
civic science 3, 5, 37, 48
climate change 95, 127, 191, 192, 359
  impact of Benguela Current on 231
coastal defense 128, 130, 176, 203, 207
  influence on development 132
  flood embankments 125
  storm surge barrier 117
coastal development 1, 239, 153
  in Philippines 222
coastal ecosystem (*see* ecosystem)
coastal management (*see* management)
Coastal Management Plan 153, 154, 159
Coastal Management Policy Programme (CMPP) 236, 239
coastal morphology 123, 125
Coastal Policy Green Paper 236
Coastal Wetlands, Planning, Protection, and Restoration Act 42
Coastwatch Report 154
coccolithophorads 74
communal governance 322
communication 135, 179, 184–187, 254, 256, 262, 267, 271, 356
  barriers to 180, 181, 187, 262
  problems in 197, 253
  role of trust in 108, 355
community awareness (*see also* public awareness) 135, 138

community empowerment 213, 219, 221, 225
community mobilization 216, 221–223
community-based management 213, 216, 220–223, 318, 321
Comprehensive Black Sea Assessment 100
Congo River 240, 241
conservation 127, 152, 169, 170, 192, 255
  awareness of 199
  in Brazil 204
  dugong conservation 143
  of fish resources 214, 223
  in Great Barrier Reef 140
  national strategy for 149
  of natural heritage 135, 137
Continuous Plankton Recorder 65
cooperation 256, 262
  Cooperative Research Centre 141, 142, 170
copepods 75
coral reefs 177, 194, 215–218, 255, 259, 269
  economic value 268
cordgrass 176
corporate governance 319
Costa Brava 151
  housing development in 152, 156
cost-benefit analysis 103, 104, 109
ctenophores 75, 76
cyanobacteria 20, 74, 104
cytoseira 73, 74

damming 95, 139, 166, 282
  in catchments 94, 103
  cost-benefit of 104
  Hoover Dam 281, 282
  riverine 275, 281, 282
Danube River 72, 76, 77, 78, 80, 285
  Danube Pollution Reduction Programme (DPRP) 89
  Danube River Protection Convention 85
  Environmental Programme for the Danube River Basin (EPDRB) 78
  N emissions 78, 79
    from human/pig populations 81, 83
  relationship of population to sewage 82
  summer hypoxia 87
  total P emissions 79
data and information systems 141, 269
DDT 61, 277, 288
deforestation 94, 177, 263, 282, 291
Delta region (NL) 113–118, 175
denitrification 20, 30, 95
devolution 213, 219, 220
diatoms 74–77
dinoflagellates 74
discounting 316, 318
dissolved inorganic nitrogen (DIN) 298, 301
dissolved oxygen (DO) 300–302
Dnieper River 72, 77, 79
Dniester River 77, 281
Don River 72, 77, 282, 283
Draft White Paper 236–238
drainage basin 1, 17, 93, 94
dredging 96, 131, 167, 193
  in Sepetiba Bay 203, 211
Drivers-Pressures-State-Impact-Response (D-P-S-I-R) framework 107, 108, 116, 159, 357, 362, 363
dugong conservation 143
dunes 123, 129, 193
Dunwich cliff erosion 125
Dutch Fifth National Policy Document 114

East Atlantic flyway 53
Ebro River 281
ecological development index (EDI) 55
ecological footprint 269
Eco-Management and Audit Scheme 261
economic factors 96, 138, 261, 317
  distribution of benefits 222
  growth in South Africa 238
  growth in Spain 153
  indicators 308, 318
  input–output models 150, 240, 269, 270
  market regulation 322
economic tools 263, 265, 316
    discounting 316, 318
    incentives for fisheries 215
economic valuation 105, 106, 110, 254, 268, 269
ecosystem 76, 121, 127, 256, 331, 343
  analysis 258, 259
  carrying capacity 253
  components 257, 258
  definition of 327
  function 259, 260
  health 54, 195, 240
    maintenance of 246, 247
  management 138, 195–200
  productivity 260
  response model 296, 297

eco-tourism 238, 239, 261
education 121, 144, 221, 222, 236, 256, 261, 262, 304, 345, 364
eelgrass 297
Elbe River 57–61, 98
energy dissipation 124, 126
energy gradients 123, 124, 128
ENVIFISH Programme 249
environmental assimilation index 116
environmental databases 261
environmental drivers 93, 99
environmental impact assessment (EIA) 267
environmental indicators 160, 161, 175, 259, 268, 318
environmental management 139, 142, 240, 301, 346
  barriers to 296
  Eco-Management and Audit Scheme 261
  quality goals 23, 161
  use of models in 102, 295–299, 304
Environmental Programme for the Danube River Basin (EPDRB) 78
EPA's "report cards" 346
erosion 123–128, 132, 222, 255, 275, 282, 363
  cliff 123, 125
  delta 103
  human-induced 193
  link with reclamation 125
  Sepetiba Bay 203
  Spanish coast 149, 153, 154
  variability in 276
Espierre River 289
estuaries 123, 124, 128, 255
  ecosystem response model 296, 297
European Monitoring and Evaluation Programme (EMEP) 102, 110
European Spatial Development Perspective 114
European Union (EU)
  membership criteria 85
  community directives 153
    EU Birds Directives 127
    EU Habitats Directive 127, 131
    EU Nitrates Directive 29
    EU Urban Wastewater Treatment Directive (UWWTD) 22, 23, 31
    EU Water Framework Directive 31
eutrophication 22, 37, 43, 55, 61, 63, 71–76, 98, 99, 101, 103, 222, 259, 270, 296, 297, 360
  Baltic Sea 15
  Black Sea 76
  Great Barrier Reef 139
  Gulf of Mexico 95
  public awareness of 89
  Wadden Sea 64
exclusive economic zones (EEZs) 169, 194, 332
Expanded Fish Production Program 215

Fenland 124
fertilizer use 40, 79, 80, 83
first-level consulting 321
fish stocks
  changes in 297
  decline of 243, 244
fisher power 321
fisheries 75, 96, 101, 140, 192, 197, 240, 262, 263, 307–309
  barachois 194
  carrying capacity for 175
  Code of Conduct for 248
  collapse of 172, 192
  coral reef 270
  exploitation of 39, 192, 213, 216–219, 239, 245, 307–310, 316, 322
  in Great Barrier Reef 137, 139
  legislation 219
  in Lingayen Gulf 214–218
  multispecies fisheries 139, 310, 311
  production function 268
  species introduction 43, 60, 239, 245
  surveys in Eastern Africa 198
  in Western Indian Ocean 193
fisheries management 43, 219, 234, 307, 313, 320, 322
  external trading pressures 96
  financial incentives 315, 316
  models for 310, 311
  Namibian 259
  in Philippines 219, 220, 224
  role of information in 309
  science/technology needs for 239
  social choice 314, 317
Fishery Sector Program 219, 220, 223, 224
fishing 71, 100
  in Great Barrier Reef 140
  illegal techniques 218
  impacts on BCLME 232
  rights 232
  in Sepetiba Bay 205
  South African 238

flooding 42, 117, 118, 123–127, 193, 280
embankments 125–128
prevention 108
protection 42, 43
risk 131, 132
food supply 260, 263, 296, 307
human 191, 192, 214
freshwater habitat, loss of 127
functional diversity 2, 8–10, 113, 114, 118
fuzzy logic 312, 314

games theory 269
Geographic Information Systems (GIS) 161, 269, 290
German Bight 55, 57–63
German Organization for Technical Cooperation (GTZ) 234
GESAMP cycle (*see also* ICM policy cycle) 6, 10, 107, 327
Global Climate Conveyor Belt 230
Global Environmental Facility (GEF) 85, 239, 240, 248, 332
Global International Waters Assessment (GIWA) 109
governance 213, 216–223, 240, 256, 308–311, 320, 327, 331–338, 343, 350
approaches to 329
capacity 266
communal 322
definition 319
description of coastal initiatives 328
fishery issues 215, 262
forms of 319, 322
international level 262–265, 322
local level 221, 262, 264
national level governance 198, 219, 262–265
in Philippines 210, 219, 221, 224
regional 159, 198, 248, 249
regulatory ladder 323
role of institutions 318
scales 262, 263, 361
science involvement in 334, 335
in Sepetiba Bay 210
State 321, 322
user participation 319, 321
gradients 117–119, 230, 299, 359
density 73, 88
energy 123, 128
salinity 41, 52
siltation 269, 270
Great Barrier Reef (GBR) 135, 170, 179, 347
catchment land use in 138
ecosystems of 136
environmental status of 145
management of 137, 140–144
Marine Park 136–137
Marine Park Authority 140–145
World Heritage Area 136, 139
gross national product (GNP) 268
ground layer 114, 116, 119, 170, 175
Guanabara Bay 177, 204, 211
Gulf of Bothnia 16
Gulf of Mexico 41–47, 95

habitat destruction 37, 43, 192, 193, 239, 244, 245, 259, 264
due to reclamation 127
North Sea 176
Philippines 214
strategy to mitigate 175, 218
heavy metals 58, 64, 131, 156, 177, 218, 276, 277, 287, 288
Helgoland 59, 62
Helsinki Commission (HELCOM) 22, 46, 57, 100, 110, 346
Himmerfjärden Bay 18, 344
map of 19
holistic approach 3, 124, 236, 249, 256, 268
Holocene 116, 124, 127
Hoover Dam 281, 282
housing development 1, 151, 152, 156, 296, 238
human behavior 2, 137, 140, 317, 346, 354, 356
belief system 317, 318, 320, 324
indicators of 276
understanding 179
Human Development Index (HDI) 268, 345, 357, 358
Humber estuary 122–128, 131
Hunstanton 128
hypoxia 42, 87, 99, 101, 300
CENR Integrated Assessment of 100
Black Sea 86
Danube summer levels 87
Gulf of Mexico 47

ICM policy cycle 328, 330, 342, 345, 347, 357
ignorance 3, 102–104, 313
IJsselmeer 61

Indian Ocean Expedition 197
indicators 55, 357, 358, 363
  development of 107, 109
  early warning 173
  of ecosystem change 345
  of environmental quality 54, 98
  need for 146
indigenous people 137–139
industrialization 2, 150, 156
industry (*see also* fisheries, shipping) 71, 78, 94, 100
  Anglian estuaries 126
  atmospheric inputs from 96
  food processing 82
  through land reclamation 124, 126
information 256, 259–262, 265, 266, 271
  availability 166
  classification of 160
  services 140–145
infrastructure layer 171, 175
input–output analysis 150, 240, 269, 270
institutions 135, 146, 166, 167–169, 265
  capacity 236, 237
  role of 318
integration 4, 13, 37, 38, 43–48, 106, 107, 114, 166, 171, 178–180, 195, 198, 220, 229, 243
Intergovernmental Oceanographic Commission (IOC) 234
Interim Benguela Current Commission (IBCC) 245, 248
International Commission for the Protection of the Danube River 72, 85, 86
International Commission for the South-east Atlantic Fisheries (ICSEAF) 232
international governance 262, 263, 265, 322
International Rhine Commission 57
International Standards Organization (ISO) 261
intertidal restoration 131
introduced species 43, 60, 239, 245
Iron Gates reservoir 79
irrigation 282, 291
Istanbul Commission for the Protection of the Black Sea (ICPBS) 72, 85, 86, 100

jellyfish 75, 76

Kattegat 16
Kenya 193, 194, 198
knowledge 101, 219, 233, 307, 329, 350
  communicating 184
  creation 182
  ontological 182
  pedigree of knowledge 313
  reliable 330, 332, 338, 342, 343
  systems 307, 311, 312
Kundryuchya River 283
Kyoto Protocol 323

lagoons 63
laissez-faire fishing regime 322
landform migration 124–128, 130
Land–Ocean Interactions in the Coastal Zone (LOICZ) 109
land reclamation 3, 52, 119, 124–127, 130, 132, 140, 193, 204, 345
layer model 113–116, 170, 171
legislation 4, 58, 64, 86, 136, 141, 144, 151, 165, 196, 203, 210, 211, 216, 219, 256, 341
  EU Directives 29, 31, 127, 131
Lincolnshire coast 125, 126
Lingayen Gulf 216, 217, 220, 270
  Coastal Area Management Commission 218, 219
Local Agenda 21 156, 159, 161
local governance 221, 262, 264
local-scale management 132, 204, 220,
Louisiana Coastal Wetlands Conservation and Restoration Task Force 46

Maas Delta 116
Madagascar 193, 194, 198
managed retreat program 131, 132
management 102, 109, 118, 151, 157, 173, 216, 232, 236, 253, 257, 258, 264, 265, 331
  (*see also* adaptive management, partnerships, sustainability, waste management)
  adaptation 37, 38, 43, 47, 48, 175
  approaches 260, 268
  beach 123, 129, 149, 151, 255
  in Brazil 203
  comanagement 221, 321, 332
  community-based 216, 223, 318–321
  devolution 213, 219, 220
  ecosystem 138, 195–200
  environmental 139, 142, 153, 301, 346
  of Great Barrier Reef 132, 140, 141, 144

management *(continued)*
integration 4, 13, 37, 38, 43–48, 106, 107, 114, 166, 171, 178–180, 195, 198, 220, 229, 243
link with science 141, 146, 166, 179, 184, 191, 195, 197, 262, 267
local 132, 204, 220
national 219
in Philippines 216, 219
reactive 117–119, 265
regional 233, 239
in Spain 153, 159
mangroves 96, 177, 211, 215, 259
MARE research program 100
mariculture 194, 199, 211, 219, 238, 263
regional policy for 248
Marine Conservation and Development Program (MCDP) 217, 221
marine protected areas 137, 152, 153, 157, 217, 220
Marne River 280
marshes 113, 114, 123
Mauritius 193, 194
Mediterranean 98
metals 277, 278, 286, 287
mining 192, 232, 239, 240
Mississippi River 43, 95, 102, 282, 287
delta 37–45, 98–101
decline in Pb levels 277
trends in nitrate concentrations 285
Mnemiopsis leidyi 75, 76
modeling 47, 48, 107, 269, 270, 298
adaptive management 342, 343
chlorophyll *a* 298, 301
climate change 109
Danube Water Quality model 78, 89
ecosystem response model 296, 297
environmental management 102, 295–299, 304
for fisheries 310, 311
input–output 150, 240, 269, 270
layer model 113–116, 170, 171
N-loading 302
river 290–292
uncertainty in 302, 304, 310, 352
for Waquoit Bay 300–303
monitoring 47, 100, 107, 237
Arctic Monitoring and Assessment Programme 110
European Monitoring and Evaluation Programme (EMEP) 102, 110
for Great Barrier Reef 145
Research Assessment Exercise 181
water quality 279
Monte Carlo simulations 302
Montreal Protocol 309, 323
Mozambique 193, 194, 198
mudflats 124, 129
multidisciplinary approach 143, 146, 184, 198, 204
multispecies fisheries 139
models of 310, 311
mussel beds 65

Namib Desert 230
Namibia 229–235, 239–242, 249, 250
fishing industry 232
National Environmental Management Council (NEMC) 198
national governance 198, 219, 262–265
National Nature Reserves 52, 126, 157
natural resources (*see* resources)
negotiation theory 269
networks layer 114
nitrates 42, 60, 61, 77, 79, 98, 286, 289, 291
from agriculture 95, 102, 285
EU directives 22, 29, 83, 86
in rivers 285
nitrogen 39, 76–79, 99, 285, 291, 297, 299
in Black Sea 77, 79, 86
to Baltic 20, 21, 109
dissolved inorganic (DIN) 298, 301
fertilizer 86
fixation 25, 30, 296
-fixing bacteria 20, 24, 25
N-loading model 302
N:P:Si ratios 95
total Kjeldhal N 278
nongovernmental organizations (NGOs) 184, 219, 249, 319, 324, 344, 347, 362
input to resource management 216
nonpoint sources 40, 41
North Atlantic Oscillation (NAO) index 65
North Norfolk coast 128, 129
North Sea 94, 96, 98, 124
shoreline transformations in 176
North Sea Quality Status Report 59
Paris Commission for 46, 57

Norwegian Agency for Development Cooperation (NORAD) 234
nuclear power station 131
nutrient loads 42, 59, 61, 71, 78, 103
  to the Baltic 20
  estimates of 77
  for European continental rivers 59
nutrients 39, 43, 61–64, 80, 131, 255, 268, 275, 296
  discharges 23, 24, 77
  N:P:Si ratios 95
  reduction goals 55, 103, 349
  trends in 284–286

occupation layer 114, 171
Odessa Declaration 85, 100
oil/gas exploration 232, 239, 240
  spills 96, 166, 169, 177, 246
    clean-up technology 239
    Great Barrier Reef 139
    Guanabara Bay 211
Oosterschelde 117
Oslo Commission 57
overfishing 37, 43, 75, 94, 100, 193, 206, 218, 307–310, 333
oxygen 297, 300
  deficiency 18
oyster 40, 41, 44

Paraíba do Sul River 204, 207
Paris Convention 22
partnerships 135, 174, 182, 185, 199, 211, 239, 249, 250, 321
  Angola–Namibia–South Africa 229, 230, 233, 235
  fundamental principles for 245
  management-user-researcher 146
PCBs 56, 61, 65
peat formation 114–117
pedigree of knowledge 313
peer review 141, 266, 316
perestroika 84
persistent organic pollutants (POPs) 275–277, 287
pesticides 80
petroleum exploitation 192
Philippines 223, 270
  characteristics of 213
  Fishery Sector Program 224
phosphate 83, 86–89, 285, 290
  in Baltic Sea 19, 24
  in Black Sea 77–79, 82, 291
  -containing detergents 58, 89, 284
  in Rhine 57, 58, 284
  in Seine 291
phosphorus 39, 76, 81, 99, 278, 284, 291
  in Baltic 20, 21, 104, 109
  in Black Sea 77
Phyllophora 73, 74, 87, 89
phytoplankton 61, 64, 66, 74, 77, 79, 94
  changes in 103, 104
  composition 55
PIREN–Seine program 291
Plan of Action 238
Plans of Excellence 158
Po River 286, 287
Polish EcoFund Foundation 110
political will 98, 104, 141, 166
polluter-pays-principle 153
pollution 22, 96, 103–105, 193, 214, 275, 360
  abatement 78, 99, 107, 223, 239, 241, 264, 275
  control 255
  freshwater pollution 151
  hotspots 288, 289
  industrial 156, 169
  land-based sources of 78, 99, 139, 223, 241
polyphosphate detergents 79, 82, 89
population pressure 1, 97, 100, 175, 176, 192
  demographic trends 79, 232, 330
  from growth 94, 260, 330
  overpopulation 253
  in Western Indian Ocean region 193, 194
Portmán 156
ports 53, 71, 131, 193
  Great Barrier Reef 139, 140
  Sepetiba Bay 211
  South Africa 236
poverty 193, 219, 253, 260, 263, 264
  alleviation 307
precautionary principle 141, 153, 171, 177, 245, 315, 317, 341, 342, 353, 362
  definition of 341
  Vorsorgeprinzip 315
prediction 251, 314, 337, 245
prisoner's dilemma 106
property/housing development 1, 151, 152, 156, 296, 238

property rights 168, 316, 317
public awareness 89, 140, 141, 145, 150, 204, 219, 236, 254, 321
public involvement 178, 236 , 245

quality of life 175, 192, 256
quality status reports 54, 57, 100
Quashnet River 297

Ramsar sites 126
rapid assessment technique 77, 268
reactive management 117–119, 265
recreational industry (*see* tourism)
red algae (*Phyllophora sp.)* 73, 74, 87, 89
Red Sea Program 262
Redfield ratio 286
regional governance 159, 198, 248, 249
regulatory ladder 323
reliable knowledge 330, 332, 338, 342, 343
Research Assessment Exercise (RAE) 181
resilience 121, 174
 definition of 54
 of the Delta area (NL) 114
 of upwelling ecosystems 259
resource and ecological assessment studies (REAs) 219
resources (*see also* fisheries) 174, 232
 exploitation of 157, 216
 human consumption of 94
 overexploitation 11, 114, 171, 175, 258
 renewable 258, 260
 resource-based economy 261
 sustainable management of 151, 153
 utilization 116, 124, 168, 259, 260
response–delay time 172, 173
reward systems 356
Rhine River 61, 98, 102, 287, 289, 362
 delta area 115, 116
 International Rhine Commission 57
 phosphate loads 57, 58, 284
 transboundary nature of 58
Rhône River 279
Rio Conference 265, 266, 315
Rio de Janeiro 204, 211
risk 108, 123, 175, 178, 307, 312–314, 355
 assessment 143, 197
rivers
 damming 275, 281, 282
 discharges 20, 57
 floodplains 127
 flux in 280
  definition 278
  evolution of 276
  modeling of 290–292
  surveys of 275–277, 288
  trends 282–289
 variability of 97
Romania 74, 76, 80
 intensive pig farming 81

Sage Lot Pond 297–300
salt marshes 52, 55, 129
 loss of 130
 vegetation of 61
sand dunes 123, 129, 193
Schelde delta 116
science 108, 141, 175, 195, 196, 216
 civic science 3, 5, 37, 48
 codes of conduct 261
 environmental aspects for 143
 fragmentation of 262
 framework for priorities 144
 key issues for 143
 link with management 141, 146, 166, 179, 184, 191, 195, 197, 262, 267
 peer review in 141, 266, 316
 role of 98, 101, 146, 254, 255, 311, 312, 332, 333, 338
 second-order 312, 321, 324
Scolt 129
Sea of Azov 73
seagrass 41, 43, 61, 66, 101, 269, 270
 Zostera marina 63
seal reproduction in Wadden Sea 65
sea-level rise 113–117, 126–131, 174, 193, 359, 360
Secchi depth 26, 27
Secchi Disk 74
sediment profiles 275
sediment transport 121, 124, 281
sedimentation 123, 125, 128, 152, 206, 270, 280
Seine River 278, 285, 287
 dilution patterns 279
 PIREN–Seine program 291
self-organizing capacity 119, 319
Sepetiba Bay 203, 204, 207
 governance of 211

Sepetiba Bay *(continued)*
map of 205
sedimentation rates 206, 207
Severn estuary 128
sewage (*see* waste management)
Seychelles 193, 194
Seychelles Conference 198
shipping 96, 100, 107, 131, 166, 192, 211, 238
in Great Barrier Reef 137, 139, 140
navigational regulations 169
shoreline development 113, 166–169, 178, 179
communication in 184, 182
definition of 165
infrastructures 170
response–delay time 173
risk mitigation 172, 178
shrimp ponds 96
silicate 77, 78, 95
Social Accountability 8000 261
Social Accounting Matrices (SAM) 269
social choice 314, 316–319
sociocybernetic system 319
socioeconomic 96, 143, 204, 210, 240
drivers 100, 106, 253, 254, 262, 263, 270, 354
indicators 160, 268, 271
and investment opportunities (SEOs) 219
Solent Science Conference 348
Somalia 194
South Africa 229–235, 239–242, 249, 250
coastal policy 237, 238
growth of 235, 236
Namaqualand 237, 238
Southern African Development Community (SADC) 234, 235, 249
South East Atlantic Fisheries Organization (SEAFO) 249
Spanish coast 150, 174
housing development in 151, 152, 156
main environmental problems 154
marine protected areas of 152
national strategy for conservation 149
Spanish Coastal Law 152, 153, 157
Spartina anglica 176
spatial design 113–115, 119
Special Protection Area 127
species diversity 74, 270
species introductions 43, 60, 239, 245
stakeholders
communicating to 188, 220, 239, 253, 262
conflicts among 3, 329
definition of 347, 348
role of 23, 31, 47, 107, 108, 160 175, 195, 240, 346, 349
NGOs 184, 219, 249, 319, 324, 344, 347, 362
State governance 321, 322
Stiffkey 128, 129
Strategic Action Programme (SAP) 240–245, 248
strategic regional plans 149, 159, 161
striped bass, recovery of 41
subsidiarity principle 331, 341
Suffolk coast 131
sulfate 283
Sumilon Island 216, 217, 220
sustainability 3, 8–12, 37, 38, 45–48, 54, 107, 143, 172–175, 193, 217, 222, 232, 240, 256, 258, 264, 308
of Anglian coast 126
Draft Whilte Paper 236–238
for human survival 192
management for 153, 242, 258
policies for 236
principles of 308
of Spanish coast 150
SUCOZOMA 16
in Western Indian Ocean region 194, 195

Tanzania 193, 194, 200
Tanzania Coastal Management Partnership (TCMP) 198–200
Thames River 122–126, 285
Thornham 128, 129
tides
amplitude 53
deltas development 124
energy 121, 123, 128
flooding 122
surge tide (1953) 126
tools 265, 268–271
development of new 220, 304
tourism 53, 96, 100, 106, 129, 140, 160, 192, 239, 261–263, 268, 270
collapse 172
eco-tourism 238, 239, 261
equity of access 139
in Brazil 203
in Eastern Africa 194
in Great Barrier Reef 137, 138
in Philippines 217

tourism *(continued)*
regulation of 145, 193
in Sepetiba Bay 205, 211
in South Africa 236, 238
in Spain 149–151, 154, 174
sustainable 156, 174, 194
trace metals in Sepetiba Bay 203, 207–210
Transboundary Diagnostic Analysis (TDA) 240–243, 248
transboundary issues 93, 102, 204, 239, 242, 244
in Benguela Current region 229
demographic trends 79, 232, 308, 330
management 102, 108, 240
in Wadden Sea 51
transdisciplinary research 179, 181, 184
trophic "dead ends" 75
turbot 75
Tuy River 279, 280

U.K. Ministry of Agriculture 128
Ukraine 74, 76
UN Conference on Environment and Development 308, 315
UN Convention on the Law of the Sea (UNCLOS III) 169
uncertainty 3, 102–104, 145, 166, 167, 177, 245, 307, 312, 313, 350, 351
communicating 178, 180, 352
in models 302, 304, 310, 352
uncoupling index 116
UNEP GEMS/WATER Programme 276
United Nations Development Programme (UNDP) 240
United States Agency for International Development (USAID) 199, 217, 221
University of Rhode Island's Coastal Resources Center (URI–CRC) 199
upwelling 95, 230, 241, 255, 259
BCLME 241, 242
urban development 1, 2, 71, 158, 192
in Great Barrier Reef 139, 140
through reclamation 124, 126
Sepetiba Bay 205
Spain 149–152, 155, 156

valuation 3, 5, 55, 138, 142, 236, 253, 268, 348, 350
variability 1, 2, 54, 65, 94, 107, 121, 177, 240, 245, 276
virtual governance 323
Vorsorgeprinzip 315

Wadden Sea 51, 57–66, 98–102
Cd concentrations 58
coastal architecture 52, 64
economic factors 53
ecosystem of 52, 60, 65
green algae in 63
resilience to perturbations 54
Waquoit Bay 297–303
Wash estuaries 125, 124, 128
waste management 88, 94, 194, 262
discharge 21, 151
EU Urban Wastewater Treatment Directive (UWWTD) 22, 23, 31
sewage treatment 80, 88, 108, 109, 156
in Himmerfjärden Bay 26
urban 156
in Western Indian Ocean region 194
water quality 96, 131, 140, 192, 217, 239, 285, 288, 297, 331, 336, 337
in BCLME 245
criteria 248, 275, 357
Danube Water Quality model 78, 89
descriptors of 275–278
drinking standards 357
in Great Barrier Reef 139
improvement of 246, 247
monitoring 279
in Philippines 218
Rhine 284, 359, 362
in Russian rivers 283
Spanish beaches 151, 155
transparency 20, 24, 74, 77, 87
wave energy 121, 123, 128, 129
Wells 128, 129
West Coast (South Africa) 238
Western Indian Ocean region 193, 194, 198–200
wetlands 37, 88, 139, 282
Chesapeake watershed 40, 42, 44
loss 43, 99–103, 333
protection 46, 47, 334
W-factors 166
World Commission on Economic Development 308

xenobiotics 61, 275–277, 287, 288

Zanzibar 194
zeaxanthin 25
Zernov's phyllophora field 74
zoobenthos 74–76